P9-DCJ-227

PHOTOGRAPHIC ATLAS OF THE BODY

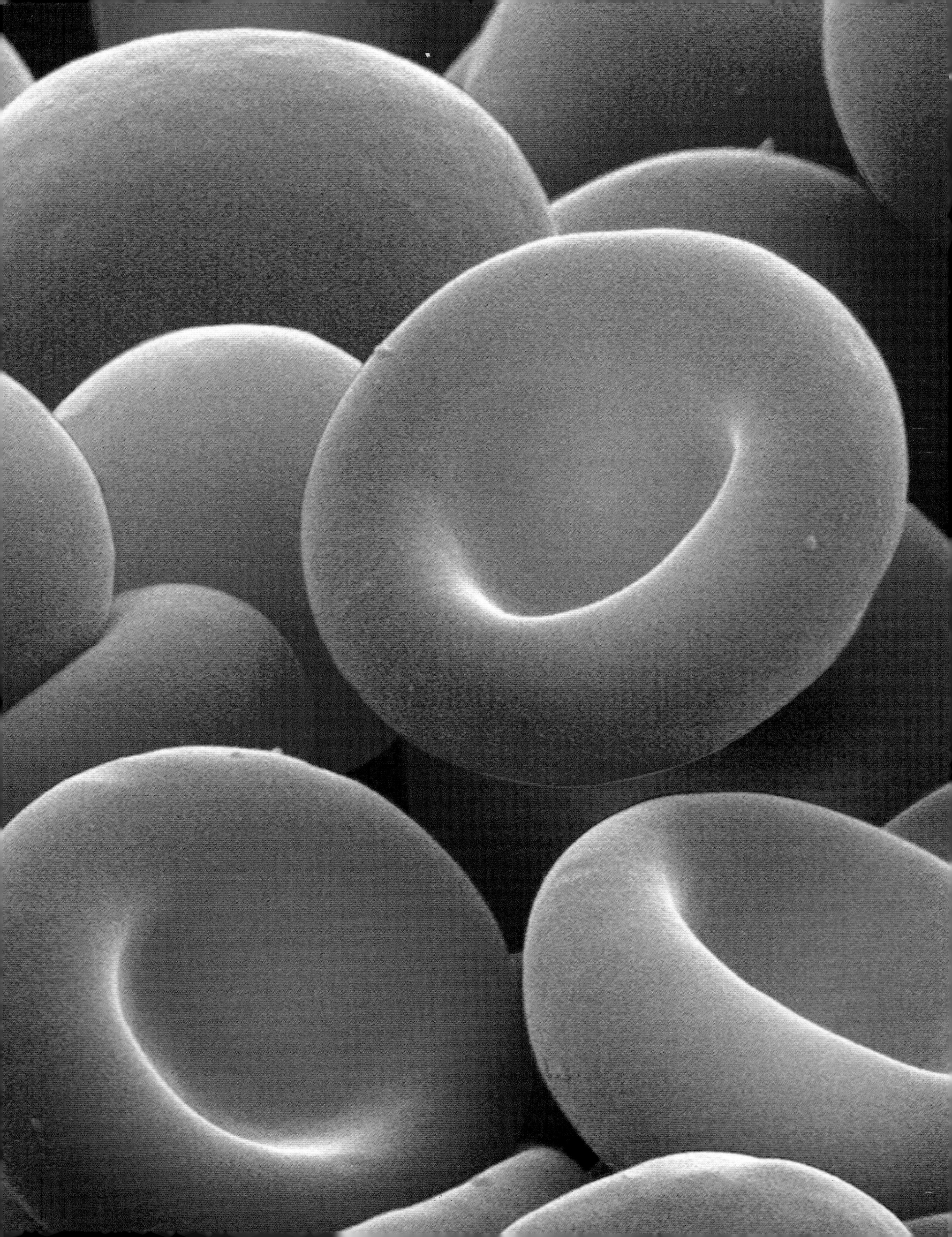

PHOTOGRAPHIC ATLAS OF THE BODY

PICTURES SUPPLIED BY
THE SCIENCE PHOTO LIBRARY

FOREWORD BY
BARONESS SUSAN GREENFIELD CBE

FIREFLY BOOKS

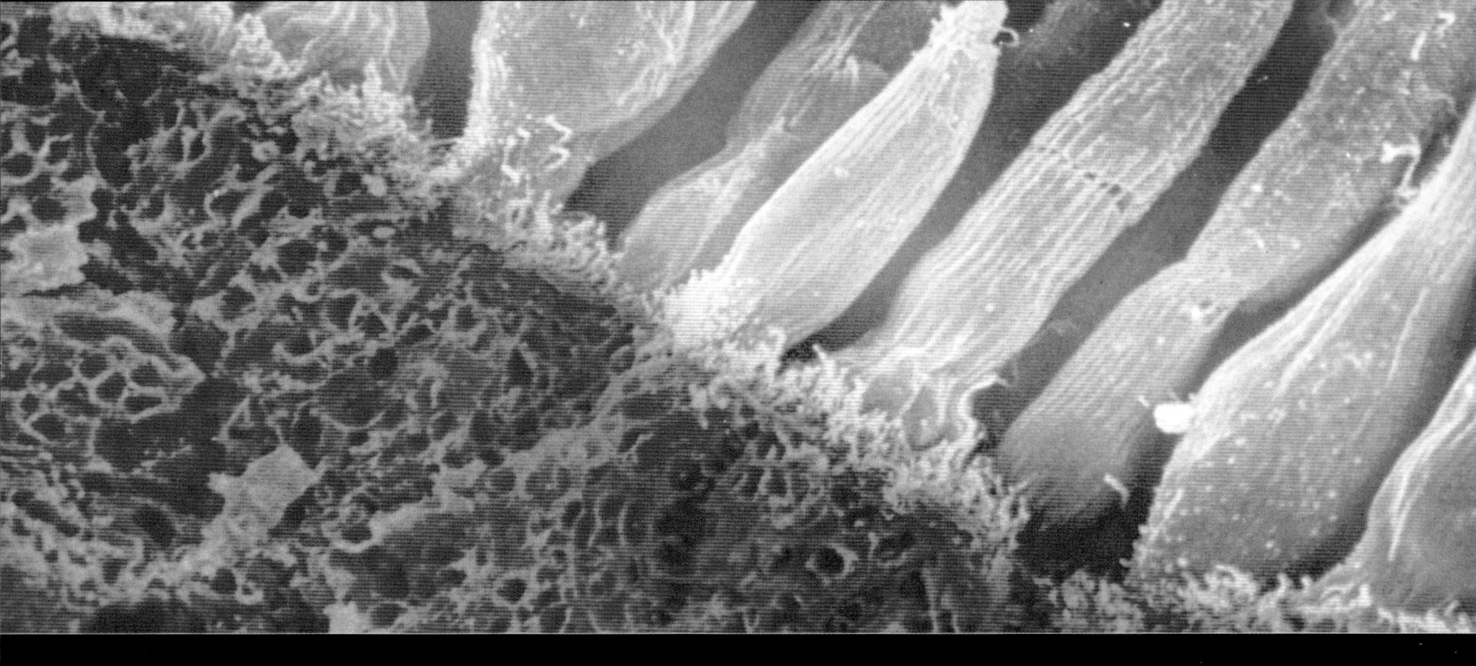

A FIREFLY BOOK

Published by Firefly Books Ltd. 2004

First printing

Publisher Cataloging-in-Publication Data (U.S.)

Photographic atlas of the body / pictures supplied by the Science Photo Library ; foreword by Baroness Susan Greenfield. _1st ed.
[288] p. : col. photos. ; cm.
Includes index.
Summary: Close-up photographs of the human anatomy taken using new imaging techniques and text explain the body's functions and inner workings.
ISBN 1-55297-973-3
1. Human physiology -- Pictorial works. 2. Human anatomy -- Pictorial works. 3. Body, Human -- Pictorial works. I. Title.
612 dc21 QP38.P56 2004

Library and Archives Canada Cataloguing in Publication

Photographic atlas of the body / The Science Photo Library ; foreword by Susan Greenfield.
Includes index.
ISBN 1-55297-973-3
1. Human anatomy--Atlases. I. Science Photo Library
QM25.P46 2004 611'.0022'2 C2004-903053-1

Published in the United States in 2004 by
Firefly Books (U.S.) Inc.
P.O. Box 1338, Ellicott Station
Buffalo, New York 14205

Published in Canada in 2004 by
Firefly Books Ltd.
66 Leek Crescent
Richmond Hill, Ontario L4B 1H1

First publishing in Great Britain in 2004 by
Cassell Illustrated,
A division of Octopus Publishing Group Limited
2–4 Heron Quays, London E14 4JP

Text by Windsor Chorlton

Editors: Victoria Alers-Hankey and Joanna Chisholm
Indexer: Sue Bosanko
Consultants: Anna Pilzer; Melissa Sayer, B.A. Hons. (Oxon), MBBS, MRCP, MRCGP; Gabriel Sayer, MBBS, M.S., FRCS; Julia Hillier, B.A. Hons. (Oxon), BM, MRCP, FRCR
Designer: Austin Taylor
Jacket designer: Jacqueline Hope Raynor

Thanks to Arran Frood for his contribution to the concept and structure of the book as well as the picture research. Thanks also to Fenella Vieceli at the Science Photo Library.

Jacket images:
Front: Artwork showing major blood vessels
Back, center: X-ray of a human skull
Back left, from top: Epithelial cells from the cheek lining; blood clot; egg cells; sperm cells; sperm production in a testis

Printed in China

contents

This book truly spans the science/art divide. Indeed, it goes one better: it shows how one can merge into the other. Science can actually be art, and in turn, in the exquisite and ultimate mechanisms and function of biology, there is an intrinsic beauty. The old images of scientists as dysfunctional nerds, and science being all about Bunsen burners, white coats, nasty smells and, above all, tedious facts, are being replaced by the excitement of gaining insights and understanding into how science is impacting on all aspects of daily life. But even if you have a complete distaste for the ways in which science can help with your daily life, at the aesthetic level alone this book should have a huge appeal.

These pages reveal your body, stripped down to its most basic components: the cells. And here you see them in huge diversity, dying, dividing or simply living, reflecting their vastly different functions. The images are startling, completely unprecedented and, above all, hugely absorbing. If you then press further and read what tissue each cell actually contributes, your insights will deepen even more. To view the banal parts of a human body – be it tooth enamel, or fat – as you have never seen them before is quite astounding. Moreover, you will come across cells and even areas of the body that you might rather not think about, such as the lining of the rectum, and find they actually look pleasingly a little like strawberries.

And then there are cells that are in the news, such as stem cells, as well as that most bizarre and secretive of organs: the human brain. Looking through the sections on the nervous system and the brain, I hope that you will be able to capture some of the excitement that I had when I was first a student, wondering how such quintessentially physical matter could generate subjective experiences such as personality.

And then there are cells that are not familiar, but are in the news, – such as stem cells, - and then that most bizarre and secretive of organs: the human brain. Just by looking through the sections on the nervous system and the brain, I hope that you will be able to capture some of the excitement that I had when I was first a student, wondering how such quintessentially physical matter could generate subjective experiences: personality.

Of course, when I was a student, we did not have access to the wonderful scanning techniques listed in the next few pages, but which now enable us to have an unprecedented intimacy and insight into what makes you the person you are. By looking at human fat or a hairshaft, anyone of any age will have their respect for and wonderment of the human body enhanced. If, in addition, younger readers are inspired to realize that science is as exciting as it is indeed beautiful, and that the old ideas are as erroneous as they are unattractive, then these pages will be even more valuable.

For anyone of any age, whether or not they are locked into the core curriculum, a few moments, or indeed hours, would be well spent looking through these pages and reflecting on how far we have come in terms of science, and how far we have to go still in understanding the most amazing machine, which is the human body. Someone once said that science is all about 'seeing what everyone else can see, but thinking what no-one else has thought'. This book will enable you to do just that.

foreword by
Baroness Susan Greenfield CBE

Imaging techniques

Light Micrograph [LM]

The microscope was invented in the seventeenth century by Antony van Leeuwenhoek, a self-taught Dutch scientist. With this new tool, English scientist Robert Hooke discovered that tissues were made of tiny compartments. He called them cells, because they looked like the rows of monks' rooms in a monastery.

Light microscopes usually have two lenses – the eyepiece lens and the objective lens – which are combined to produce a much greater magnification than is possible with a single lens. An image produced from a light microscope is called a Light Micrograph (LM). Light microscopes cannot show up structures smaller than the wavelength of light, so their powers of magnification are limited to x1250 in ordinary light. This is good enough to see cells and some of their larger internal structures, such as nuclei.

The latter are more clearly seen if they are stained with dyes – a discovery made in 1872 by a young Italian medical graduate, Camillo Golgi, who accidentally knocked a piece of brain tissue into a solution of silver nitrate and found that the chemical had made the nerve cells stand out in stark contrast to the rest of the tissue.

Electron Microscope [EM]

By the 1930s, the theoretical limitations of the light microscope had been reached, and the electron microscope then began to have a huge impact on biology, enabling researchers to investigate tiny cell features that are only faint blurs when viewed with a light microscope. Instead of using lenses to focus a beam of light, the EM uses electromagnets to focus a beam of electrons in a vacuum.

Because the wavelength of electrons is much smaller than that of light, they have much higher magnifying and resolving power. A modern EM can magnify up to one million times, while its resolving power is about 250,000 times finer than that of the human eye.

The drawback of EMs is that, because living tissues cannot survive in a vacuum, they cannot show the ever-changing movements of a living cell.

Transmission Electron Microscope [TEM]

The TEM was the first type of electron microscope to be developed. A TEM works rather like a slide projector: it transmits a beam of electrons through a thin specimen stained with substances that deflect some of the electrons. The remaining electrons pass through the specimen, and the pattern they produce is projected on to a viewing screen, forming an enlarged image. A TEM is particularly useful for examining a two-dimensional section of a tissue or cell.

Scanning Electron Microscope [SEM]

SEMs first became widely used in the 1960s. They are used to study three-dimensional objects. Electrons bounce off the surface of an object rather than passing through it. By moving the specimen about, the operator can build up a detailed picture of its surface structure – a bit like taking aerial views of the countryside.

X-ray

In 1895 the German physicist Wilhelm Röntgen discovered that a stream of electromagnetic radiation shorter in wavelength than light could penetrate living tissue and form an image on a photographic plate on the other side. Such an X-ray photograph resembles a negative of an ordinary photograph, with dense tissues such as bones showing up as white shapes.

Ordinary X-rays show different density of tissue. For

example, the border of the heart shows clearly, in sharp contrast to their surroundings. This is because the lung predominantly contains air, whereas the heart is predominantly soft tissue.

Contrast medium is used to outline structures that wouldn't otherwise be 'seen' by X-rays. Barium sulphate, which is opaque to X-rays, is used for the X-ray examination of the digestive tract. It is administered by mouth, and hence is known as barium meal. Different types of contrast media are used to examine the liver, kidneys and some other organs.

Angiogram

An angiogram is a special type of X-ray that produces images of blood vessels, including those of the head, heart and lungs. A dye s injected into the part of the circulatory system to be investigated. In the case of a coronary angiogram, the dye is introduced through a tiny tube (catheter) inserted into a main artery at the groin and then threaded through blood vessels to the heart. X-rays produce real-time images of the dye as it travels through the blood vessels, outlining them and showing any problems such as blockages or constrictions

Computed Tomography [CT]

Invented in 1972, computed tomography machines represent a major improvement over conventional X-ray radiography. X-rays are used, but images with far more detail can be produced. Technology in this field is rapidly advancing. Nowadays, multiple images may be taken using an X-ray tube that rotates rapidly around the body. Digital processing can manipulate this data to show an image of the body in any plane or even 3-D.

Positron Emission Tomography [PET]

Whereas a CT scan shows an organ's shape and structure, a PET scan provides images that reveal how the organ is functioning. PET imaging involves tracing the action of radioactive substances inserted into the body. The radioactive material emits positrons, which collide with electrons, releasing energy in the form of gamma rays. The gamma-ray flashes are plotted, and the information is processed by a computer to provide images of blood flow and metabolic processes within the tissue.

Although the equipment is complex and costly, PET scans are particularly useful for brain research, where they help diagnose brain damage and map brain functions.

Magnetic Resonance Imaging [MRI]

One of the greatest advances in diagnostic medicine since the invention of X-ray, MRI involves placing the human body in a strong magnetic field that excites atoms within the body to emit signals. A computer then translates these signals into images depicting the body parts being examined. Widely regarded as the most versatile and sensitive imaging technique available, MRI has been generally available since the 1970s and has proved to be an invaluable tool for the detection of some cancers that are too small to show up on X-rays.

Electron Beam Tomography [EBT]

Because EBT scanners work much faster than CT scanners, they produce clearer images of moving tissues and organs. An EBT scanner fires an electron beam against tungsten targets arranged in an arc under the patient. These produce X-rays, which pass through the patient, and the resulting data is processed by a computer to produce cross-sectional images that can be combined into a three-dimensional image.

Thermogram

Cameras sensitive to infrared heat radiation produce images called thermograms, which are 'maps' showing the different temperature zones of all or part of the body surface. Abnormal heat patterns can help specialists to identify subsurface disorders. A 'hot spot', for example, may indicate the presence of a tumour, where cells are more active than usual, while an abnormally cool area may point to a blocked blood vessel.

Ultrasound

Ultrasound uses high-frequency sound waves to 'see' within the body. As it does not invlove ionizing radiation (such as X-rays) it is of particular value in examining pregnant women, being safe for the developing fetus. It is also used for diagnosing problems in soft tissues such as the heart, breast, liver and gall bladder.

Ultrasound is also used to destroy diseased tissue, drill teeth and shatter kidney stones. The principle involved is resonance, which occurs when an object is put into motion by sound waves having the same frequency as the object. Resonance makes the target object vibrate with greater and greater amplitude until it shatters.

Endoscope

An endoscope is a narrow tube with a light and camera attached used to take images inside a body cavity or organ. The light is usually carried into the body by an optic fibre. The tube is passed into the body through a natural orifice, such as the mouth or ear, or through an incision in the skin. Endoscopes are given different names according to the parts of the body they are designed to probe. For example, a bronchoscope is an endoscope used to examine the lungs.

Gamma Camera Scan

Gamma camera scans involve injecting a small quantity of radioactive material into the body. The radioactive material accumulates in the organ or tissue under examination and the radiation emitted is recorded by a gamma camera that builds up a composite image of the target organ. Gamma scans are not as detailed as conventional X-ray images but are more sensitive in detecting abnormalities such as tumours or infections.

Resin cast

The network of blood vessels in an organ can be mapped by first injecting a low viscosity resin into the organ and then dissolving the surrounding tissues with chemicals. Obviously, this imaging technique is not used on living human tissue.

Fundus Camera Image

A fundus camera is a low-power microscope attached to a special camera designed to take images of the inside of the eyes, or fundus.

Macrophoto

A macrophoto is a close-up image usually taken by a conventional camera fitted with a macro lens.

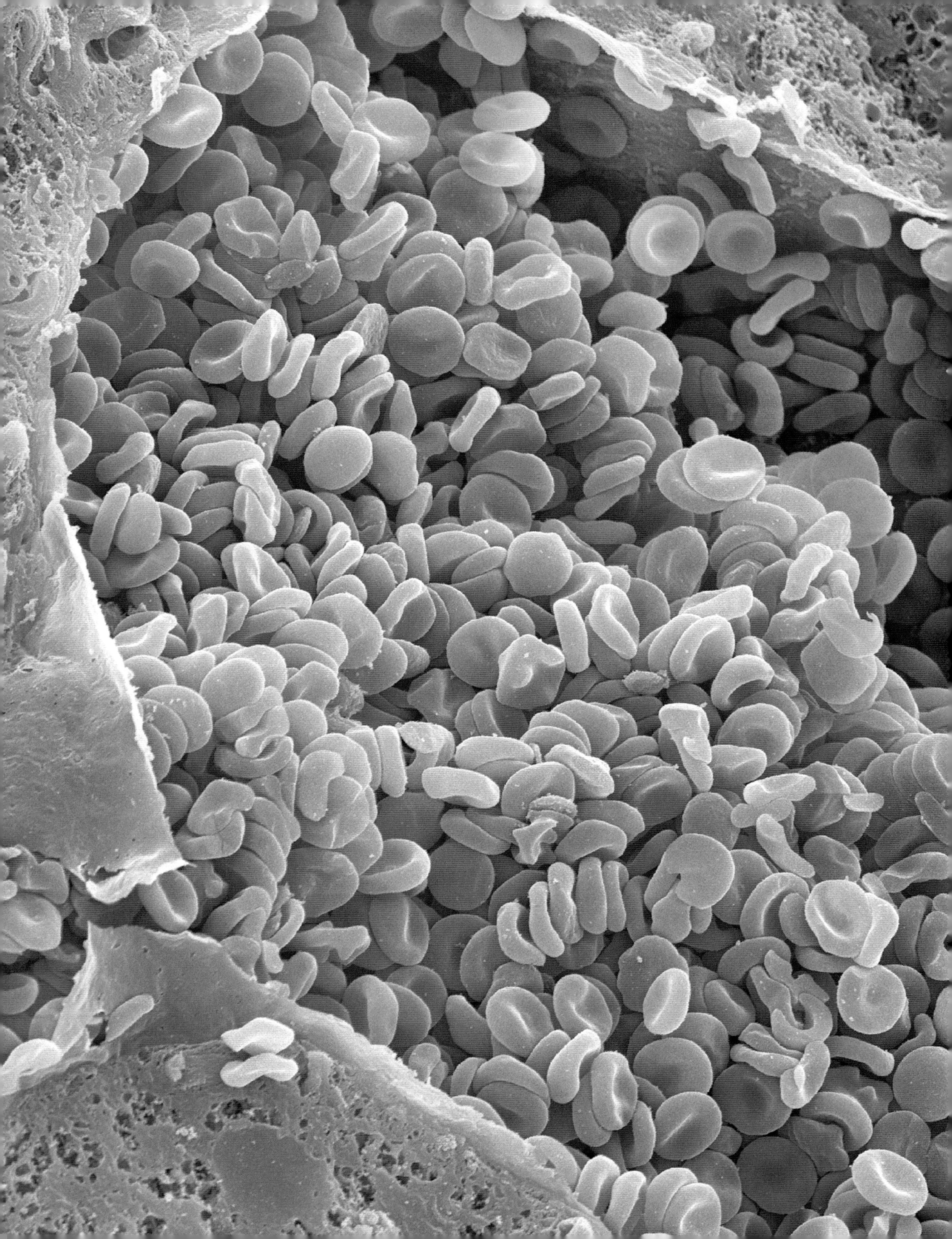

CELLS

MILLIONS OF RED BLOOD CELLS pack a vein in the liver. Produced in the bone marrow at the rate of about 200 billion a day, red blood cells carry oxygen and carbon dioxide to and from body tissues. Unlike most cells they do not possess nuclei and cannot reproduce. The life span of a red blood cell is about four months.

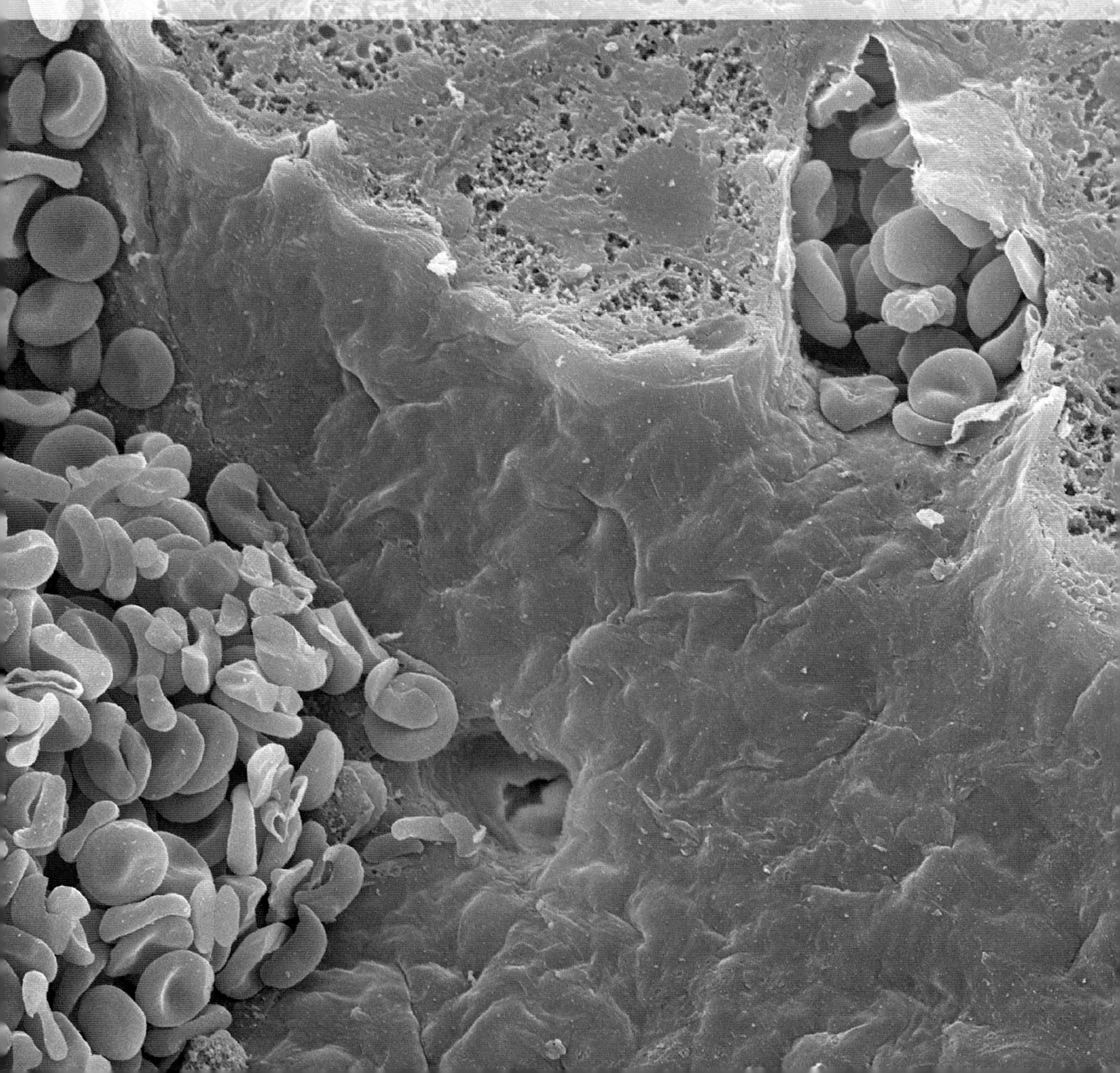

As the basic units of life, cells are the building blocks of all living things. While the simplest organisms get by with just one cell, the human body contains billions of cells assembled into tissues and organs. Most human cells are microscopic. The largest, the ovum or egg, is narrower than a human hair, while one of the smallest, the sperm cell, is less than three millionths of 1m/3½ft across.

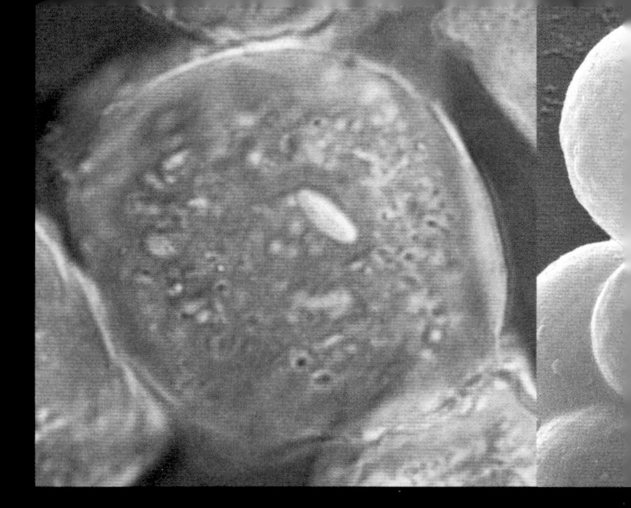

At the cellular level, the human body has been compared to a vast city. But outside the pages of science fiction, no city is as large, complex or efficiently organized as the human body. The thousand or so different types of cell we possess work in superb harmony to regulate all the processes essential for life – food processing, production and storage; repair, transport and waste disposal; surveillance and defence; communication and administration.

Cells come in weird and wonderful shapes. In many cases their form is related to function. For example, the red blood cells that carry oxygen and carbon dioxide to and from our tissues are tiny, elastic-walled discs designed to slip through the narrowest capillaries. Some of the white blood cells that form part of our defences against disease have 'feet' that enable them to move to sites of infection. Nerve cells have long extensions that link with other nerve cells to carry tiny electrical

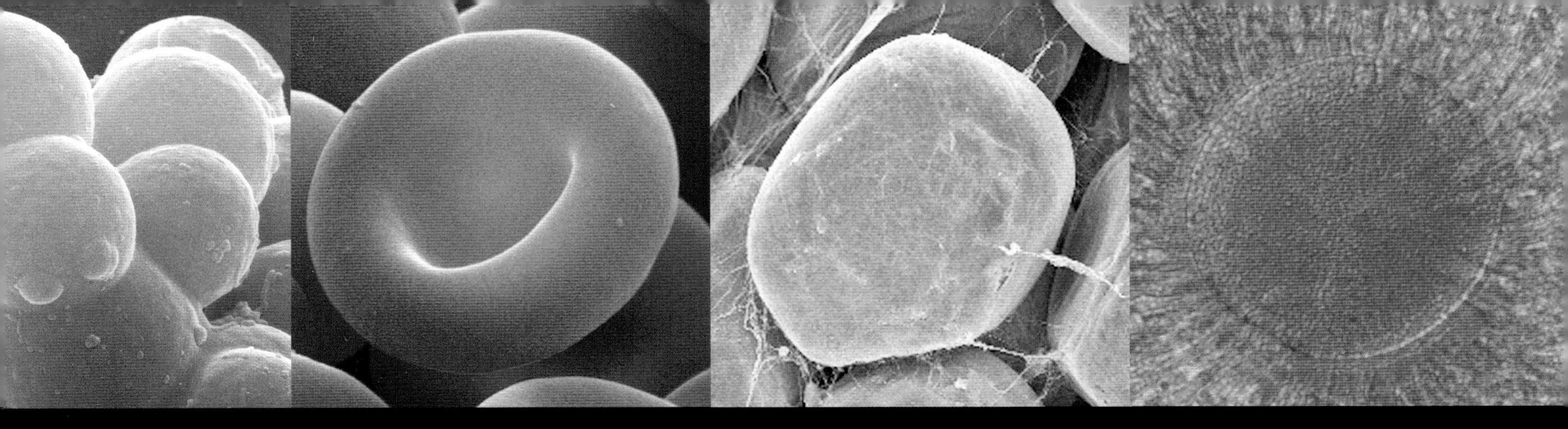

Despite their incredible diversity, all the cells in an individual's body have come from the division of a single fertilized egg and contain the same genetic material. This means that in theory every cell has the capacity to become any other type of cell. In fact, as cells differentiate, most of the genes they contain are 'switched off', and those that remain active ensure that, when cells divide, they replicate themselves exactly.

The exception to this rule is a class of cells called stem cells. Stem cells from the bone marrow give rise to all the different types of red and white blood cells, and some stem cells from the older embryo have the potential to become all other tissue types – or even a new individual. Experiments to produce babies from stem cells are banned in most countries, but there is intense research into the possibilities of using stem cells as a source of new, healthy tissue to repair or replace damaged organs.

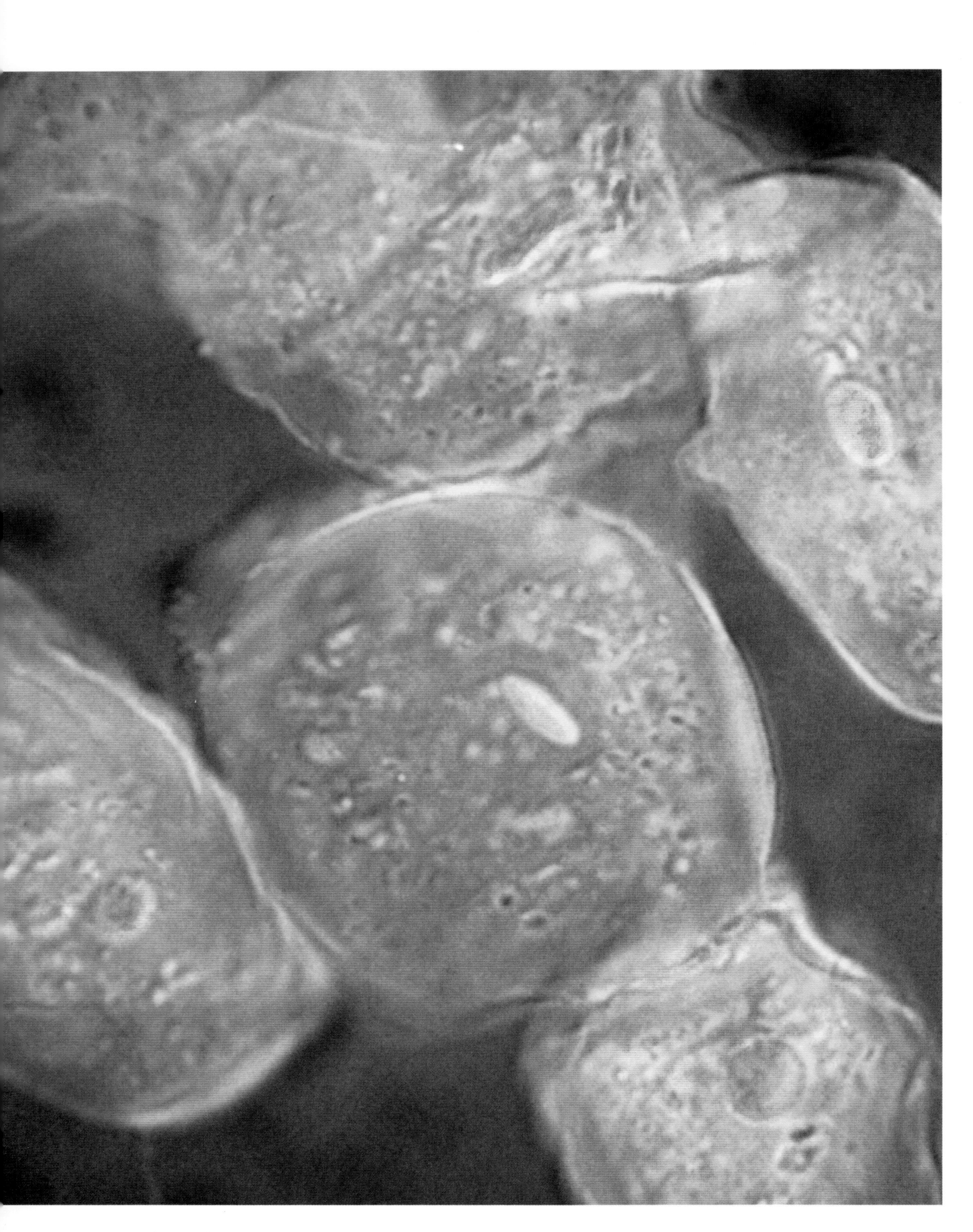

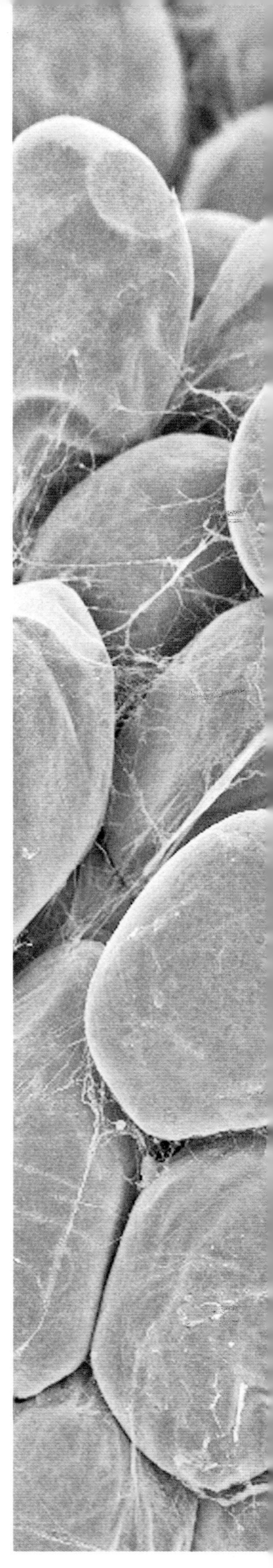

Epithelial cells [LM]

SCALE-LIKE EPITHELIAL CELLS from the human cheek lining are shown here with the cell contents stained green and the nuclei in yellow. These cells form part of the epithelium, the lining and covering that protects the body's internal and external surfaces. Epithelium in the cheek is simple in arrangement, forming a layer only one cell thick that allows the rapid exchange of nutrients and other substances.

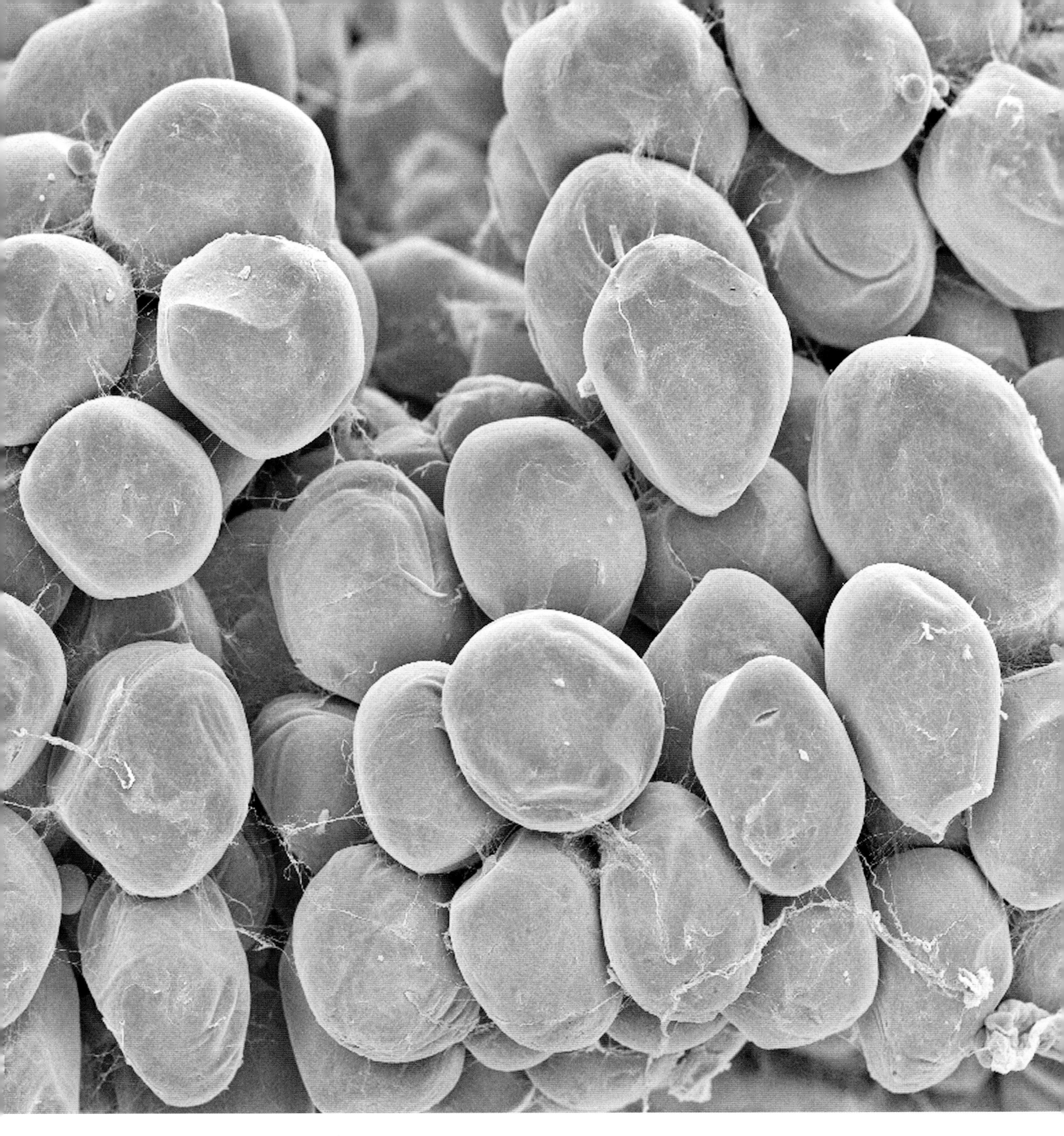

Fat cells and tissue [SEM]

FAT CONSISTS OF ROUND, FAT-STORING CELLS supported by strands of connective tissue. Almost the entire volume of each fat cell consists of a single fat or oil droplet. Fatty tissue provides a source of energy and insulates the body against heat and cold. About half is under the skin, and most of the rest is in the abdominal cavity or around internal organs.

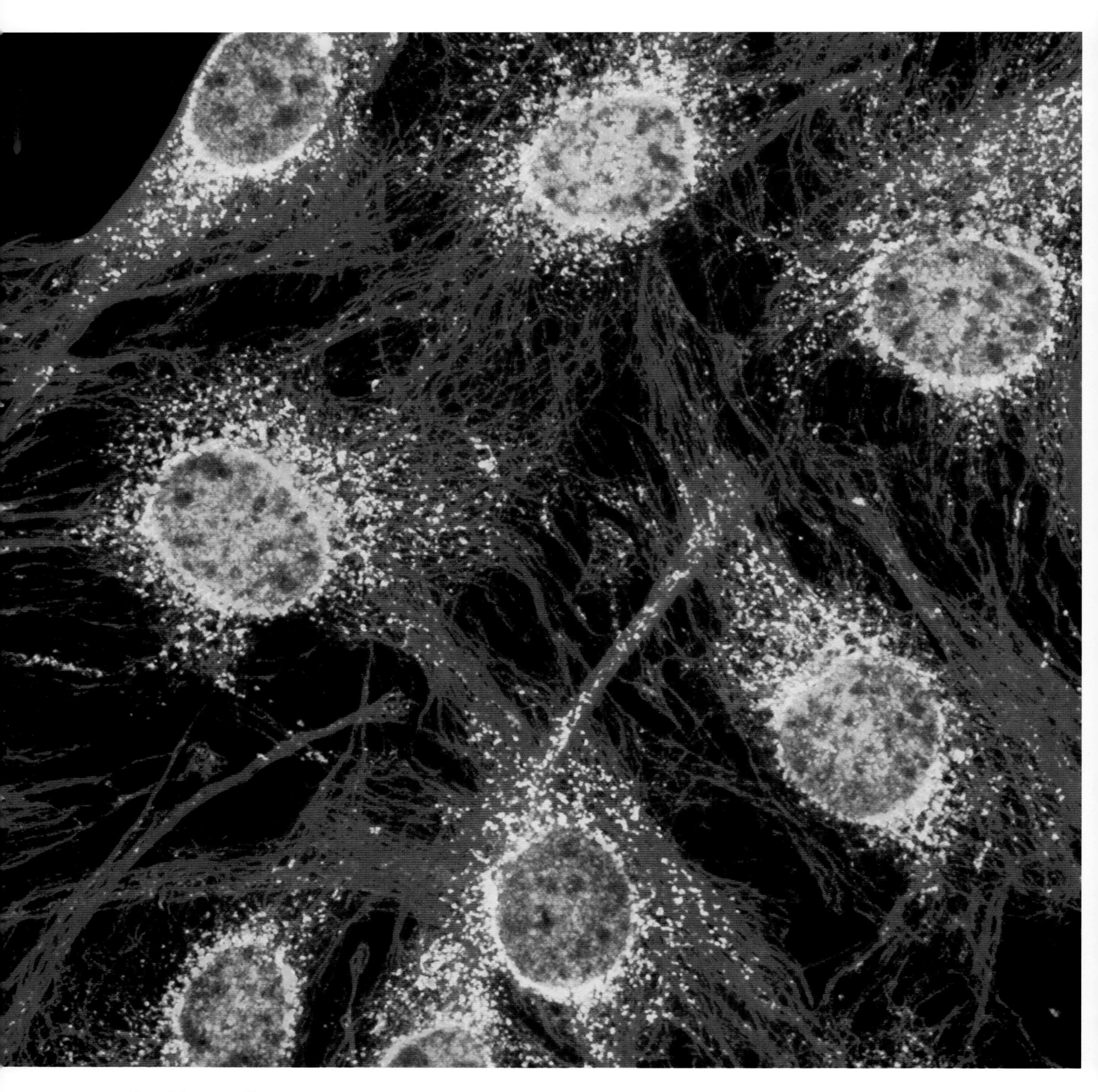

Fibroblast cells [LM]

THE FIBRES THAT STRENGTHEN connective tissues such as skin, bone and cartilage are manufactured by fibroblast cells, here shown with blue nuclei. The orange threads make up the cytoskeleton, the scaffolding that supports each cell. Fibroblasts produce collagen, a protein that forms long fibres with the tensile strength of steel.

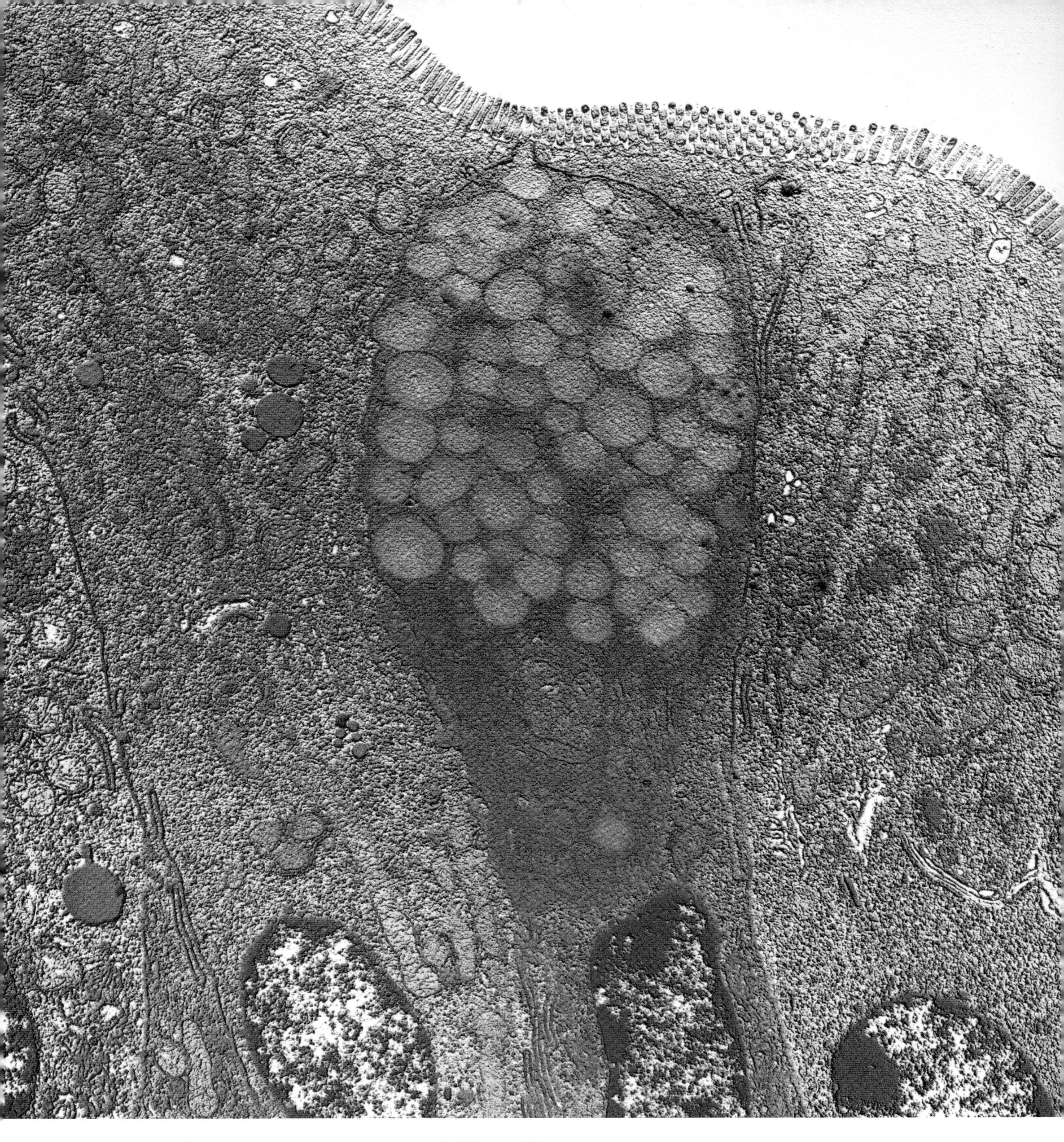

Goblet cells [TEM]

IN A SECTION THROUGH part of the small intestine, a goblet cell (pink and blue) is flanked by secretory cells. Goblet cells secrete mucus, which protects the stomach lining from acid, enzymes and mechanical abrasion during food digestion. The mucus is formed from mucigen granules (circular objects at upper centre) combined with water. The adjacent secretory cells also help in the digestive process. Their secretory capacity is increased by the green hair-like structures – microvilli – on their outer surfaces.

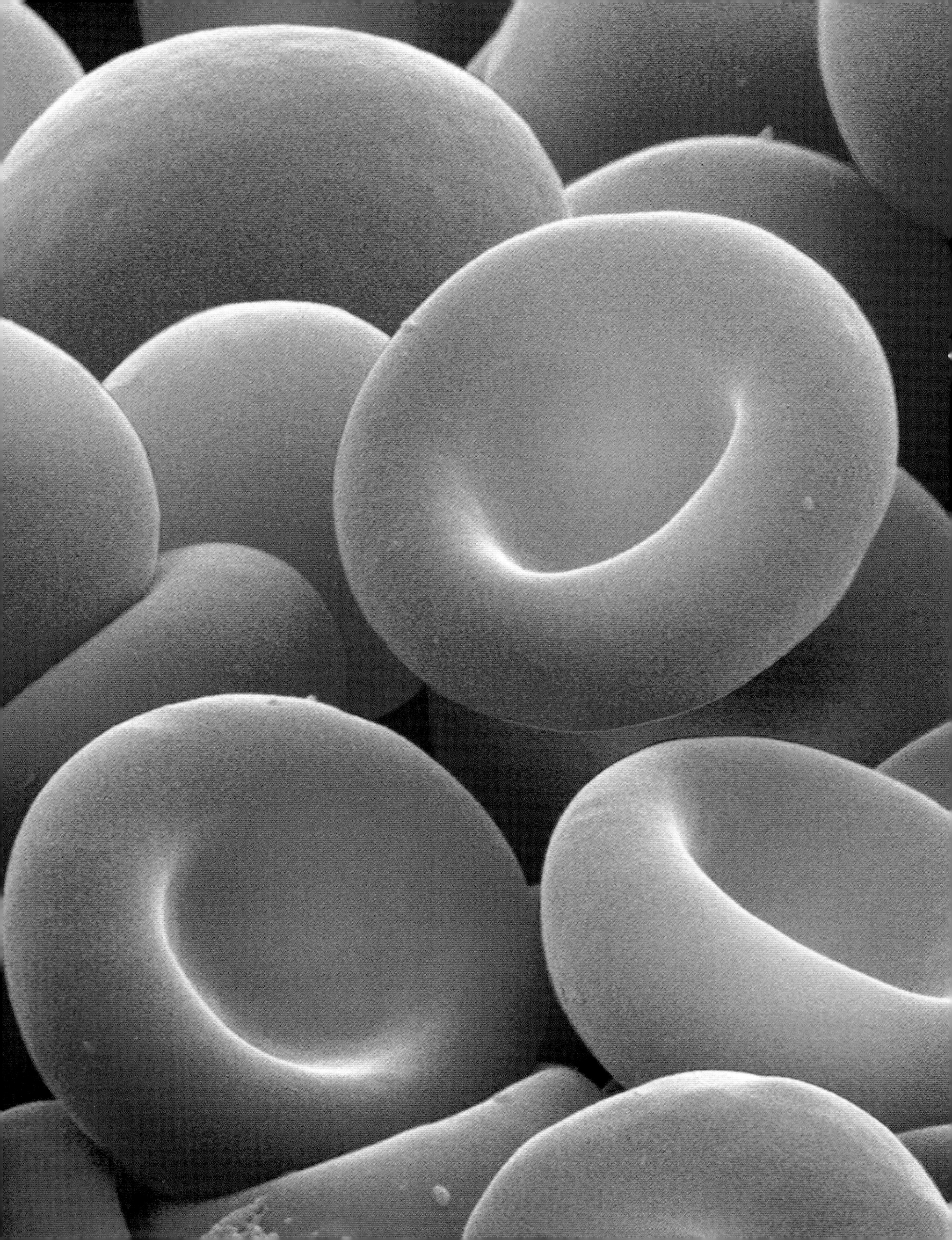

Red blood cells [SEM]

THE MOST ABUNDANT CELLS in vertebrate blood, red blood cells are biconcave, disc-shaped cells that transport oxygen from the lungs to all the body cells. They also remove carbon dioxide produced by cells in respiration and transport it back to the lungs, where it is exhaled. The red colour is due to haemoglobin, a protein compound that binds reversibly with oxygen.

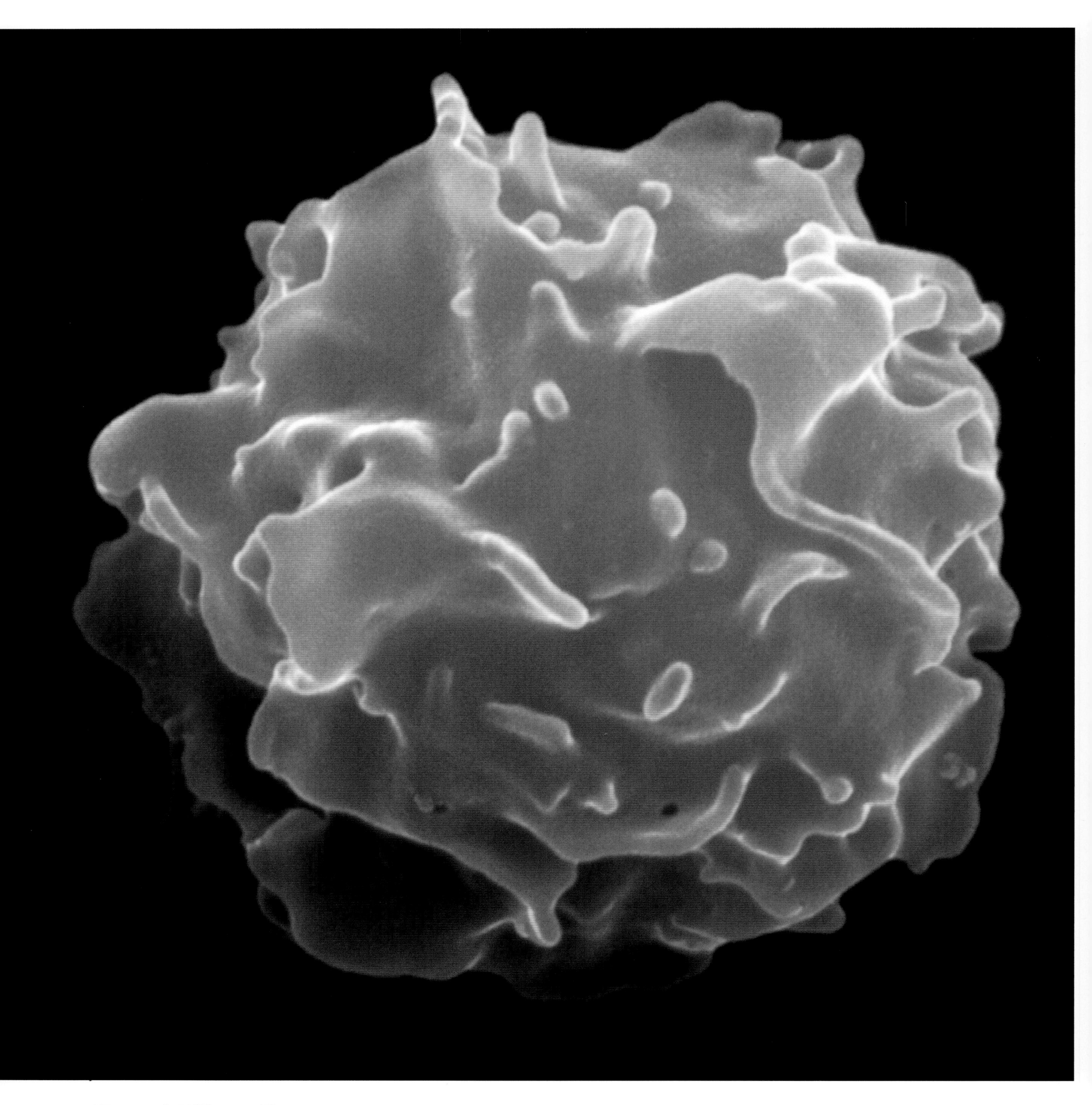

Natural Killer cells [SEM]

A VITAL PART OF OUR IMMUNE SYSTEM, this Natural Killer (NK) cell is a type of white blood cell that attacks virus-infected cells and some tumour cells. When an NK cell contacts the surface of a tumour cell, it recognizes certain proteins called antigens, which activate its cell-killing mechanism. The NK cell binds to the tumour cell and releases toxic chemicals that make it burst.

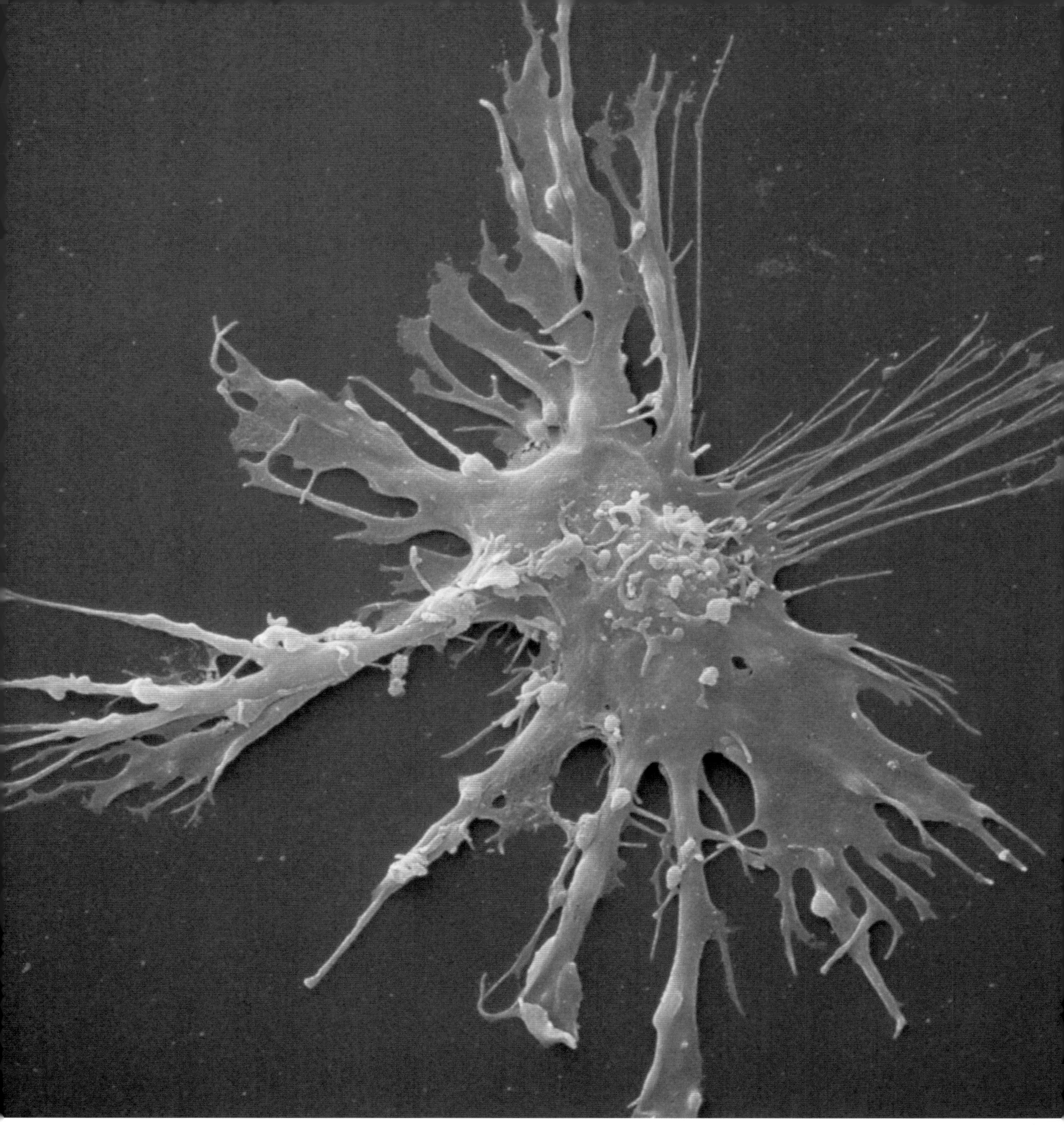

Dendritic immune cells [SEM]

THIS DENDRITIC CELL is found in the upper layer of the skin, where it helps protect the body against infection. The long projections on the cell's surface are 'feet' that help it to move to sites of infection. When the cell encounters a foreign protein (antigen), the cell engulfs it and processes it before releasing it once again into the skin. The processed antigen then acts like an alarm signal, alerting other immune cells of the body to the infection.

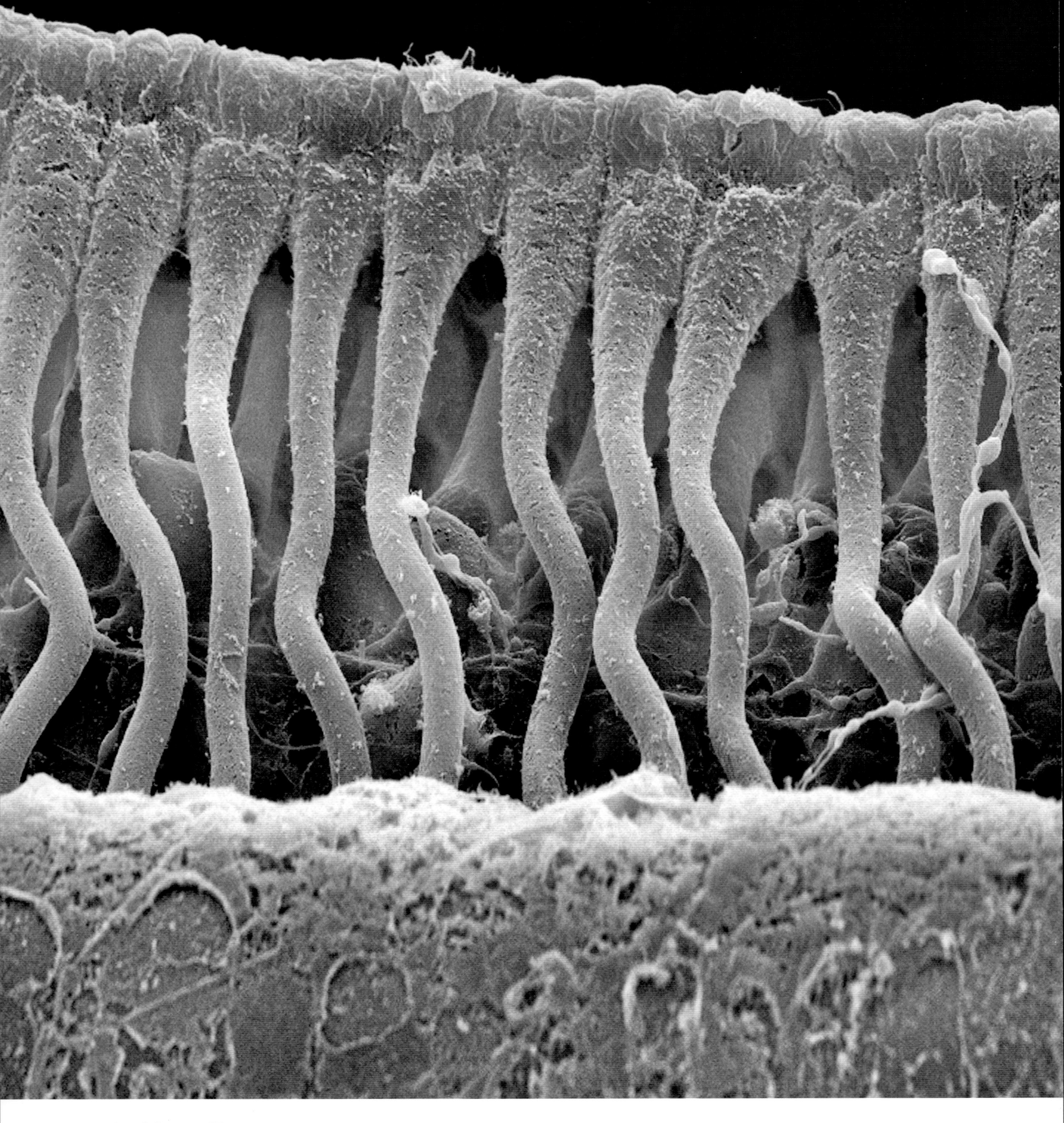

Cochlea cells [SEM]

A SECTION THROUGH PART of the cochlea inside a human ear shows a row of pillar cells on the organ of Corti, the structure that transforms sound waves into auditory signals. The cells arise from a flexible membrane (along the bottom). Sound waves distort the membrane, moving the pillar cells, which in turn flex hair cells (not seen) that trigger auditory nerve impulses.

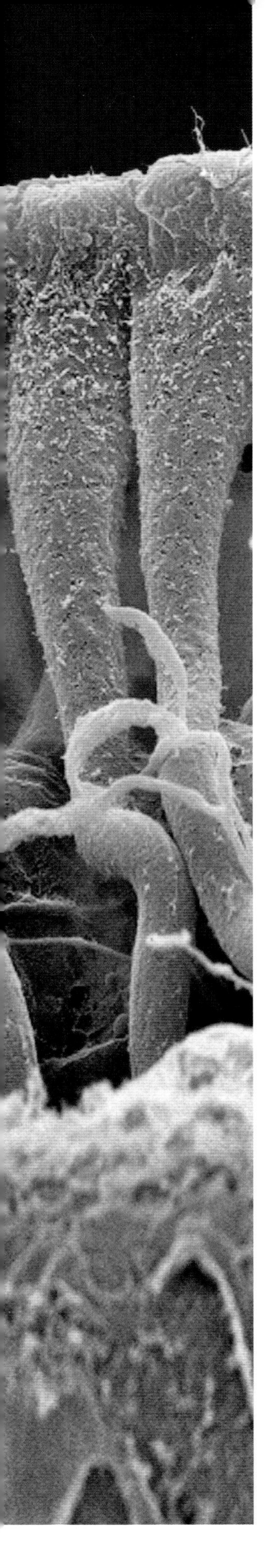

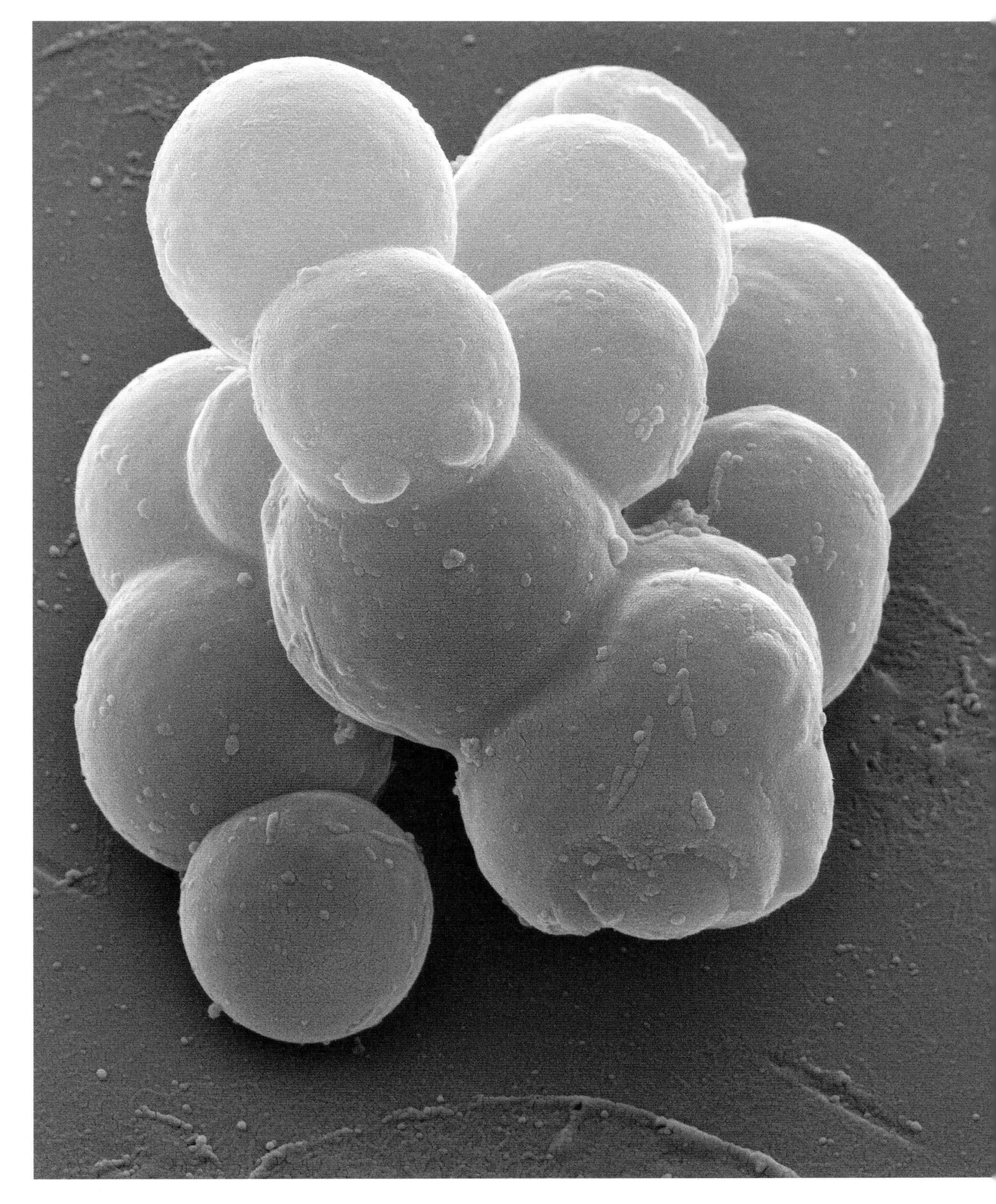

Stem cells [SEM]

STEM CELLS ARE CELLS THAT CAN DEVELOP into different tissue types. These stem cells, taken from umbilical cord blood, give rise to all the body's specialized blood cells. In this process the stem cells develop either into red blood cells or one of several types of white blood cells that make up the immune system. The purification of stem cells from umbilical cord blood allows scientists to research the function of the immune system and to develop treatments for diseases such as AIDS and leukaemia.

Nerve cells [SEM]

THIS NERVE CELL is from the outer layer of the cerebellum, a part of the brain that regulates muscle activity and co-ordination. Its spherical nucleus has several branching processes that carry nerve signals to and from adjacent cells (not seen). The nervous system is built up of millions of such interconnected nerves, or neurons.

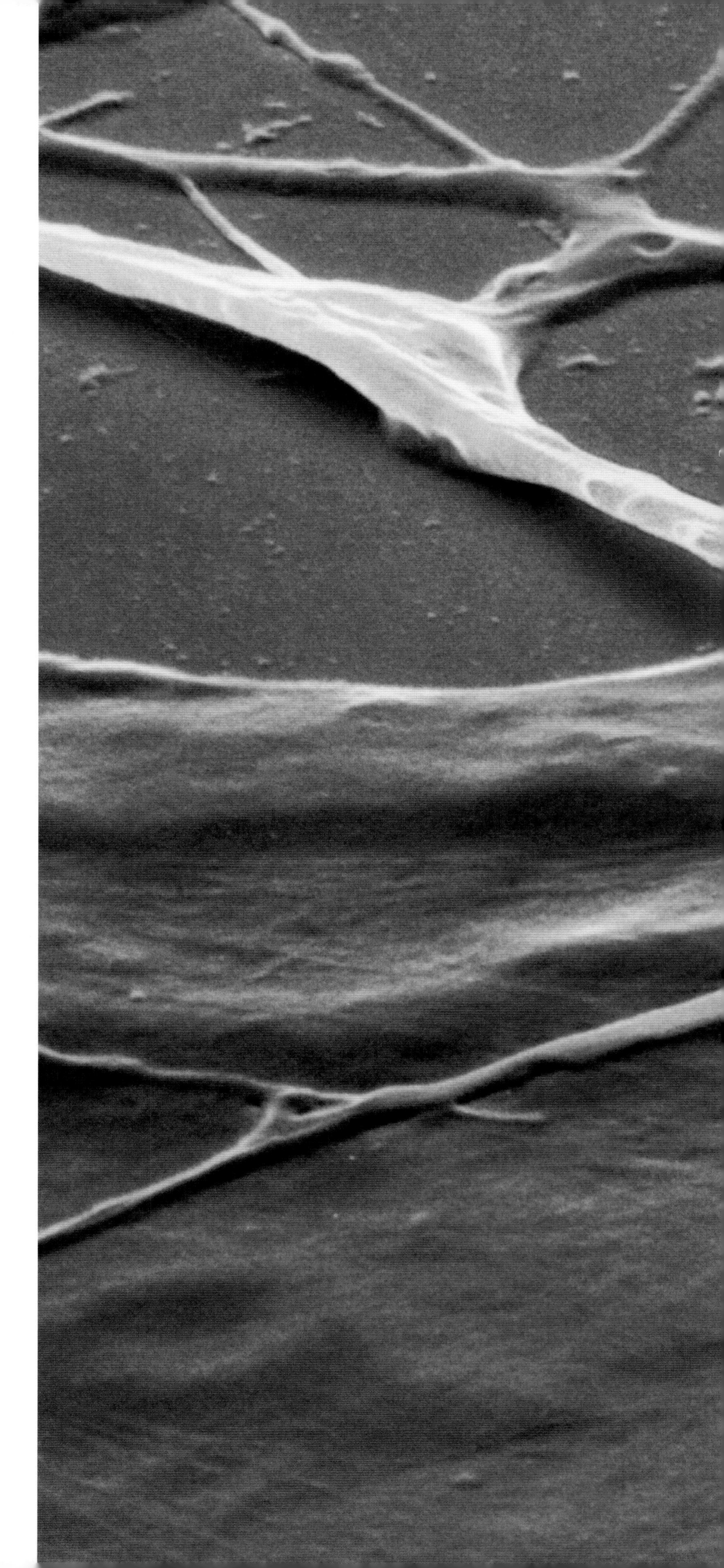

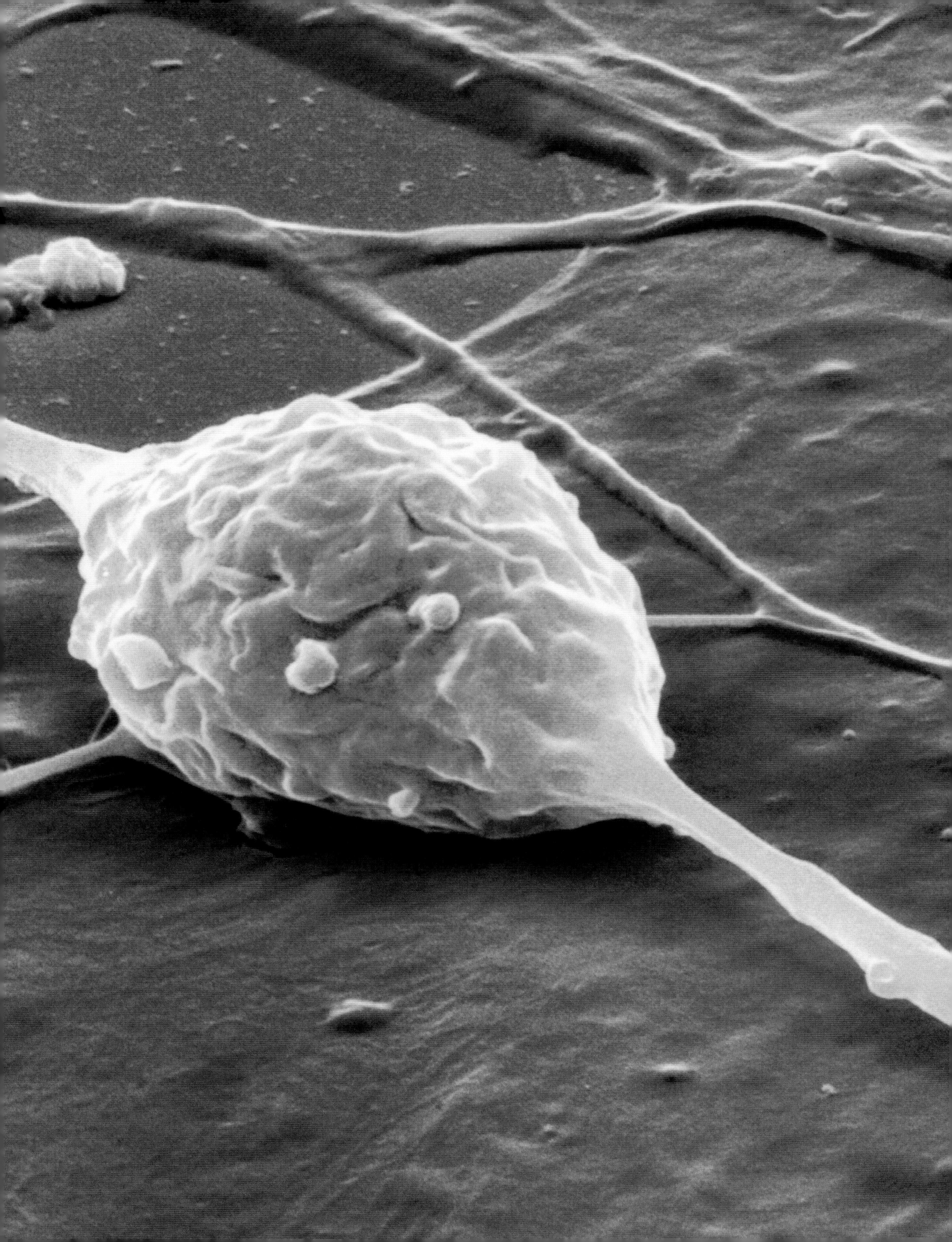

Sperm cells [SEM]

TADPOLE-SHAPED SPERMATAZOA are the male sex cells produced by the testes. They deliver genetic material, contained in their oval heads, to the female sex cell (ovum), using their tails to swim through the female reproductive system. Human males ejaculate, on average, about 100 million sperm, but normally only one sperm will fertilize the ovum. Each sex cell has twenty-three chromosomes that fuse at conception to make a cell containing forty-six chromosomes, half from each partner. This mixing of genetic material creates the diversity of human life.

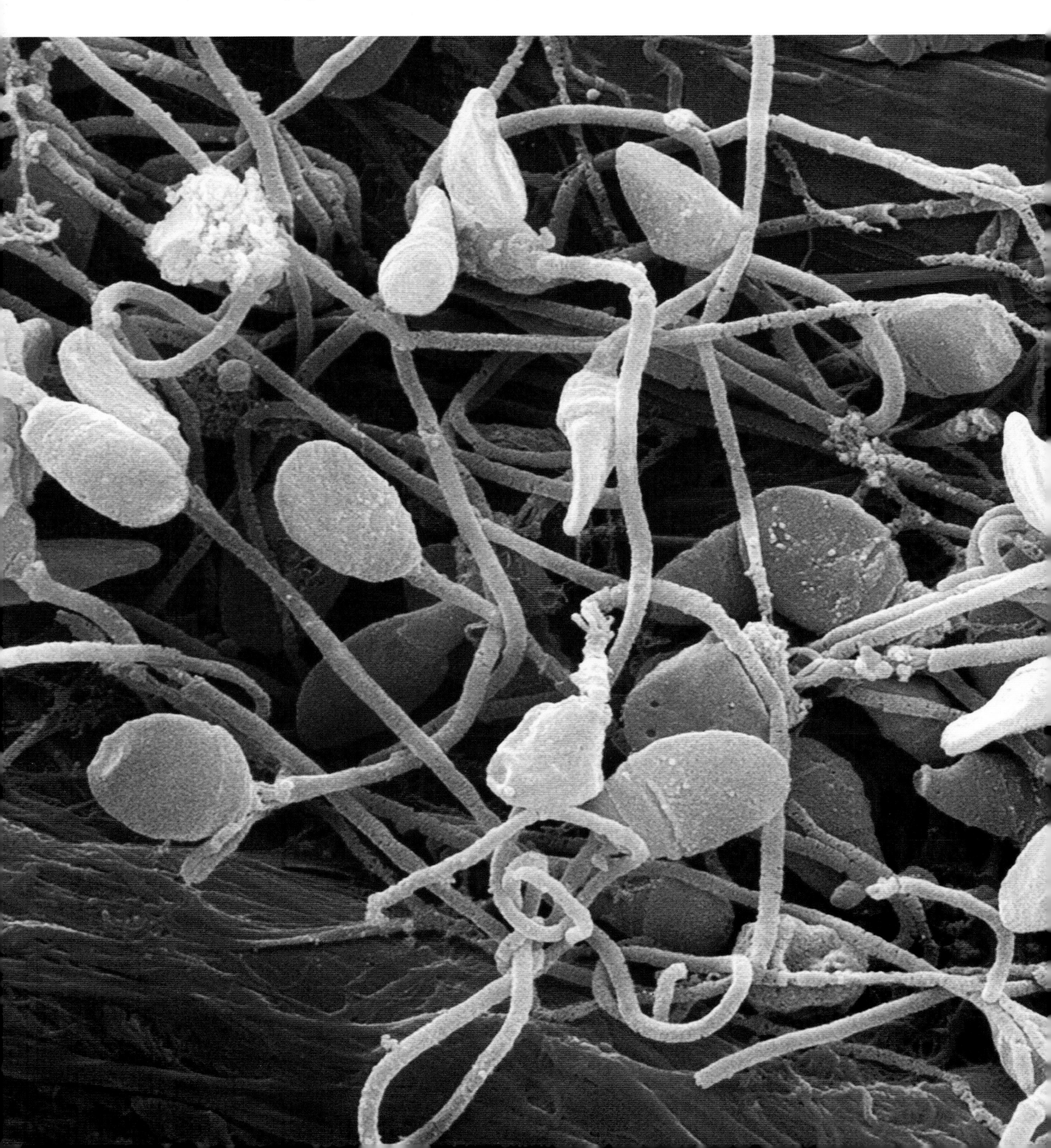

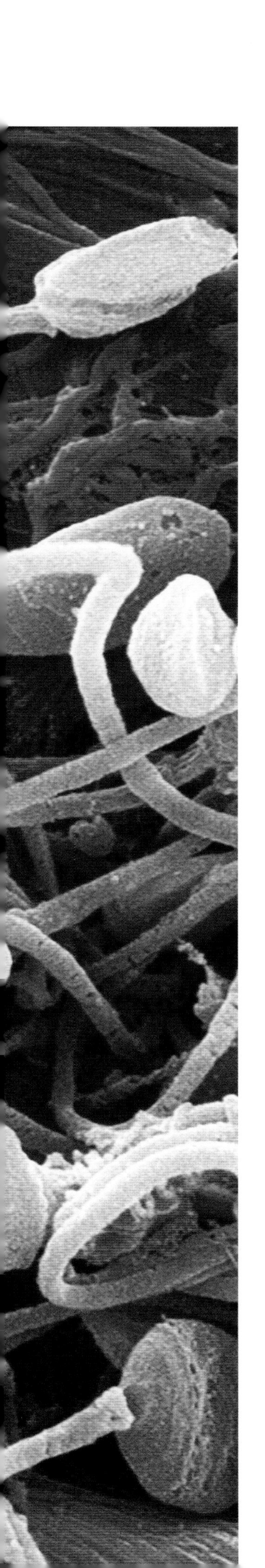

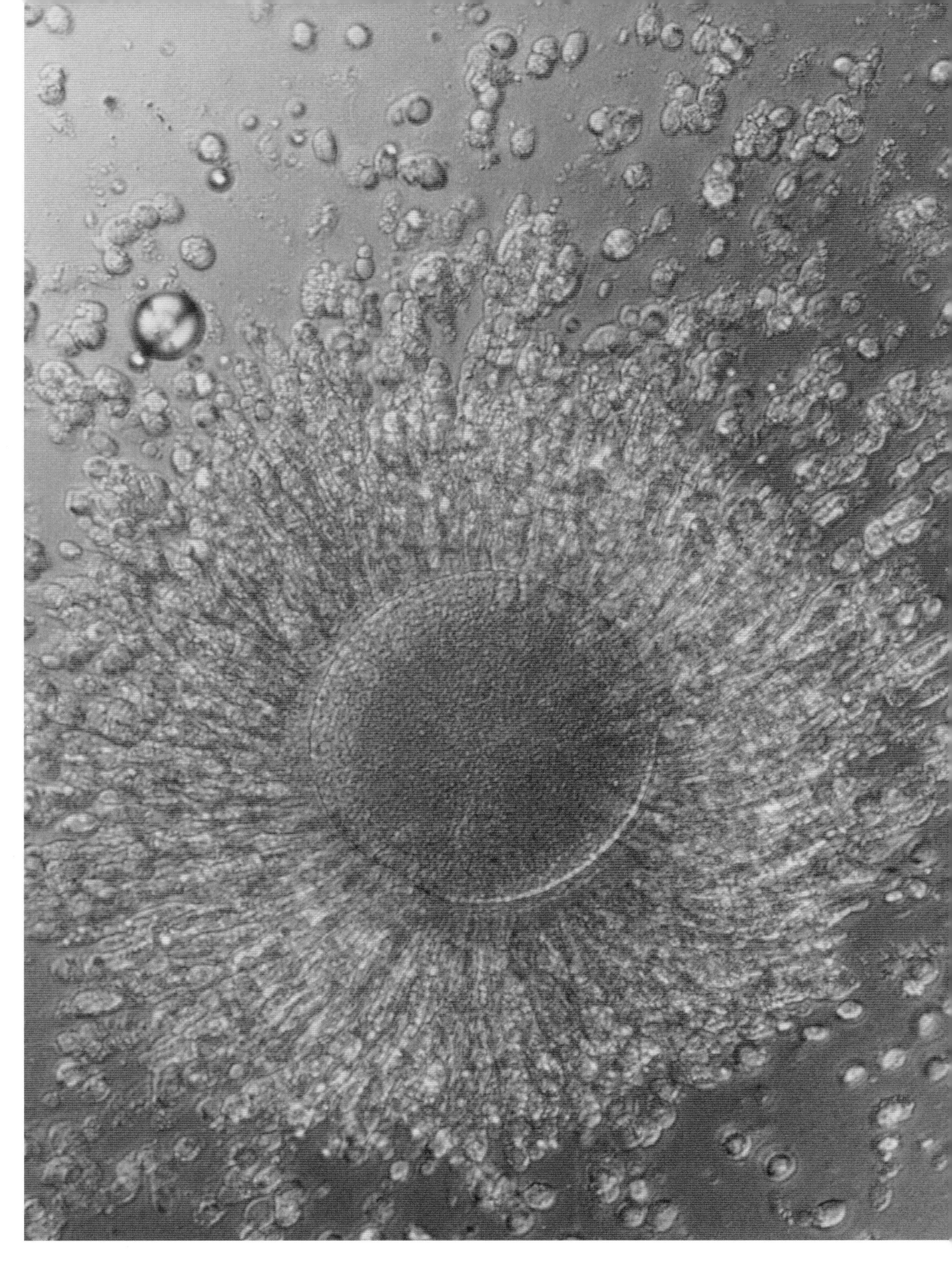

Egg cells [LM]

A MATURE HUMAN EGG to be used in in-vitro fertilization (IVF) is surrounded by layers of much smaller cells that support and nourish the growing egg. At birth, a girl has more than a million egg cells, but most of these degenerate during childhood. Of those that are left, only about 450 mature fully – one per month throughout the female's reproductive life. In IVF treatment, the egg is removed from the ovary and fertilized in culture using sperm (which may be from a donor). The resulting embryo is implanted directly into the uterus (womb).

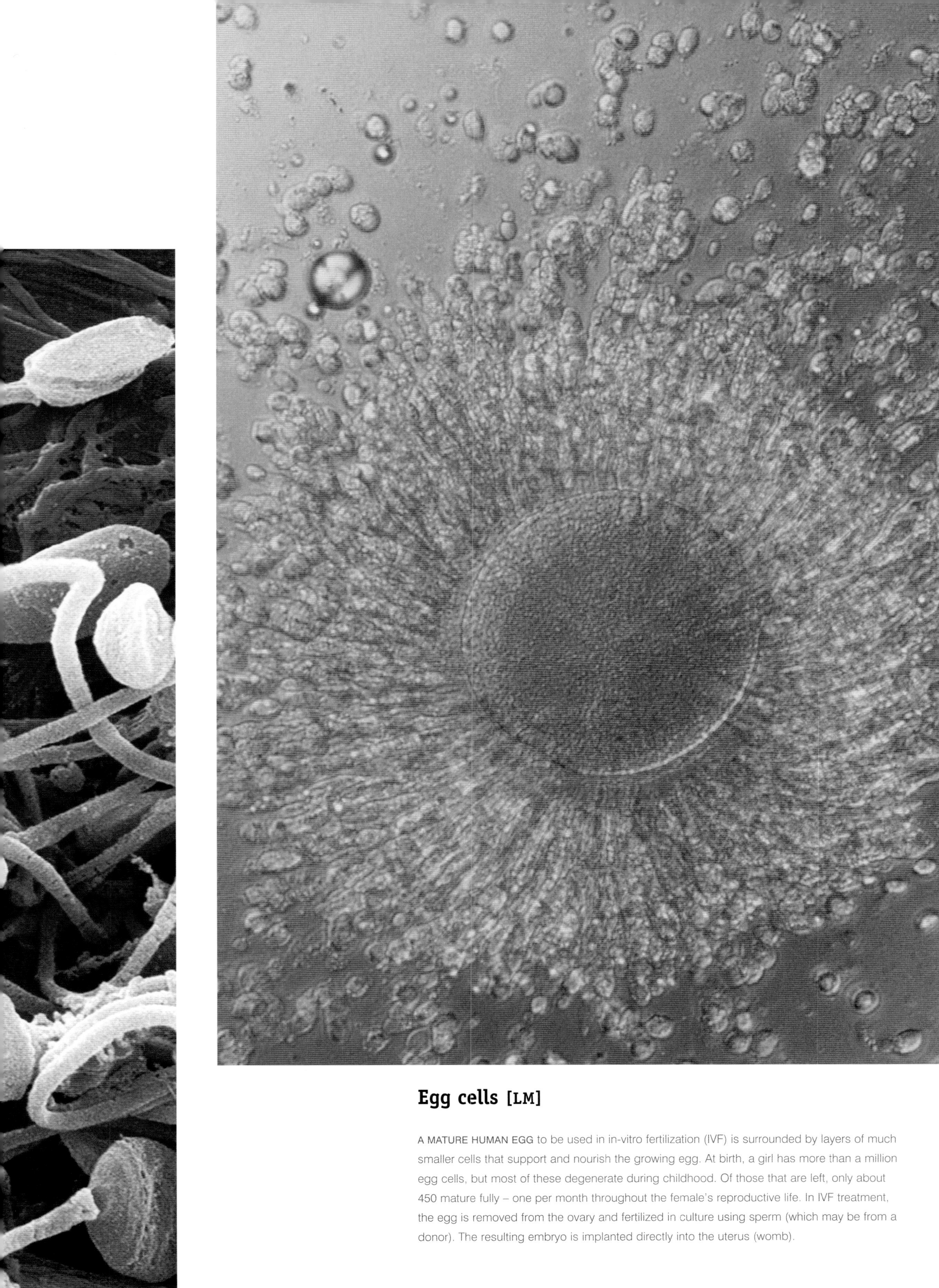

Egg cells [LM]

A MATURE HUMAN EGG to be used in in-vitro fertilization (IVF) is surrounded by layers of much smaller cells that support and nourish the growing egg. At birth, a girl has more than a million egg cells, but most of these degenerate during childhood. Of those that are left, only about 450 mature fully – one per month throughout the female's reproductive life. In IVF treatment, the egg is removed from the ovary and fertilized in culture using sperm (which may be from a donor). The resulting embryo is implanted directly into the uterus (womb).

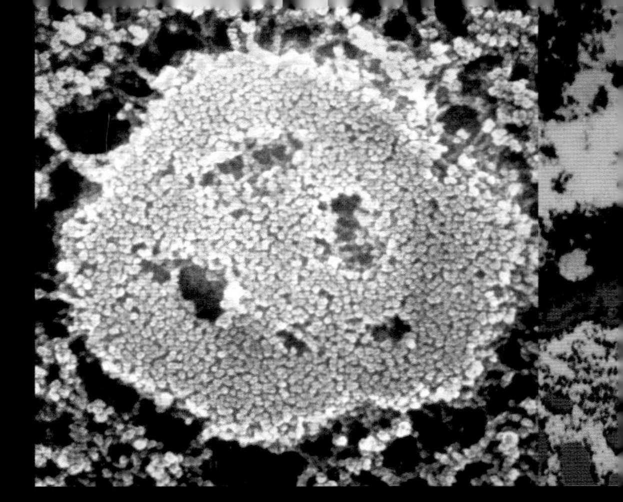

When scientists first examined cells with the electron microscope, they were amazed by their internal complexity. At the same time they discovered that most cells, no matter how specialized their function, have many structural features in common.

To understand the workings of a cell, it helps to think of it as a factory: a cell takes in raw materials, processes them and then either uses the finished products itself or exports them, together with waste. Instead of walls and gates, a cell has an outer membrane with pores that selectively block the movement of substances. In many cells – those involved in digestion, for example – the external membrane is folded to form tiny extensions called microvilli that increase the surface area for absorption and secretion.

The machines inside this cell factory are called organelles, each with its own specialized job. They float in a watery gel called cytoplasm, but are surrounded by their own membranes so that they do not interfere with each other.

The most conspicuous organelle is the nucleus, a compartment containing the instructions that control the cell's activities and enables it to reproduce. The instructions are stored in genes, units of a chemical called deoxyribonucleic acid (DNA), which is packed in structures called chromosomes.

Inside the nucleus there is usually at least one nucleolus, which assembles particles containing ribonucleic acid (RNA), a compound involved in the synthesis of proteins. The RNA leaves the nucleus and attaches itself to tiny organelles called ribosomes, which it codes for the manufacture of specific proteins.

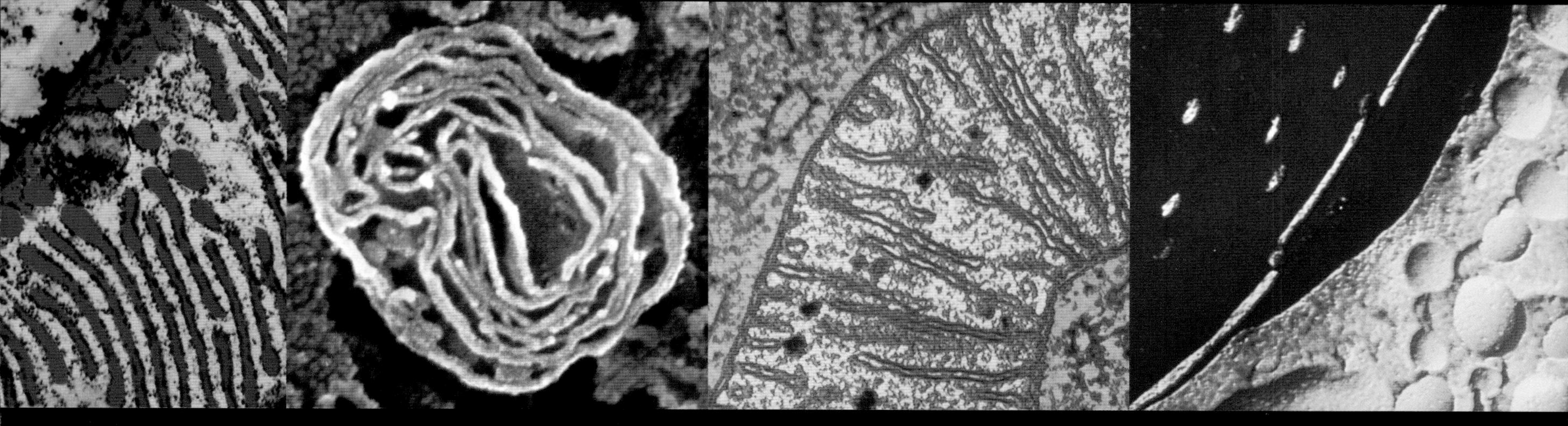

The ribosomes can be seen as granules covering rough endoplasmic reticulum (ER), a complex membrane system that acts like a conveyor belt, carrying proteins to different parts of the cell. Some of these proteins are used by the cell itself; others, such as digestive enzymes and hormones, are exported.

Like factory machines, organelles become worn out or redundant. When that happens, they are destroyed by organelles called lysosomes – bags of powerful enzymes that also digest foreign bodies such as bacteria.

The power to drive the cell factory comes from mitochondria. These sausage-shaped or spherical organelles convert molecules of food such as glucose into energy. The more energetic a cell, the more ribosomes it needs, so they are particularly numerous in very active cells, such as muscle and sperm cells. Mitochondria are thought to be derived from free-living bacteria that, at an early stage of evolution, invaded larger cells and adapted to a mutually beneficial way of life.

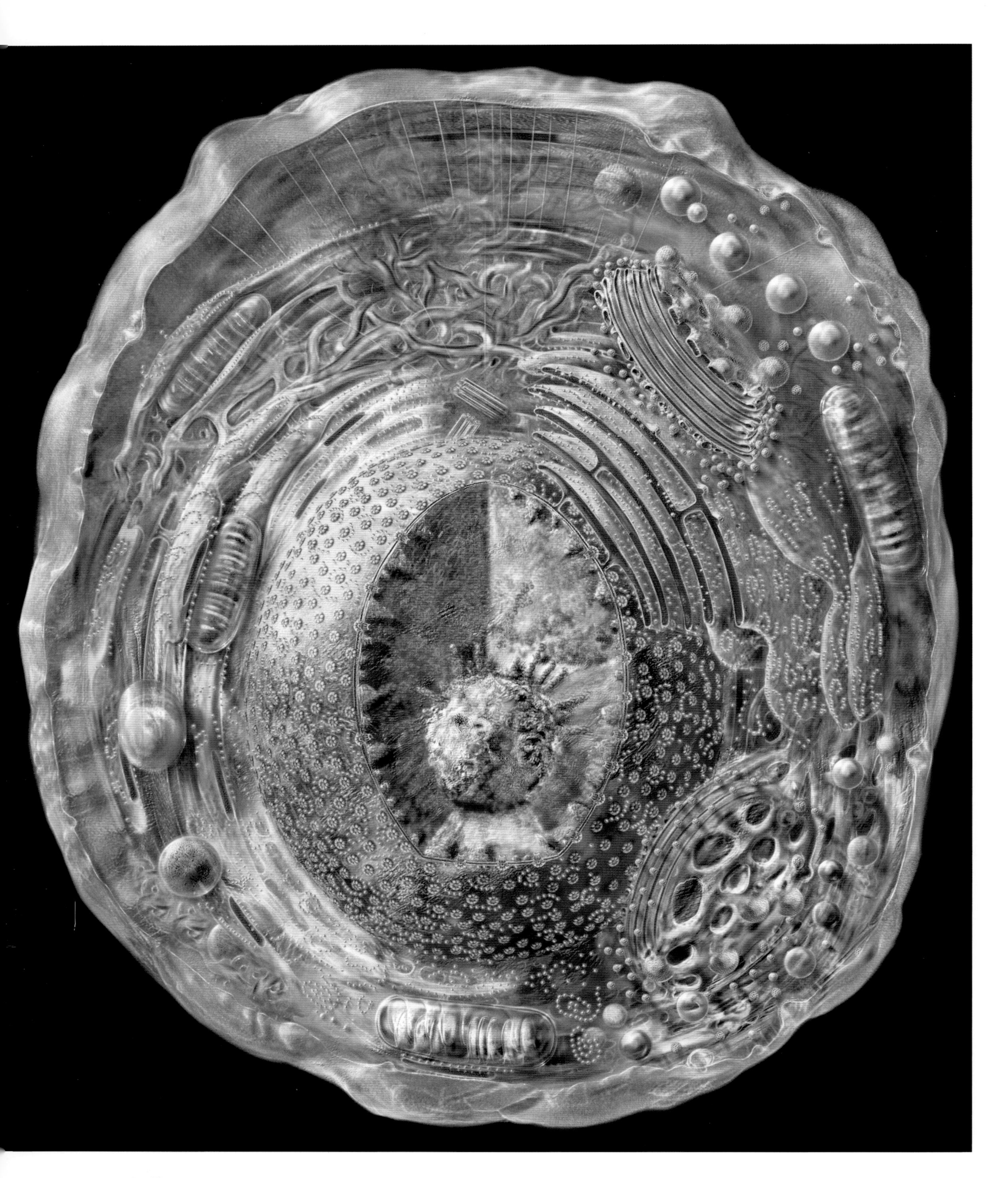

Cell structure [Artwork]

DESPITE THEIR DIFFERENCES in form and function, most cells possess similar internal structures. The nucleus (large purple oval) houses the cell's genetic material and contains at least one nucleolus (sphere, lower centre), which controls protein synthesis on ribosomes (chains of small yellow spheres) in the endoplasmic reticulum (folded structures above and at right of nucleus). The proteins are stored in the flattened membranes of the Golgi complex (above and at right of ER). Energy for the cell's activities is generated by mitochondria (cylinders, one at bottom).

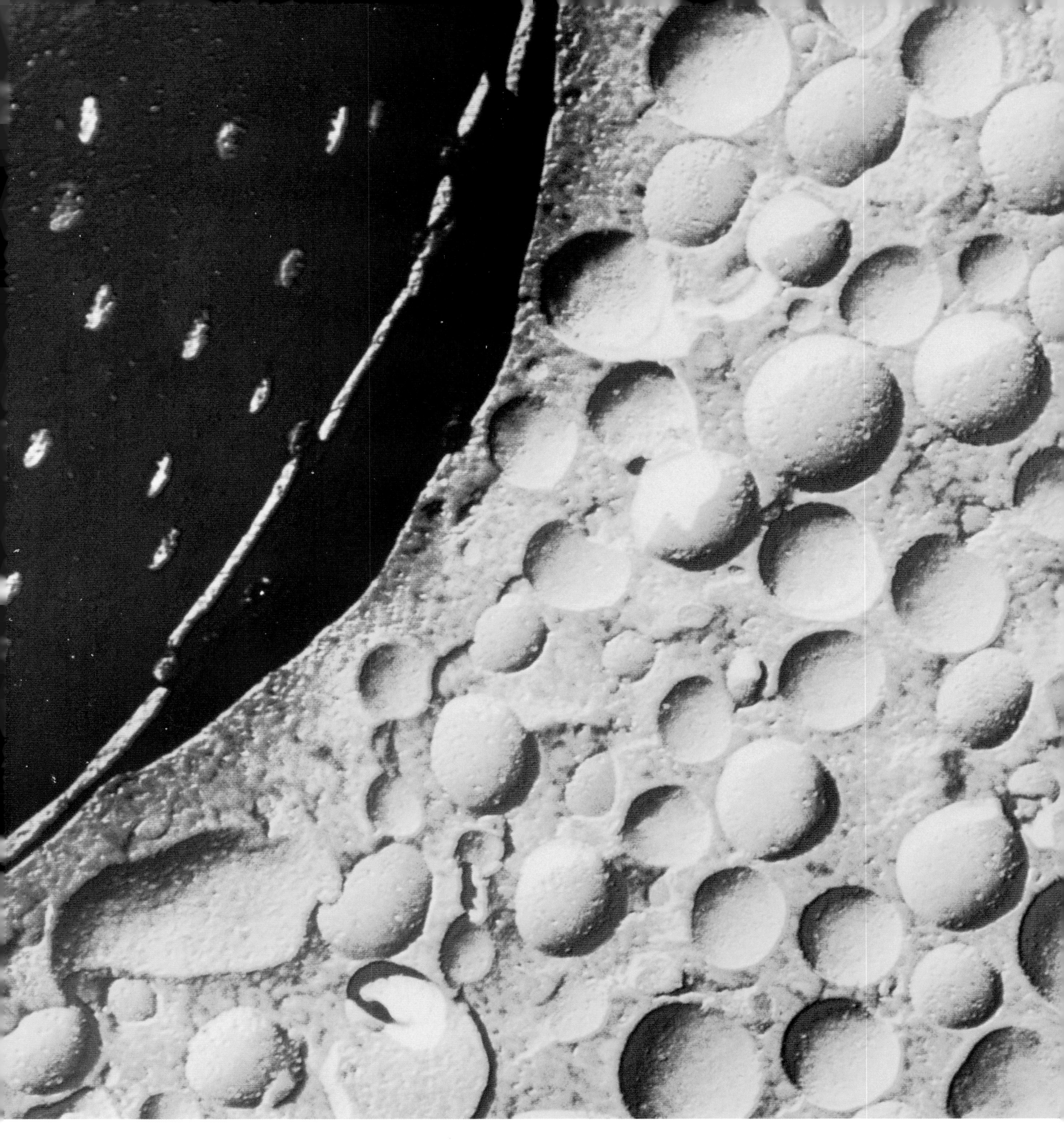

Nuclear membranes [SEM]

A MEMBRANE SEPARATES the nucleus (left) from the cytoplasm (right) containing the rest of the cell contents. Pores (pink) in the membrane allow large molecules to pass out of the nucleus. The nucleus contains a complete set of the organism's genes, which are encoded into DNA molecules on chromosomes.

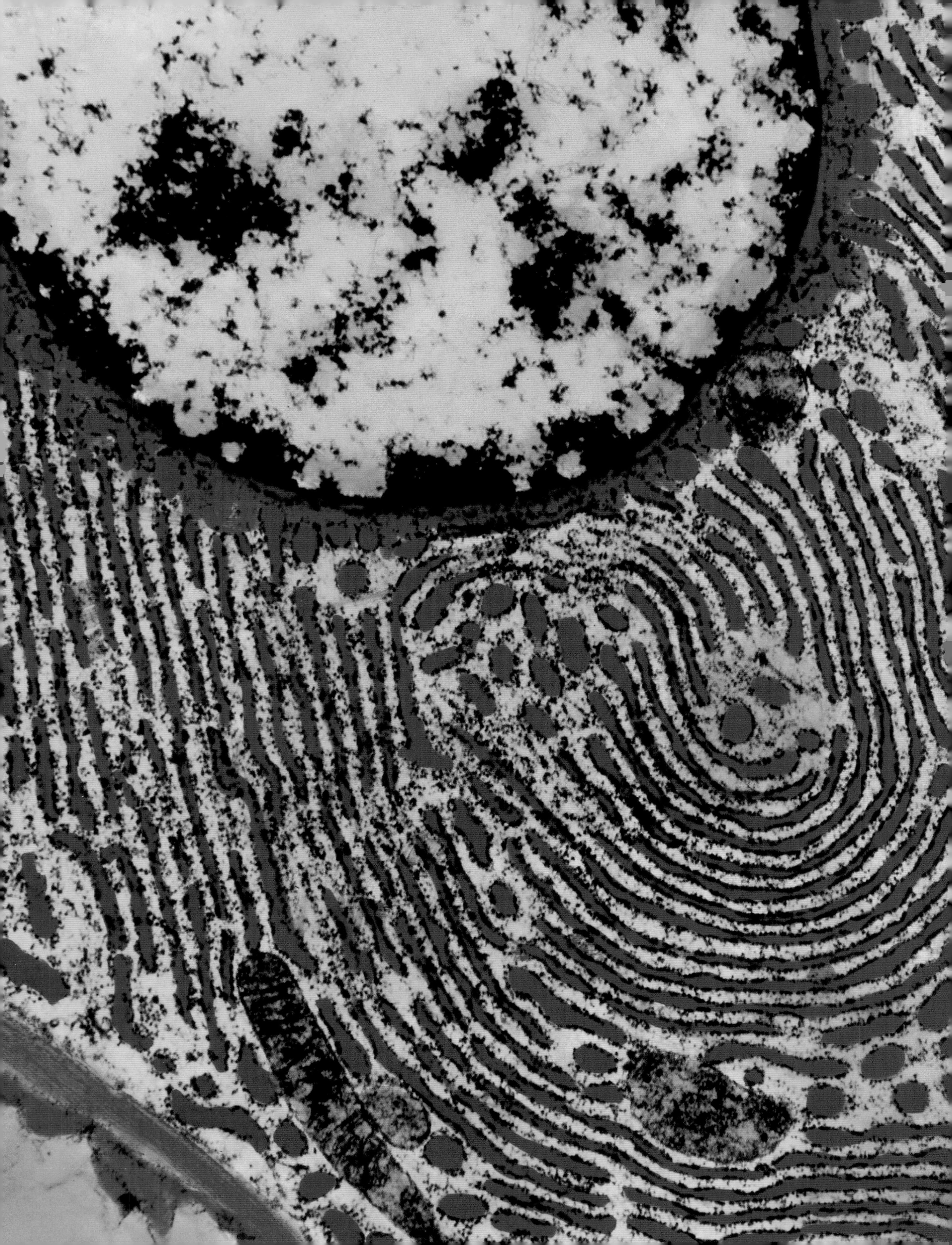

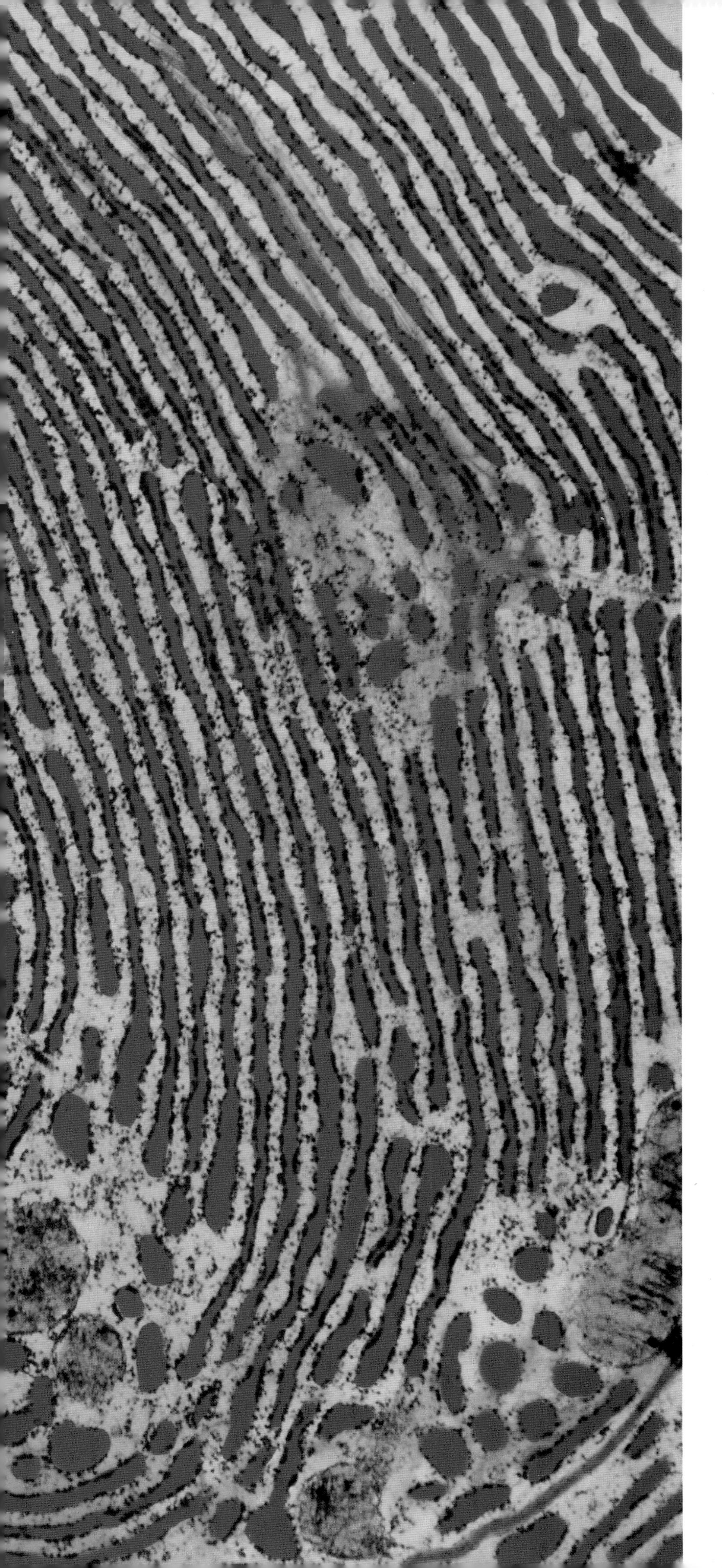

Endoplasmic reticulum and nucleus [TEM]

THE NUCLEAR MEMBRANE around the nucleus at top left joins with rough endoplasmic reticulum (ER), coloured red, a complex membrane system that divides the cell cytoplasm into cavities. The surface of rough ER is covered with ribosomes (black dots), which produce proteins, such as enzymes. The proteins are transported through the cavities, called cisternae, and secreted out of the cell.

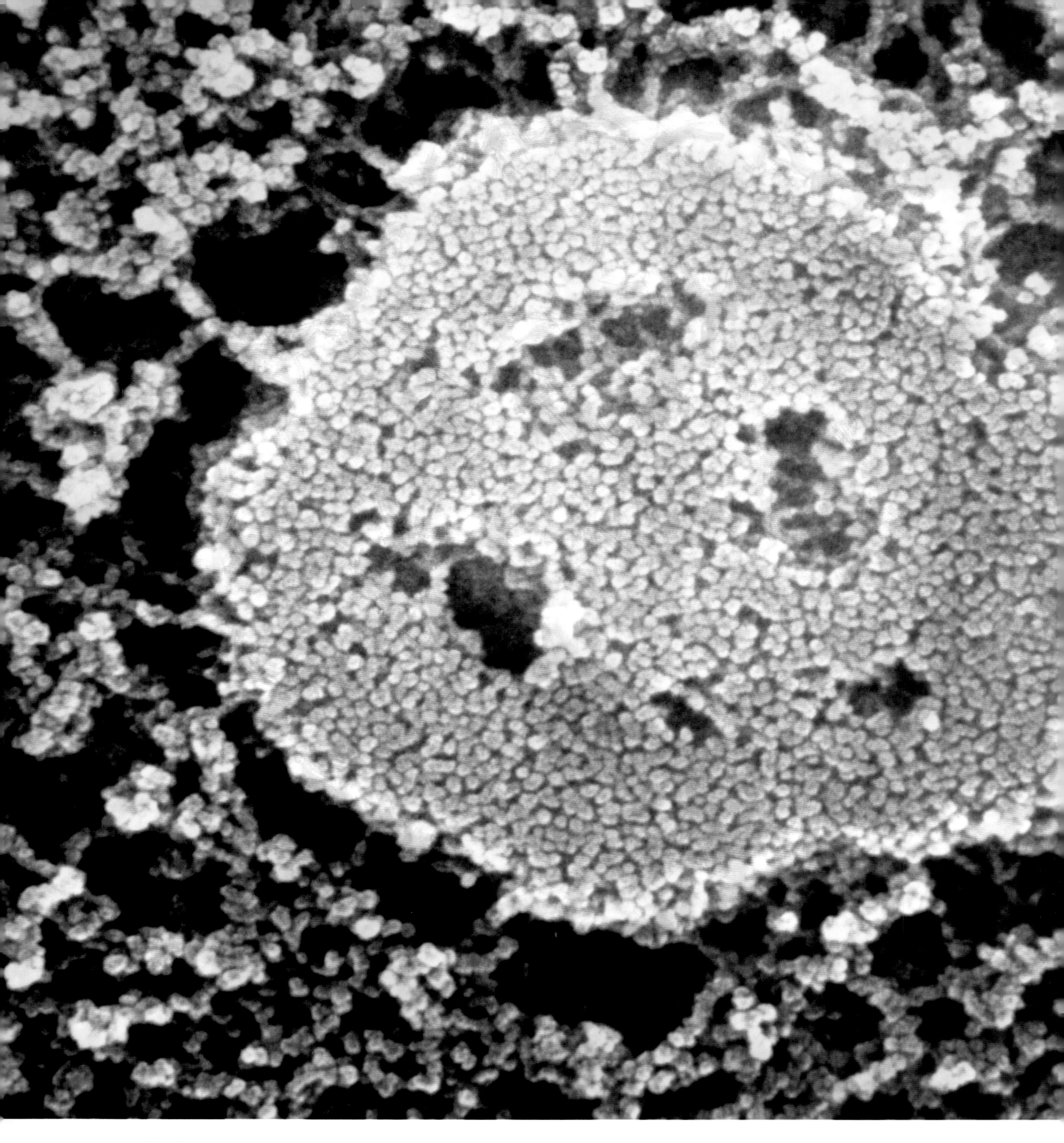

Nucleolus [SEM]

THE NUCLEOLUS (centre) is a specialized part of the nucleus that assembles particles containing ribonucleic acid (RNA), a compound involved in the synthesis of proteins outside the cell nucleus. Surrounding the nucleolus are chromatin fibres, which contain the cell's DNA, the genetic material that governs all the cellular processes, including cell division. When the cell divides, the DNA becomes visible as rod-like structures called chromosomes.

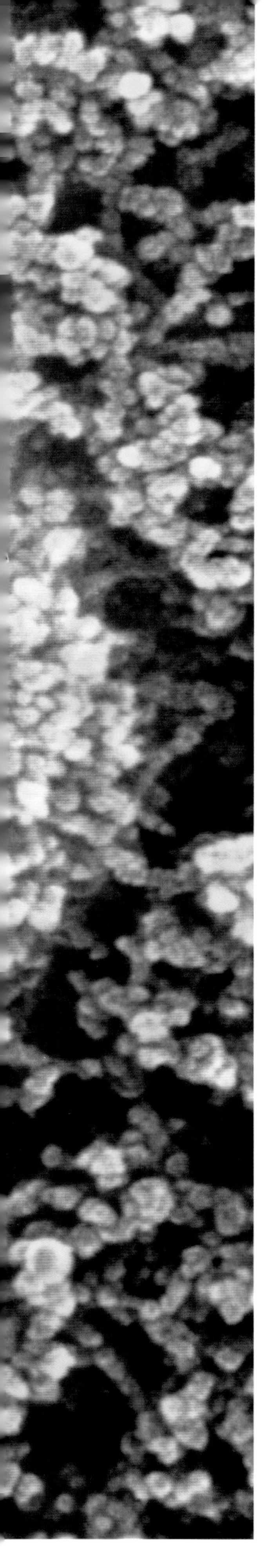

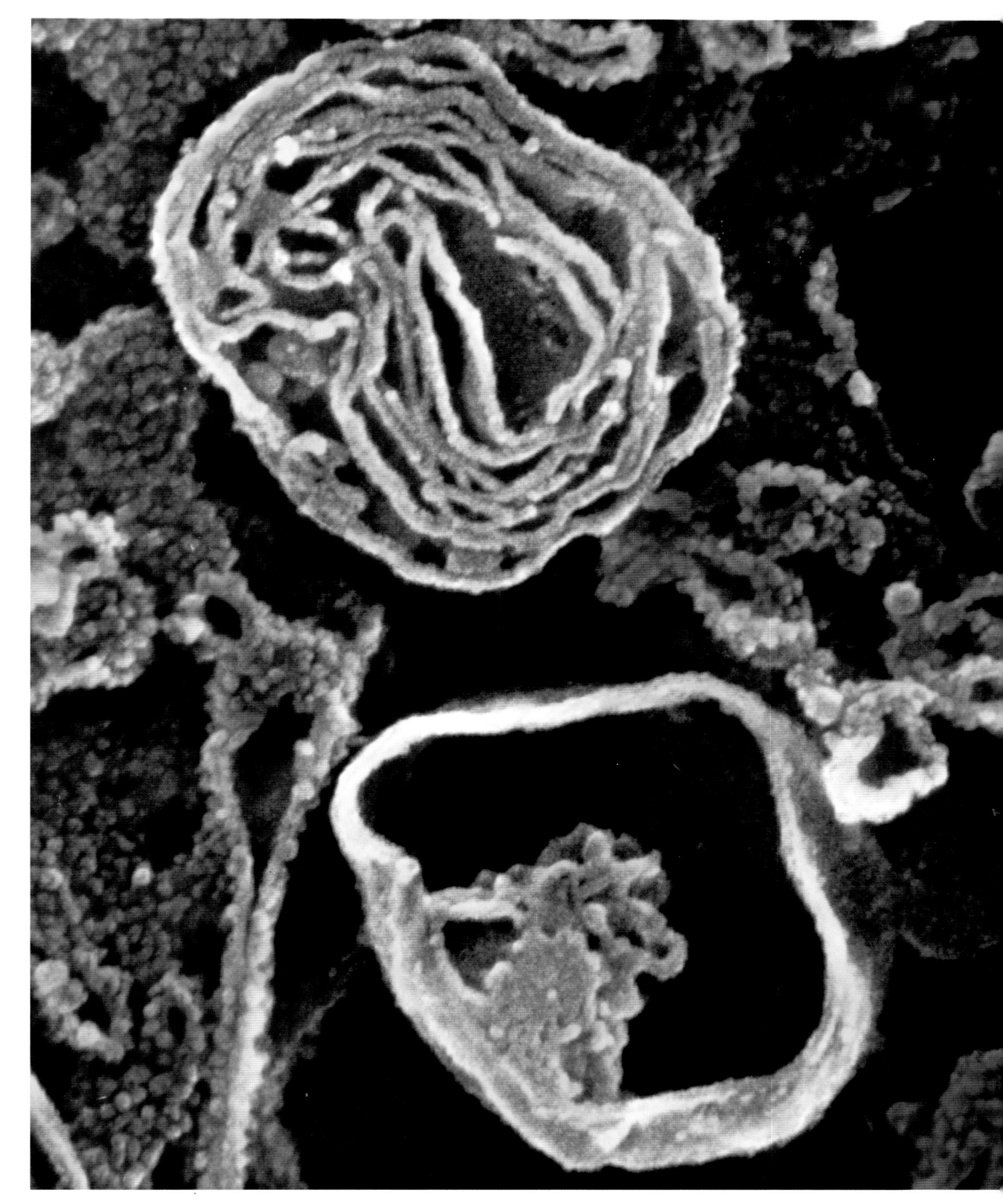

Lysosomes [TEM]

TWO LYSOSOMES (green) are simply bags of digestive enzymes bound by single membranes. Although simple in structure, lysosomes have several functions. The enzymes they contain are used to destroy redundant organelles (digested material absorbed by the cell), as well as cells and tissues that are no longer needed, such as milk-producing tissue after weaning.

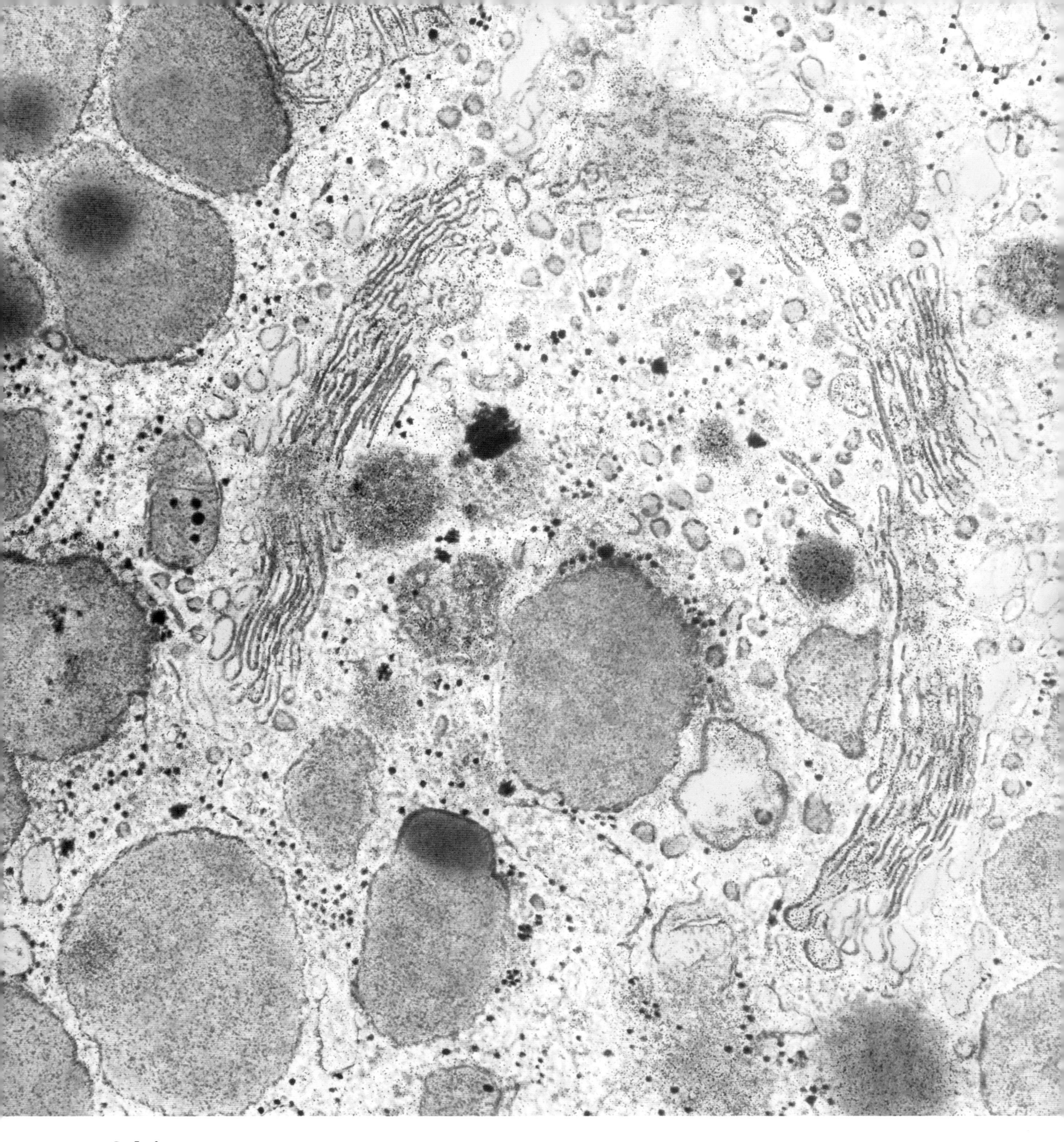

Golgi apparatus [TEM]

THE BLUE HORSESHOE-SHAPED STRUCTURE in the cytoplasm of this intestinal cell is called the Golgi apparatus. Variable in shape but usually located near to the nucleus, the Golgi apparatus is made up of flattened cavities that resemble a stack of plates The Golgi apparatus stores and modifies proteins before they are secreted out of the cell.

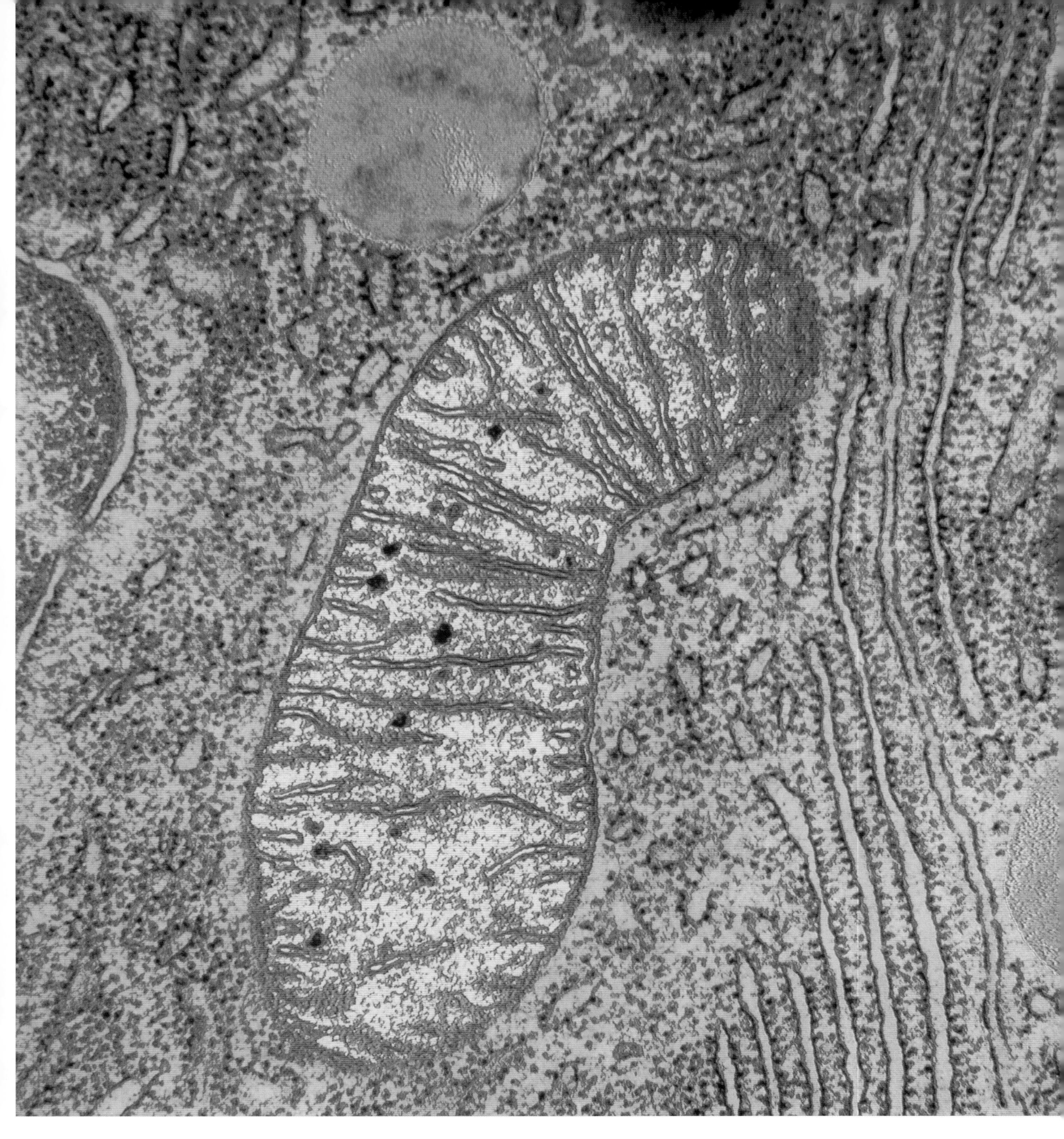

Mitochondria [TEM]

MITOCHONDRIA ARE THE POWERHOUSES of cells, responsible for releasing energy from food. They are particularly numerous in very active tissues such as heart muscle. Their sausage-shaped structure has a double membrane with folded ingrowths called cristae, in which chemical reactions occur. The more metabolically active the cell, the more cristae it has. The long turquoise tubules outside each mitochondrion are part of the endoplasmic reticulum, which stores and transports proteins synthesized by ribosomes (dots).

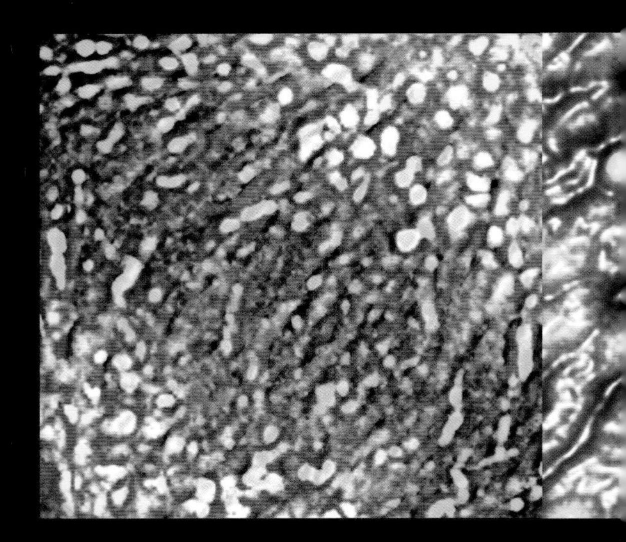

A cell lives and functions until it reproduces or dies. Every day millions of new cells are born; every day millions of cells die. Children grow because cell production outnumbers cell death. In the womb, for example, new nerve cells of a fetus are produced at the rate of about 250,000 a minute.

Once we reach adulthood, the cell population remains relatively constant. Some types of mature cells lose the capacity to replicate. Brain cells, for example, cannot be replaced if they are destroyed. Fortunately, they can live for decades. Liver cells divide only rarely, although, unlike brain cells, they can regenerate if the liver is damaged. Many cells, however, divide fairly frequently, and some are specialized for rapid turnover. Each year about 4kg/9lb of skin cells are born to replace those that die and flake off. Our intestinal lining is replaced every few days. A man's testes manufacture enough sperm in a year to populate the entire world.

All but a few highly specialized cells duplicate themselves by a process called mitosis. In mitosis, all the forty-six chromosomes in a cell nucleus make copies of themselves. The nucleus divides and the two

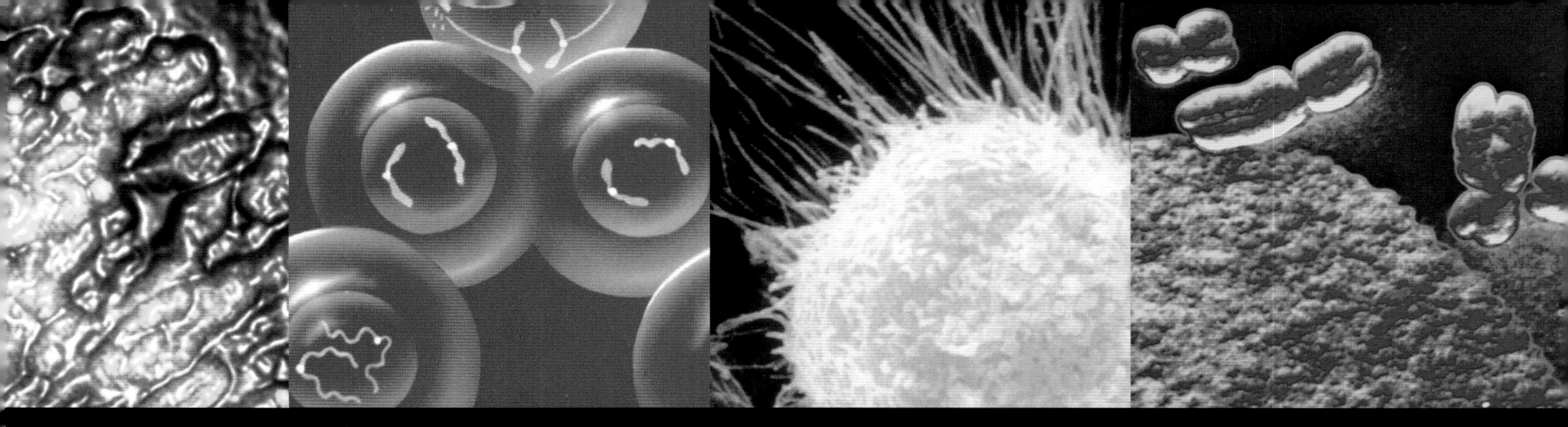

sets of forty-six chromosomes separate, so that two genetically identical cells now exist where previously there was only one. The other type of cell division, called meiosis, involves the germ cells in the ovaries and testes, which produce eggs and sperm. Instead of producing cells with forty-six chromosomes, meiosis halves the genetic material so that eggs and sperms have twenty-three chromosomes. When a sperm fertilizes an egg, the chromosomes come together to make a cell with forty-six chromosomes, half from each parent. During the halving process, the genes get shuffled, so that each egg or sperm is genetically different. This explains why children born to the same parents are not identical.

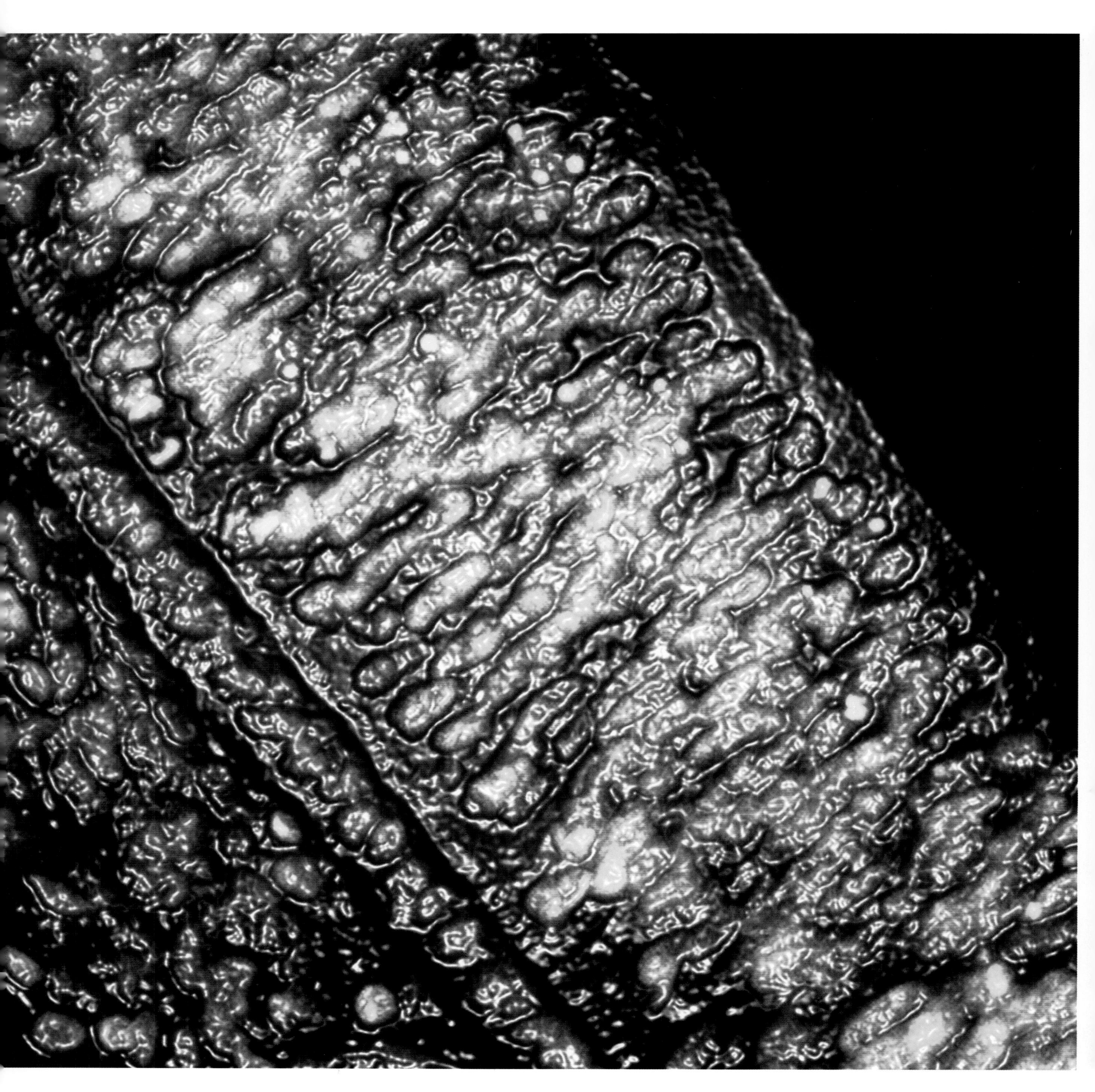

Cell death [LM]

THE BRIGHT YELLOW DOTS are the nuclei of cells in the retina of :he eye undergoing programmed cell death. Cell suicide, or apoptosis, is a normal part of the development of tissue. More cells are produced than are required, and the redundant cells are instructed to kill themselves by chemical messengers. After their death, their remains, which contain valuable cellular material, are digested by other cells.

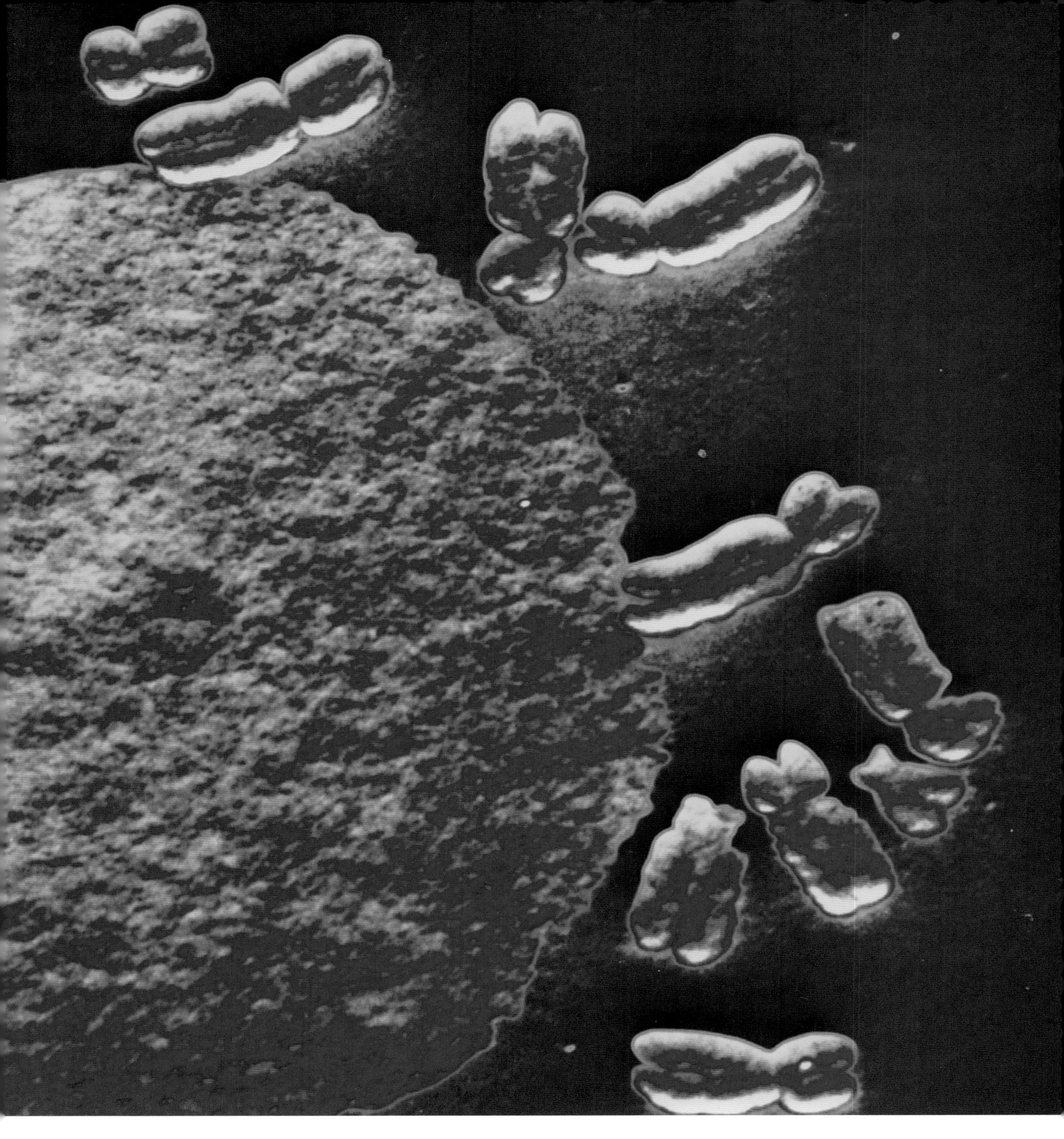

Chromosomes [SEM]

SOME OF THE TWENTY-THREE PAIRS of human chromosomes are shown beside a cell nucleus. Chromosomes carry the genetic information that determines how a person will grow – whether he or she will be dark or fair, short or tall, blue-eyed or brown. Chromosomes are only visible when the cell is about to divide, as in this image. The rest of the time, the genetic material is dispersed inside the nucleus.

Cell division [LM]

THIS SEQUENCE OF IMAGES shows stages in the process of mitosis, the replication of a cell into two identical new cells. The cell's genetic material (blue) begins to separate, drawn apart along spindles (red). The process is aided by a protein called dynein (green). As the two move to opposite poles, a dark line of cleavage forms in the centre. Some cells, such as muscles cells, divide infrequently, while others have a very rapid turnover. The lining of the intestine, for example, is replaced every two or three days.

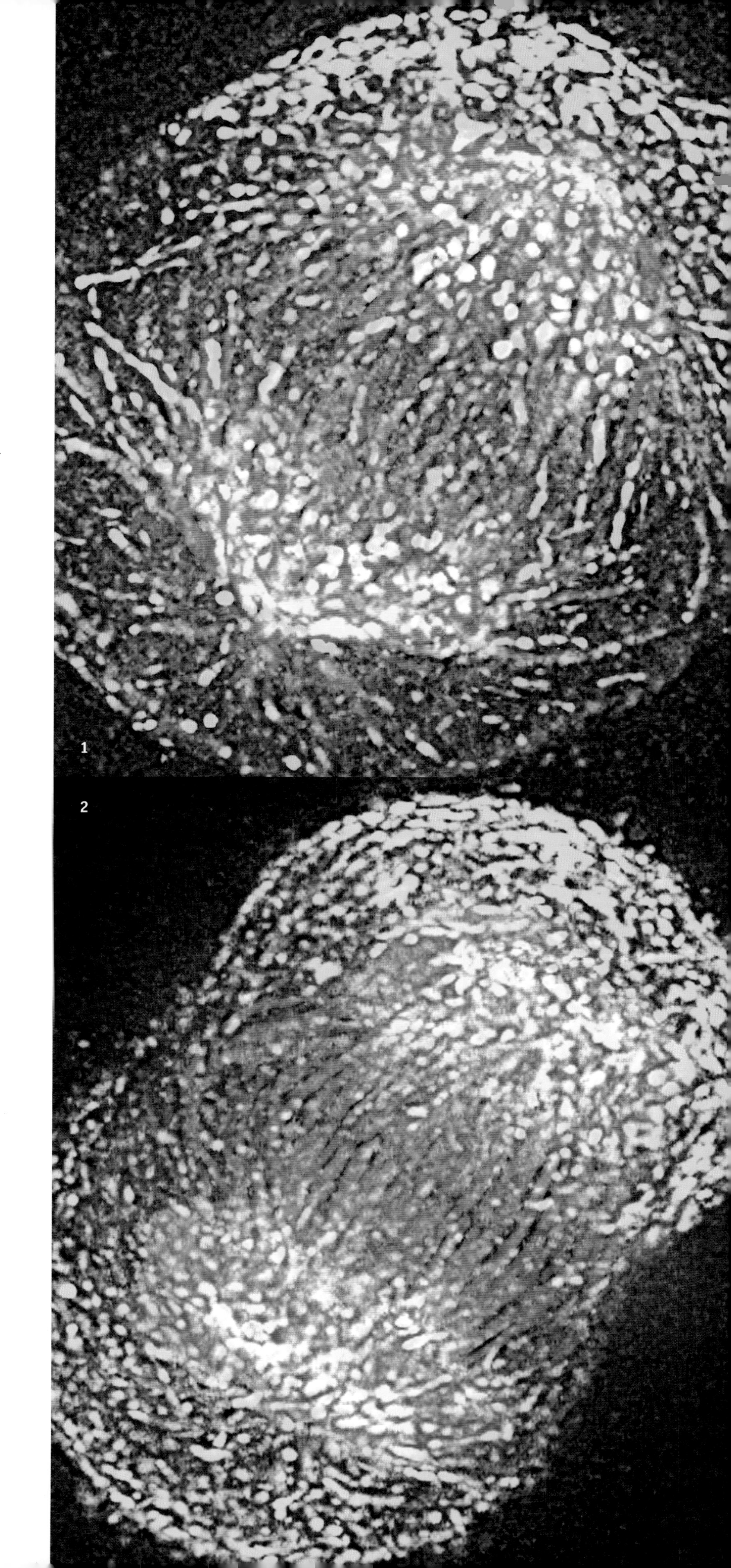

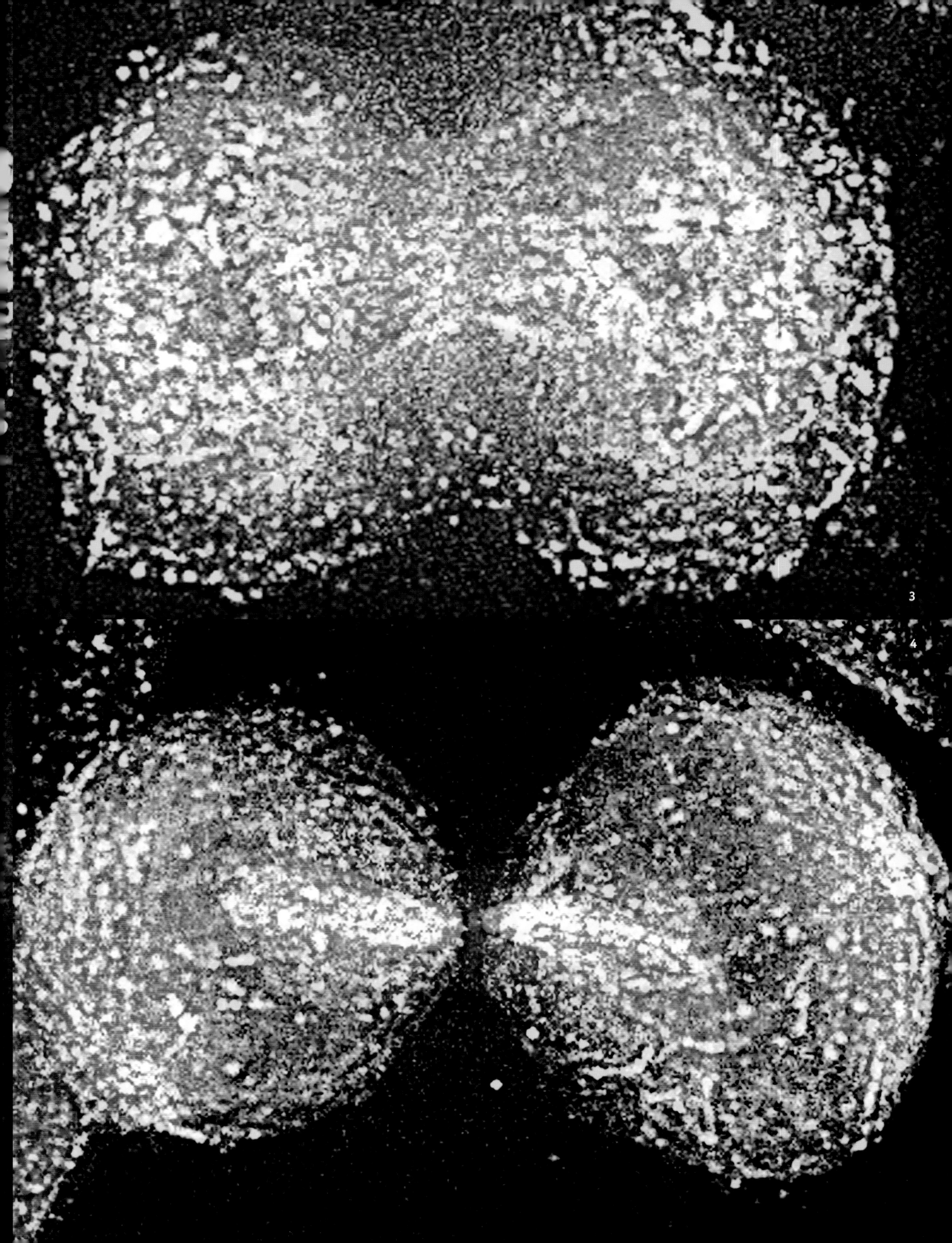

3
4

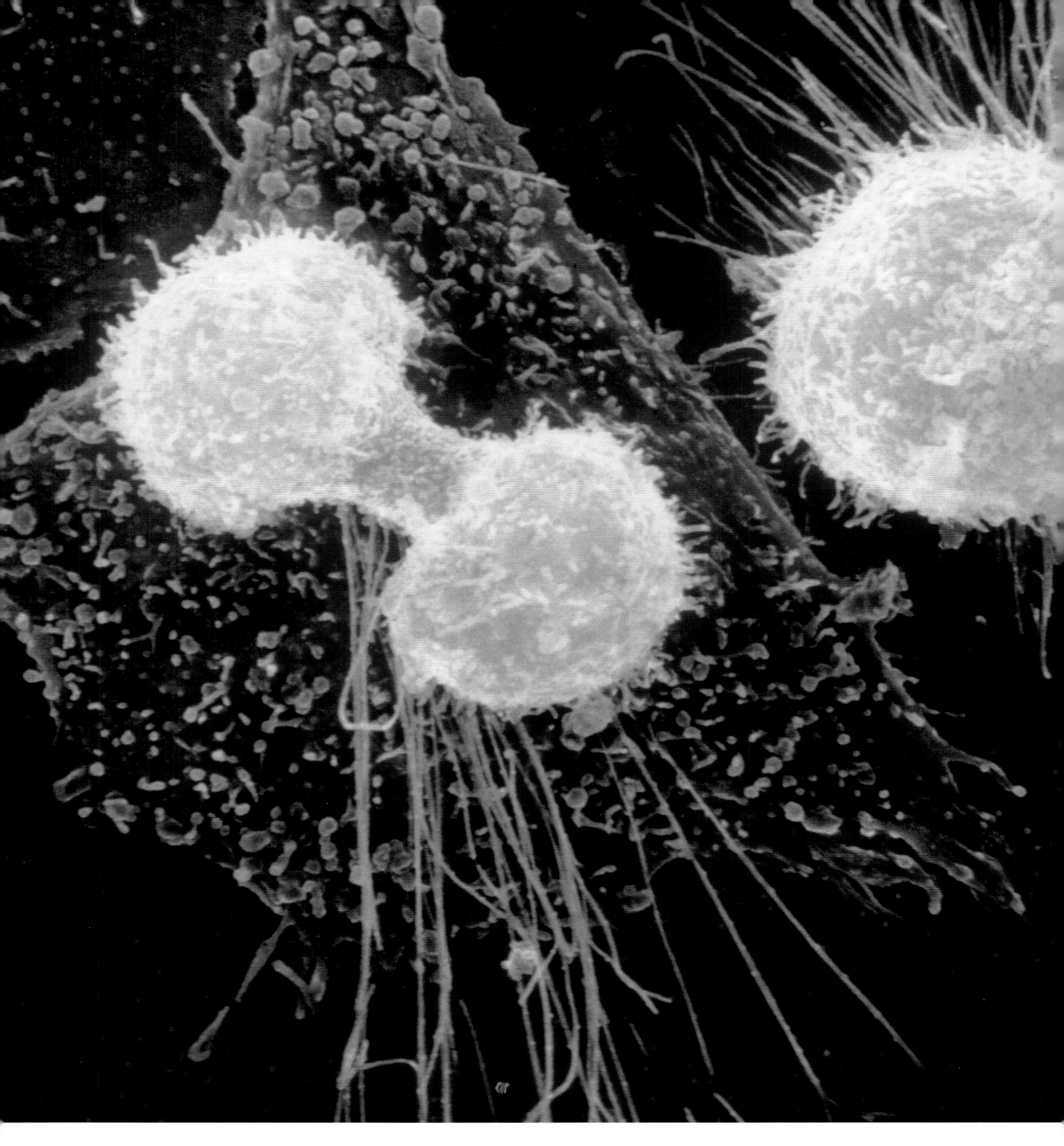

Mitotic cell division [SEM]

DURING THE LAST STAGE OF MITOSIS, two daughter cells containing identical sets of chromosones are connected by a narrow cytoplasmic bridge that will soon split. The long filaments protruding from the cells' surfaces provide mechanical support and supply the cells with nutrients from the surrounding environment.

Stages of meiosis [Artwork]

IN MEIOSIS A CELL in the testes or ovaries containing forty-six chromosomes divides to form sperm or egg cells, each one with twenty-three chromosomes. It involves two divisions. The first one (top centre) produces two daughter cells with twenty-three chromosomes. The second division (bottom right and left) is a mitotic type of division that produces four identical sperm or egg cells. During the first division there is exchange of genetic material between the paired chromosomes so that each of the final sperm or egg cells receives a unique mix of the parental genes.

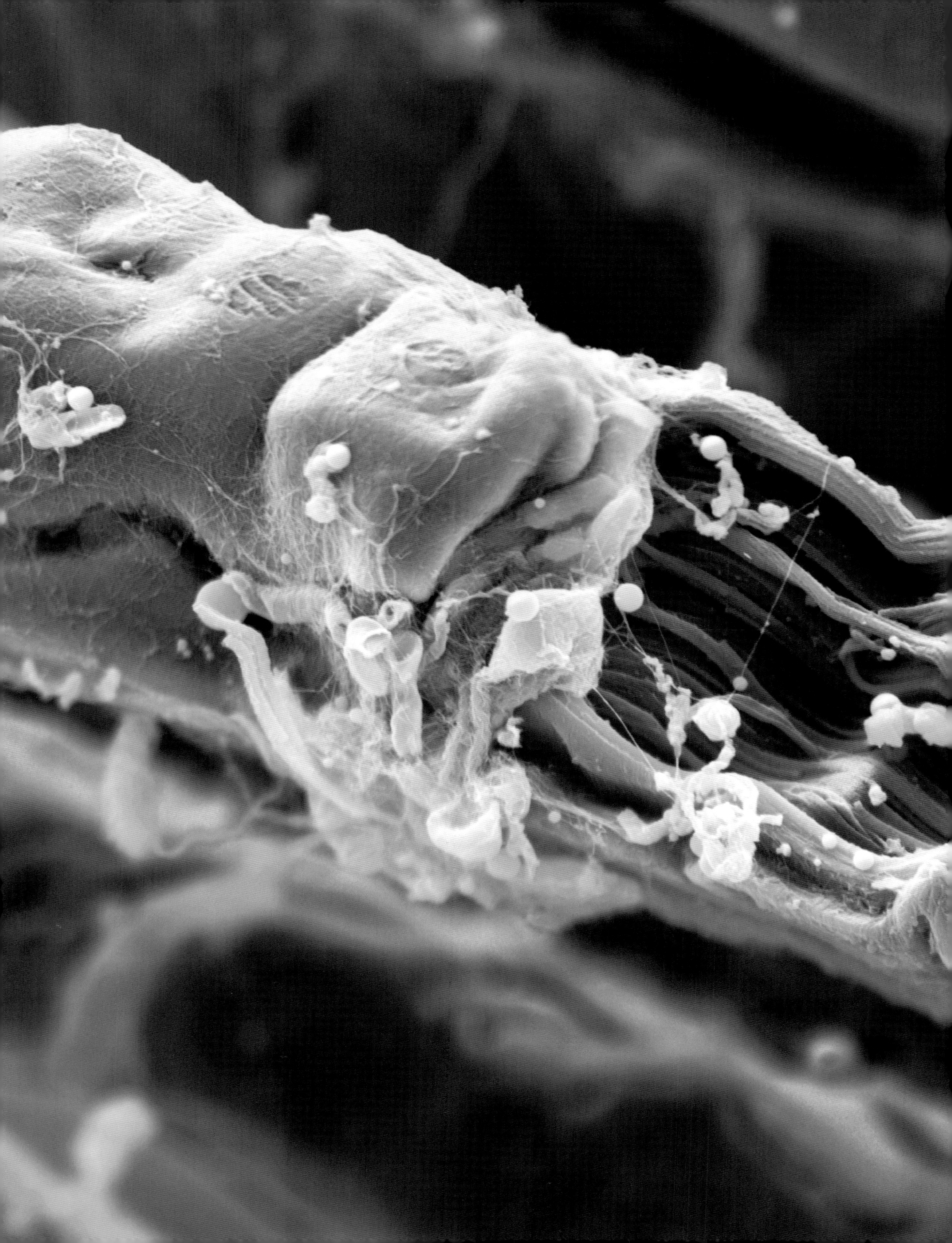

TISSUES

A SKELETAL MUSCLE FIBRE has been torn open to expose the thin threads called myofibrils that enable muscles to contract. Myofibrils are formed of alternating filaments of the proteins actin and myosin. When the muscle is stimulated, 'heads' on the myosin filaments attach themselves to 'hooks' on the actin filament and pull the actin filaments past them, causing the muscle fibre to shorten.

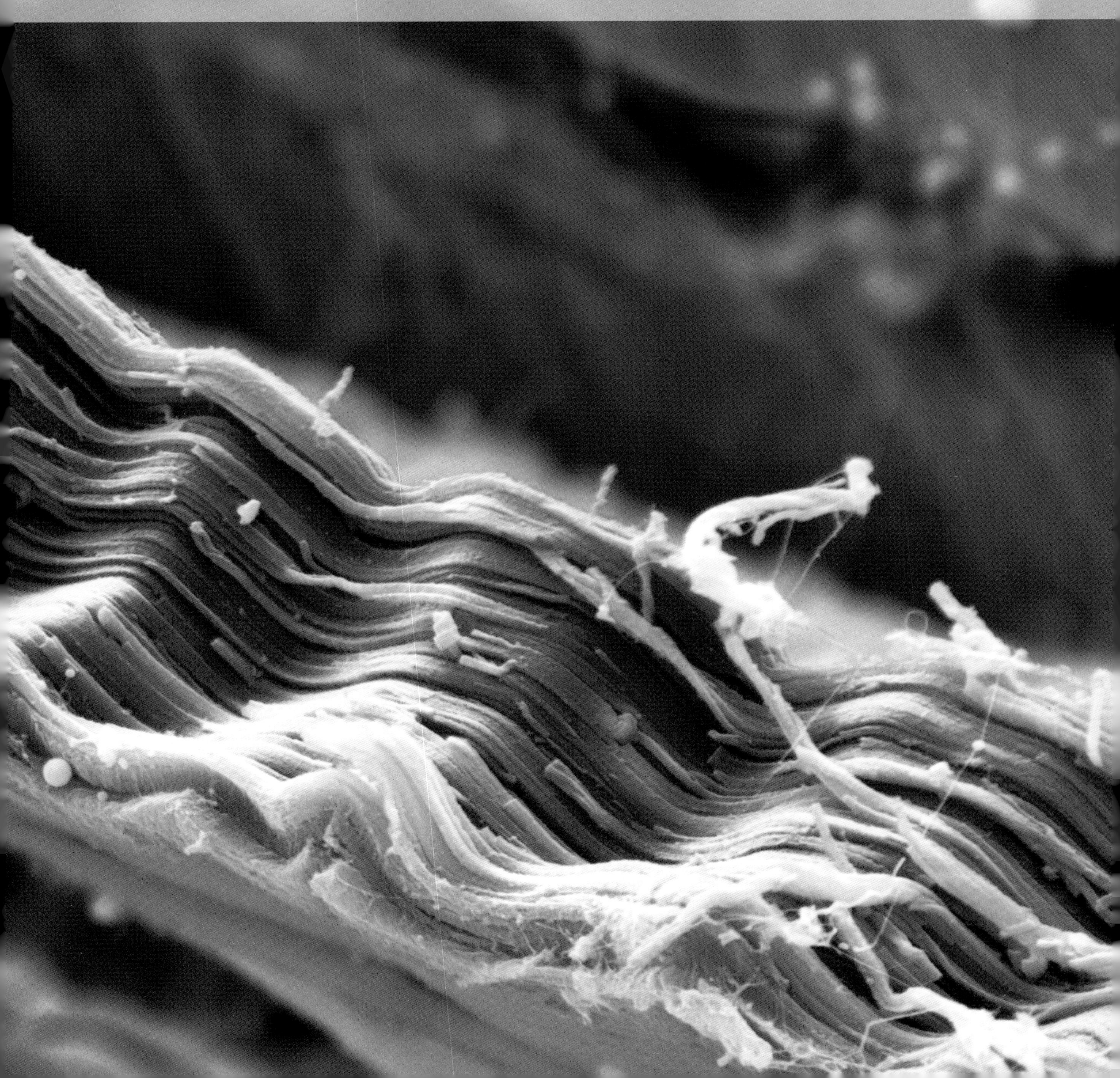

Each day our bodies are exposed to fluctuating temperatures, infectious agents and harmful light rays. Epithelial tissue is the barrier against the world, forming continuous protective surface layers both outside and inside the body.

The skin is the first line of defence against invasion. The largest organ of the body, it protects the tissues beneath from mechanical injury, ultraviolet radiation, germs and dessication. As well as being a physical barrier, it secretes chemical defences. Sweat and tears contain lactic acid and the enzyme lysozyme, which slow down bacterial growth. Skin cells also produce our primitive natural weapons – teeth, fingernails and toenails.

But skin is more than just a protective covering. As the part of the body in direct contact with the external environment, it contains numerous sense organs that make us aware of changes in our surroundings. It plays a major role in regulating body temperature. Much of the heat produced by the body is lost through the pores of the skin by sweating, a process controlled by nerve signals to the sweat glands and to small blood vessels just under the skin. While we no longer depend on body hair to keep us warm, the oily sebum secreted by glands in the hair follicles helps waterproof our bodies.

Tough skin that is several cells thick merges into thin epithelial tissue at the mouth, nose and other orifices. The epithelial linings of the digestive and respiratory tracts lack structural strength but have their own defences. Microorganisms that enter the windpipe are trapped in mucus and then expelled by the rapid beating of tiny hair-like extensions called cilia. Coughing and sneezing achieve the same effect but more explosively, blowing out particles and mucus at speeds of up to 160kph/100mph. Any bacteria that reach the stomach are usually destroyed by protein-digesting enzymes and a hydrochloric acid solution strong enough to dissolve zinc.

Epithelial tissue

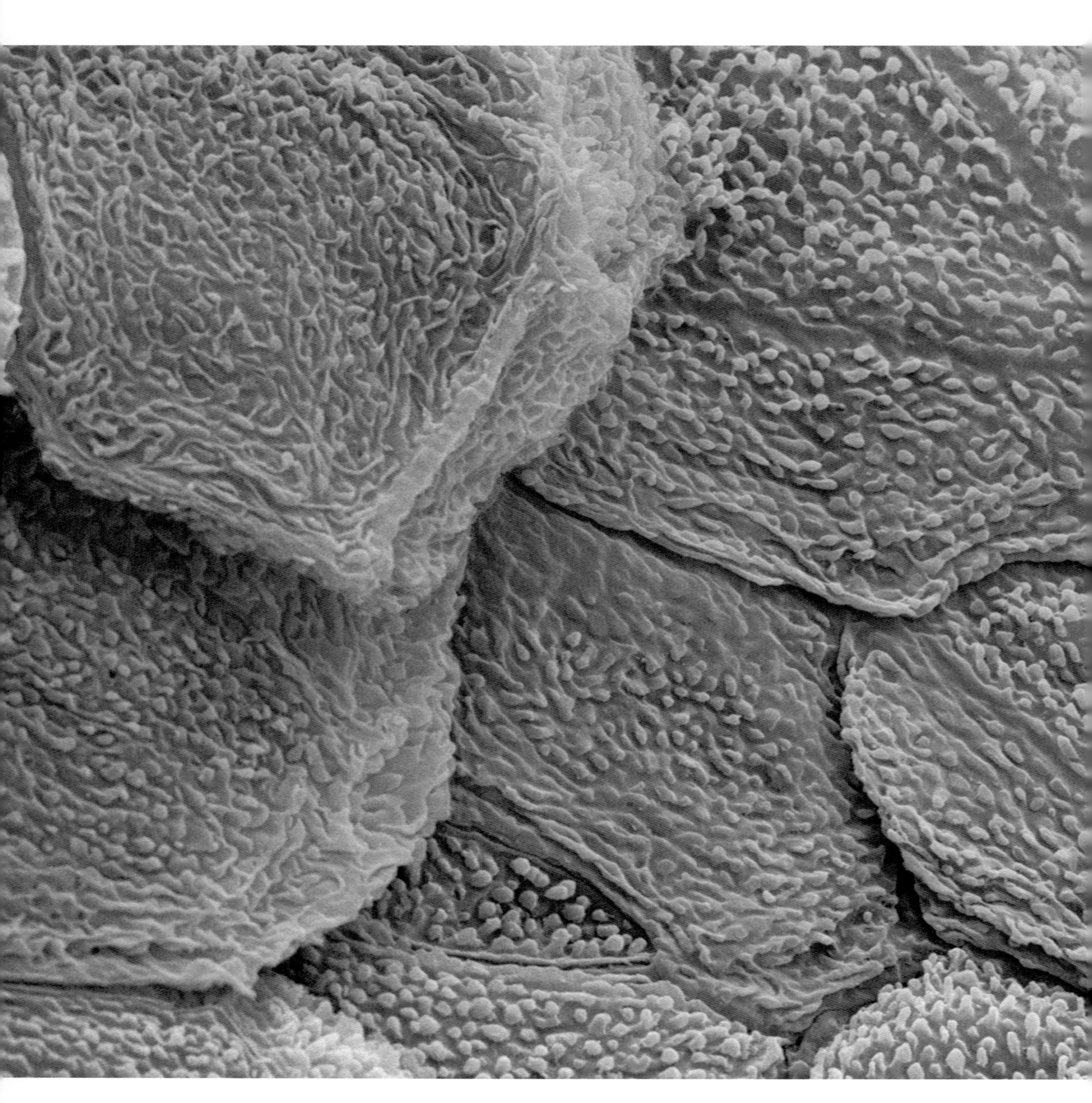

Outer layer of human skin [SEM]

THE OUTER LAYER of human skin is a tough protective coating formed from overlapping layers of dead, flattened cells that are continually shed and replaced with cells from the living layers below. As the cells move upwards, they become filled with a fibrous protein called keratin, which also makes up hair and nails.

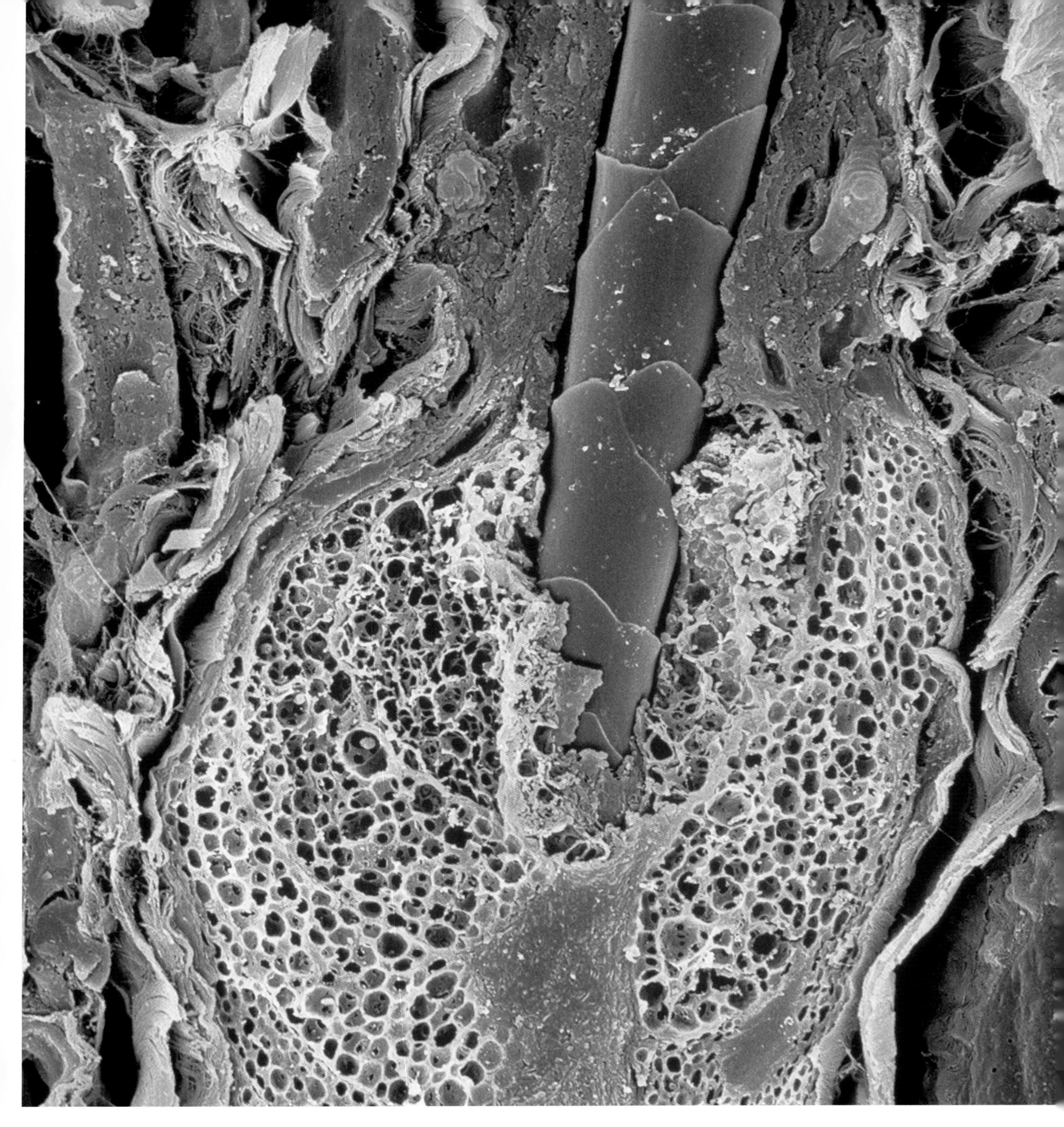

Hair follicle [SEM]

A HAIR SHAFT (purple) rises through the skin from its follicle (not clearly seen). The pale brown honeycomb structure surrounding the follicle is a sebaceous gland, which produces the sebum that lubricates the hair and skin. The green-brown structures at top left are parts of the muscle that causes 'goose bumps' – the erection of hairs in cold or fear.

Tooth enamel [SEM]

A SECTION THROUGH a tooth shows the specialized epithelial cell layer (green) that produces enamel (yellow, at bottom). The hardest substance made by mammels, enamel covers and protects the teeth.

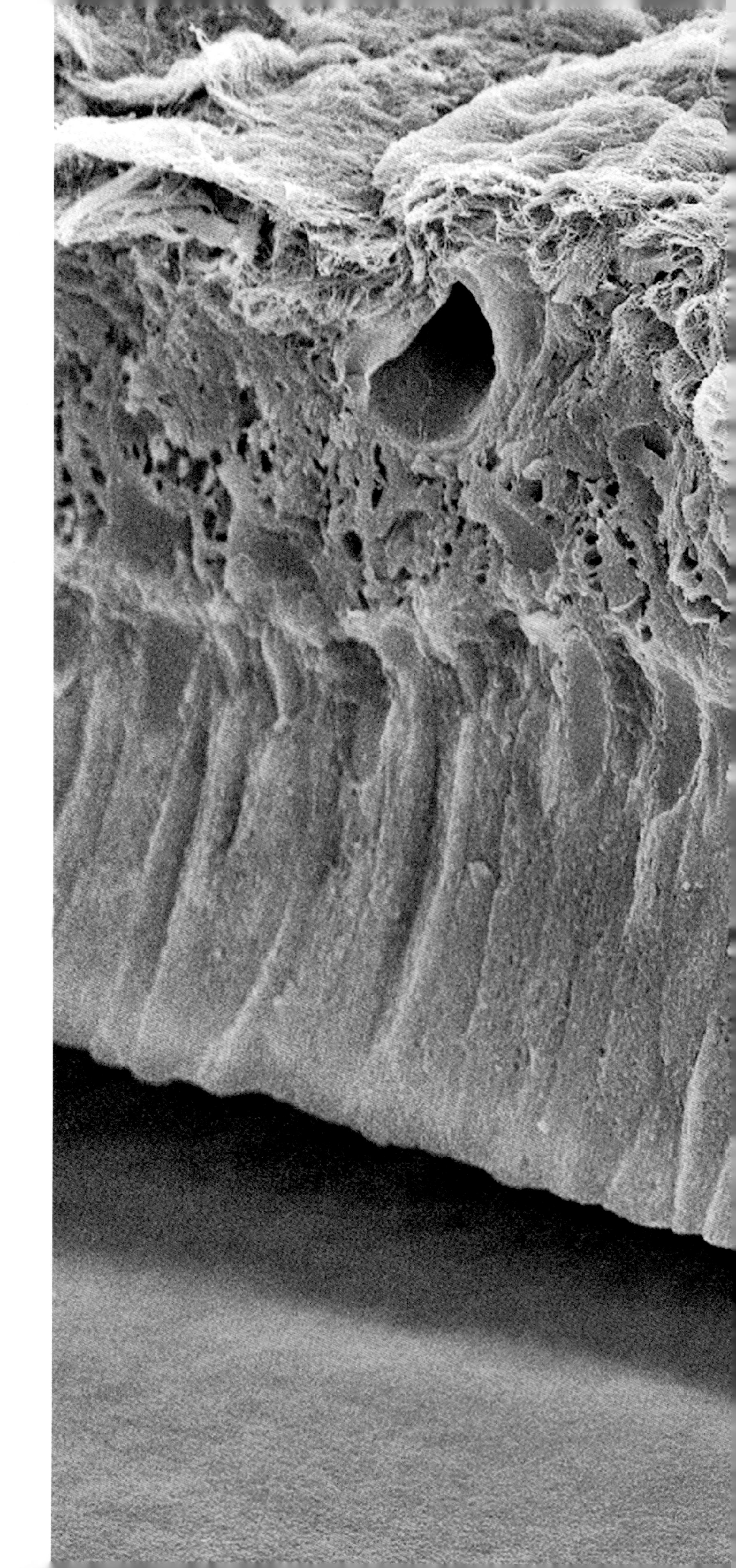

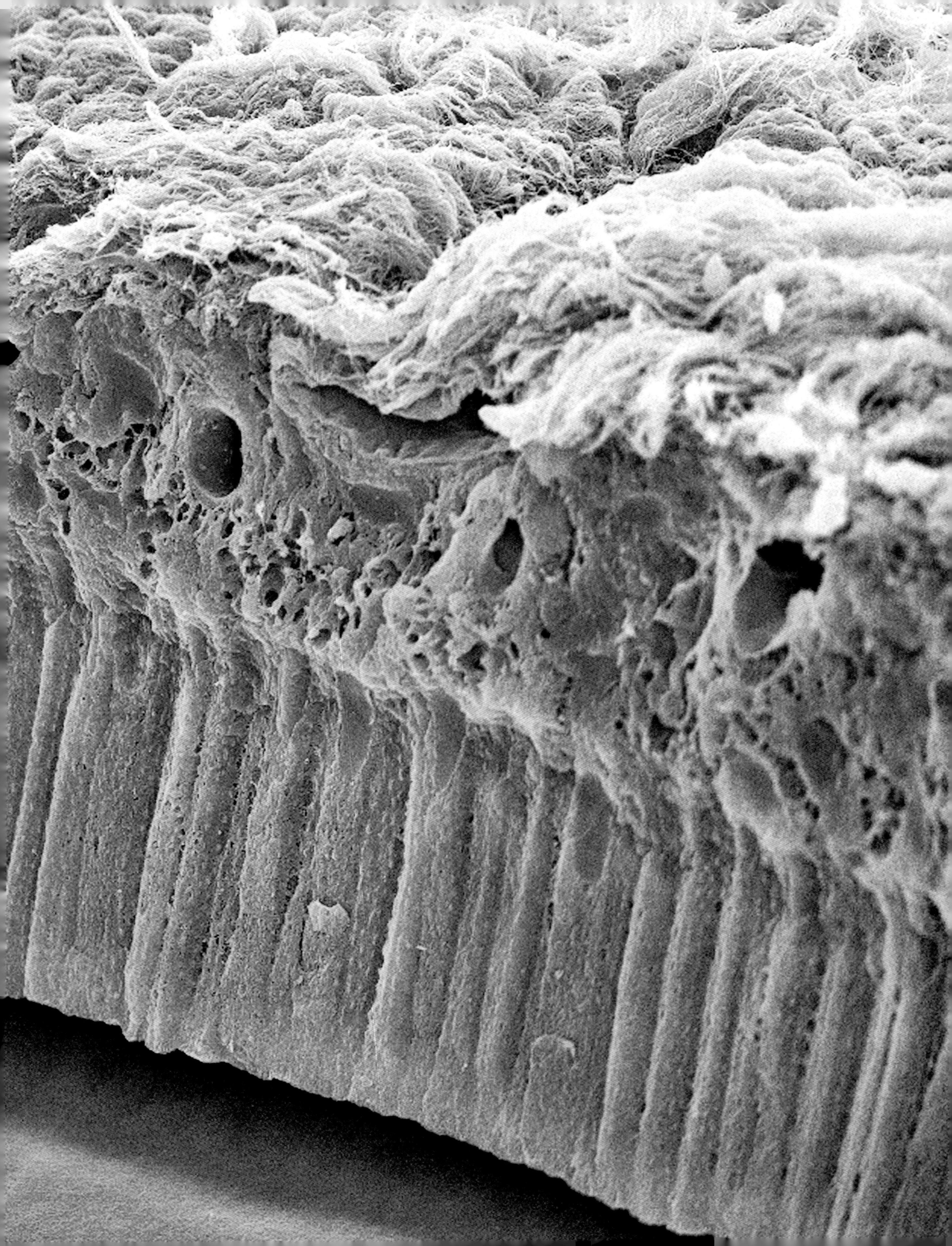

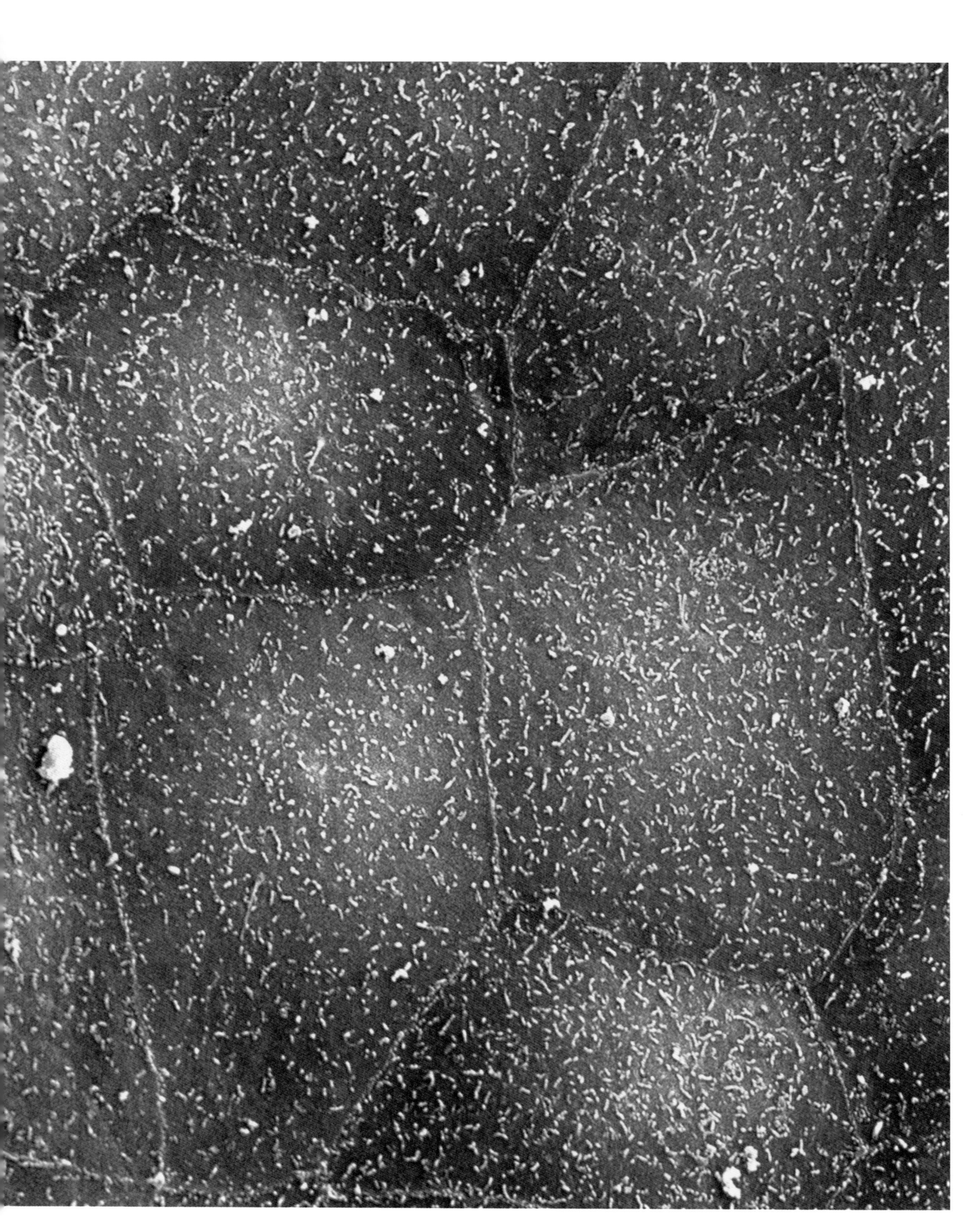

Epithelium of palate [SEM]

THE PALATE IS LINED with a layer of cells resembling paving stones. The cells are covered with white, hair-like structures that increase the cell's surface area and aid the exchange of substances. This simple epithelium lines most of the body's hollow internal structures. Stratified squamous epithelium, several cells thick, covers the body's external surfaces.

Bladder epithelium [SEM]

THE HUMAN BLADDER is lined with a highly folded epithelium formed from cells that have tiny plaques (green and blue dots) on their surface. The plaques and folds enable the bladder to deal with the large changes in its surface area as it fills and empties.

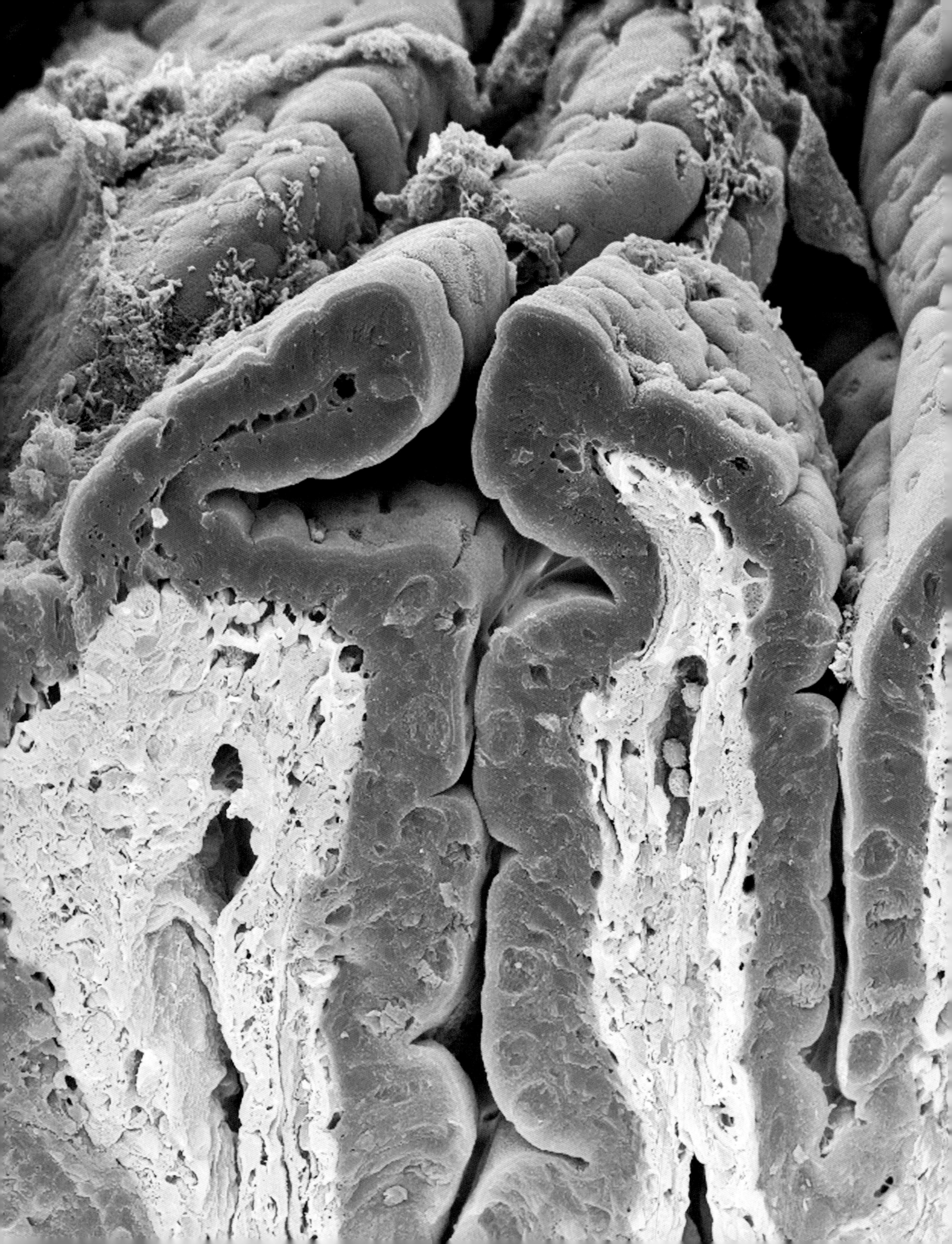

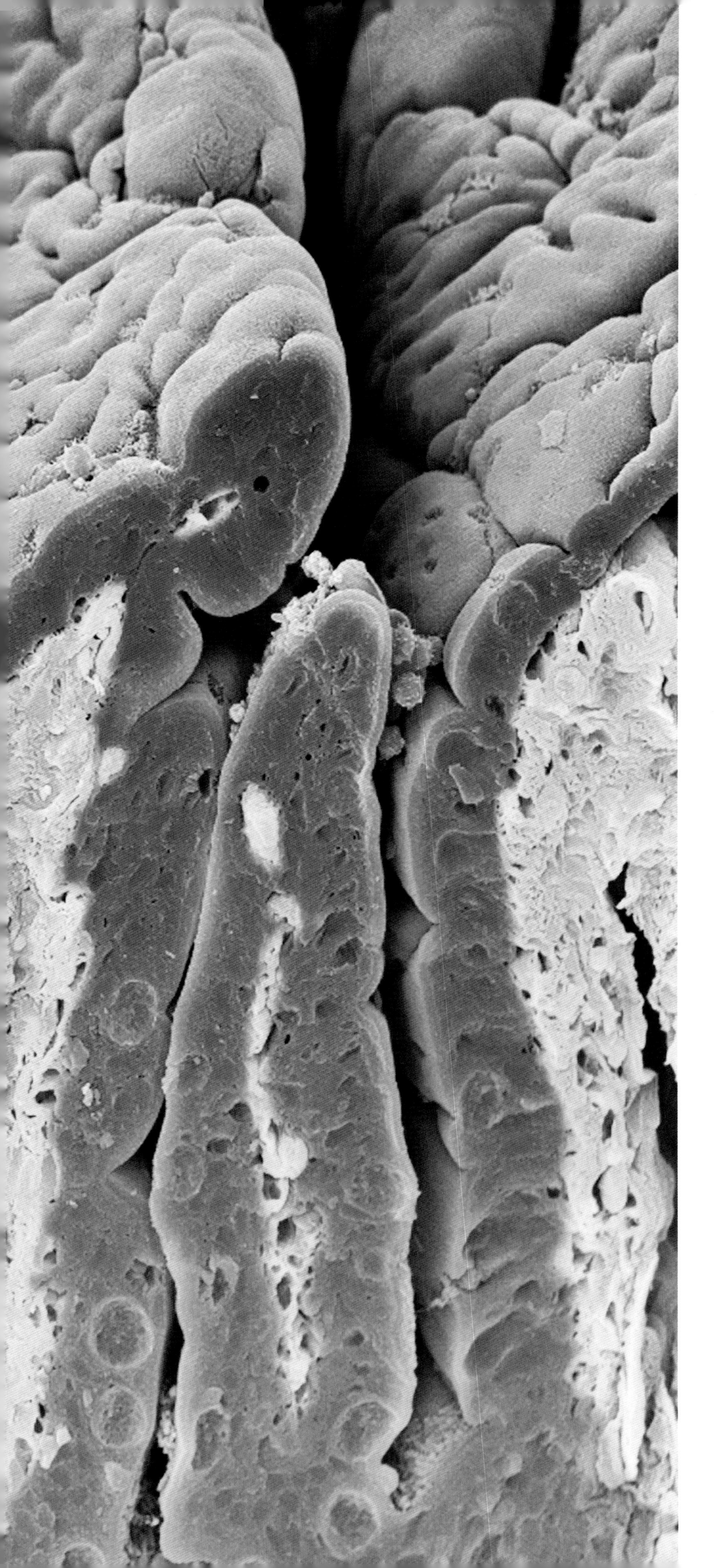

Lining of the small intestine [SEM]

THIS SECTION THROUGH the surface of the small intestine shows the deep folds, called villi, which increase the area for the absorption of nutrients from food. In the sectioned area, the epithelial cells (red) are supported by connective tissue (light brown) that forms the core of each fold. The height of a villus in the small intestine varies from 0.3mm/1⁄80in to 0.8mm/1⁄32in.

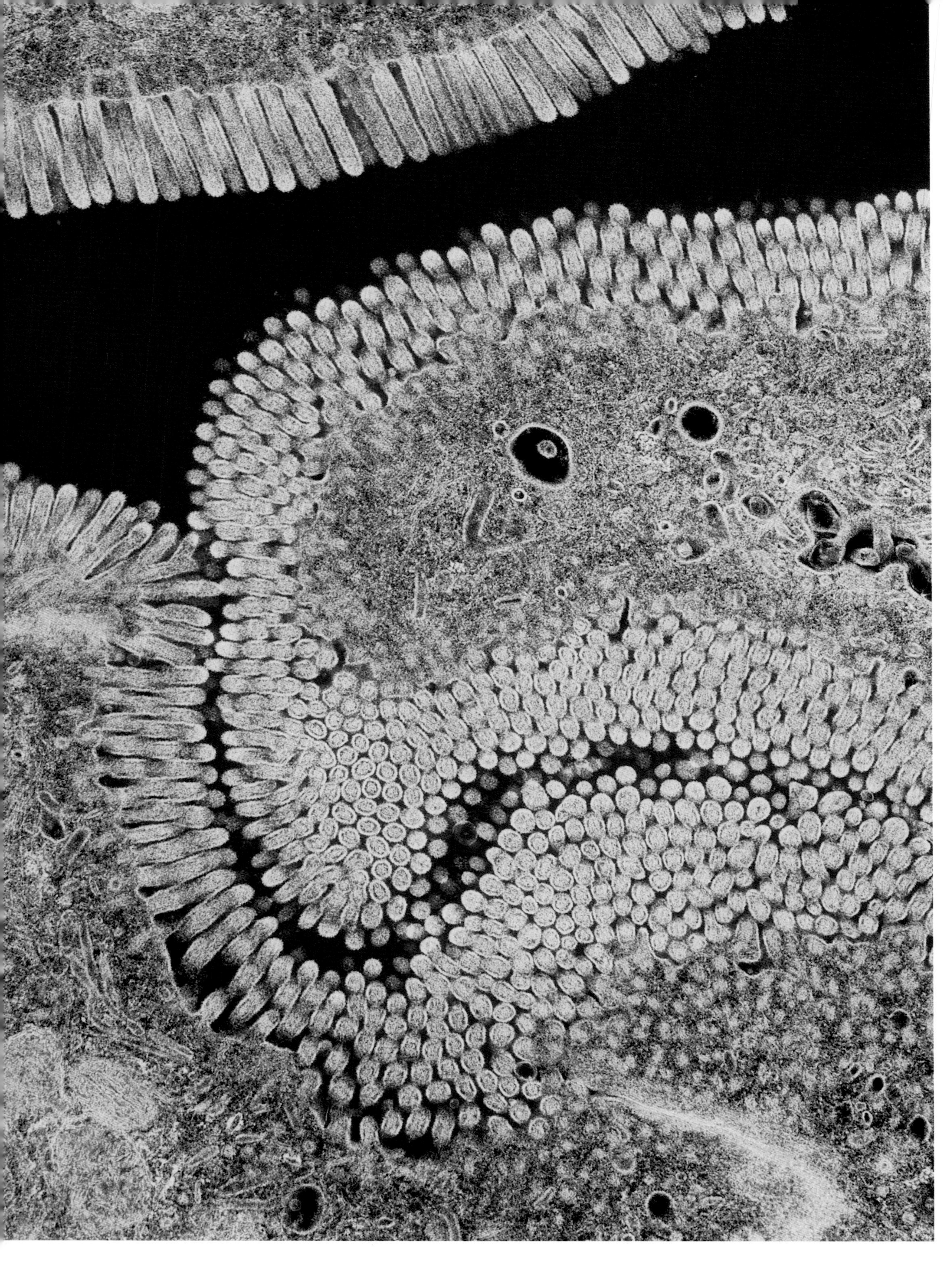

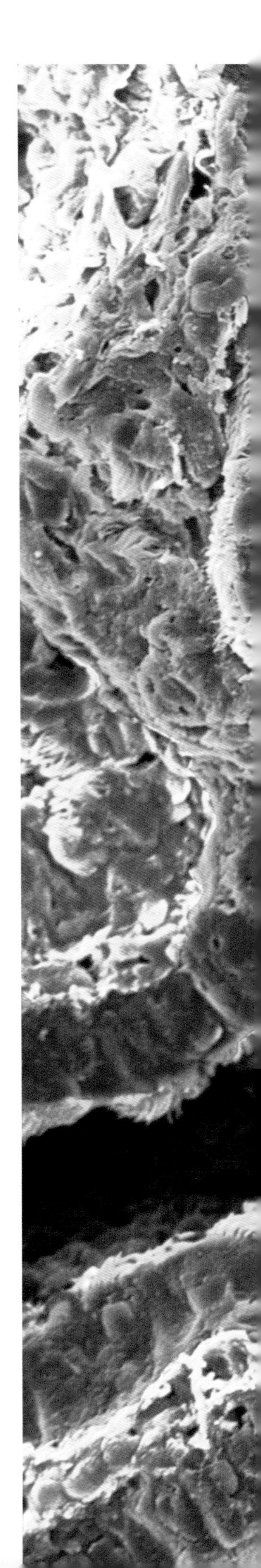

Large intestine microvilli [TEM]

THE BLUE AREAS IN THIS SECTION through a large intestine are microvilli. These tiny, hair-like structures increase the surface area through which water and any remaining nutrients are absorbed. Reabsorption of water used in the digestive process takes place mainly in the small intestine, but of the 1 litre/2 pints or so that reaches the large intestine each day, about 90 per cent is reabsorbed.

Fallopian tubes [SEM]

FALLOPIAN TUBES ARE THE PAIR of muscular ducts through which eggs pass from the ovaries to the uterus (womb). Each tube is lined with a highly folded mucus membrane (pale purple). The mucus secreted by the membrane helps transport the egg, aided by contractions of the tube's muscular walls and the rhythmic movement of tiny, hair-like structures on its surface.

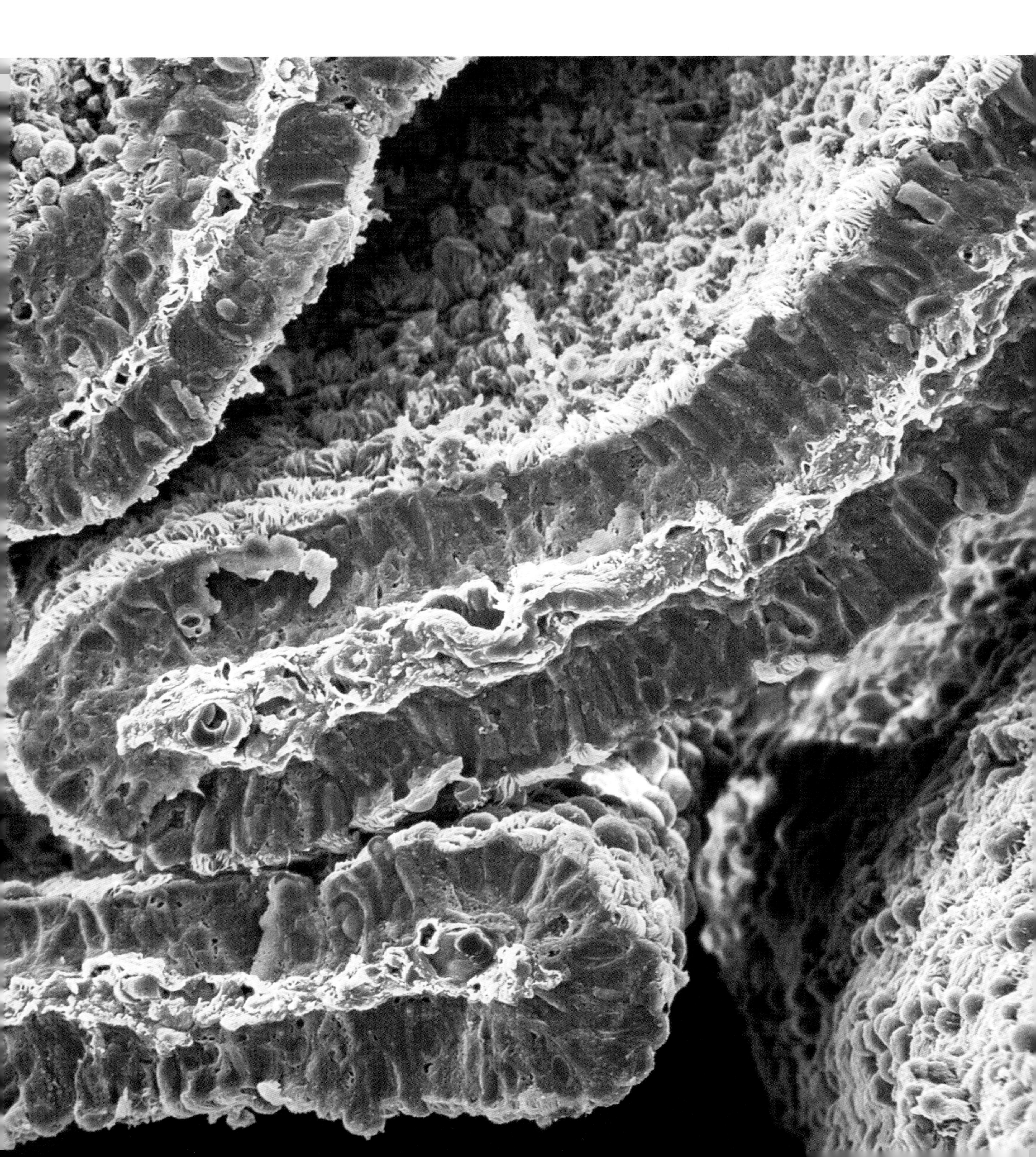

Our organs are supported and held in position by connective tissues. Unlike epithelial tissue, where the cells are packed together, connective tissue consists of a matrix of protein fibres secreted by a scattered population of cells. They are the fibres that give connective tissue its strength.

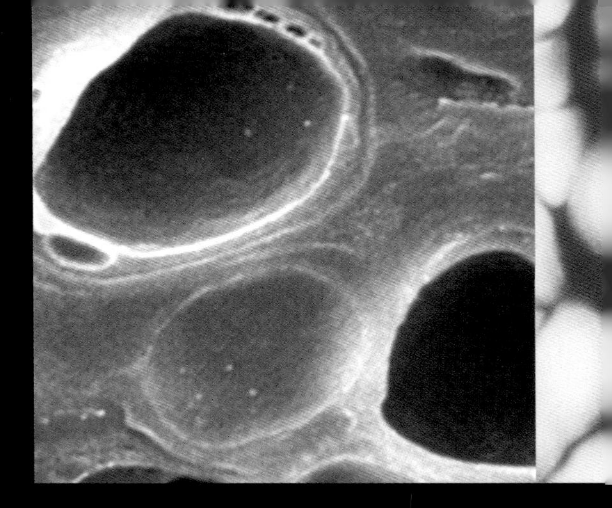

Collagen is the most widespread protein fibre. In fact, it is the most abundant protein in the body, representing 25 per cent of total body protein. Its structure, resembling that of a plaited, three-strand rope, gives it high tensile strength but relatively low elasticity. The loose connective tissue that fills spaces between organs has a low density of collagen. Tendons, which attach muscles to bones, and ligaments, which bind bones together, are composed mainly of collagen. Bones are also made up of collagen, with the mineral calcium phosphate providing rigidity. In cartilage, the collagen fibres are embedded in a flexible and compressible matrix.

The other type of structural protein is elastin. As its name suggests, elastin forms a flexible fibre with great resilience. Under tension it can be stretched to several times its resting length, then snap back when the tension is released. Elastin is found in the walls of the lungs and large arteries, but we

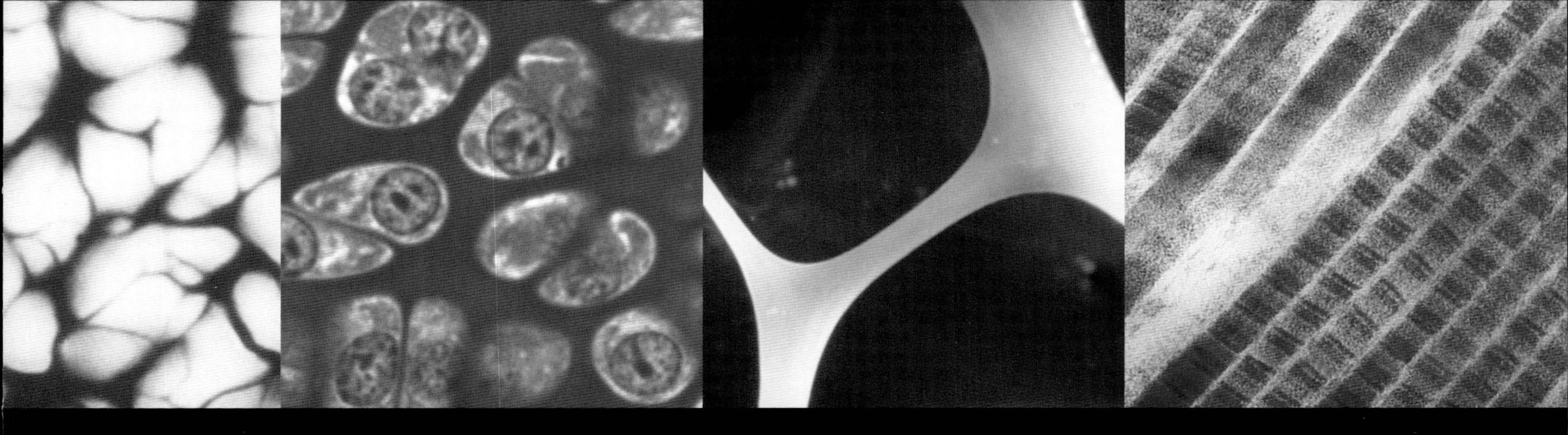

are most familiar with its elastic properties in the skin. It is the loss of elastin fibres with age that causes skin to wrinkle.

Fat is also classed as a connective tissue, although it acts more as a cushion and insulator than as a support. Fat cells, as well as being a source of energy, also produce the hormone leptin, which acts on centres in the brain that tell us to stop eating. Injections of leptin can help curb the appetites of overweight people who are deficient in this hormone. However, most obese individuals have more than adequate leptin levels, suggesting that their brain receptors have reduced sensitivity to this hormone – or that they simply enjoy food too much and override the brain signals telling them to stop eating.

Supporting and connecting tissue

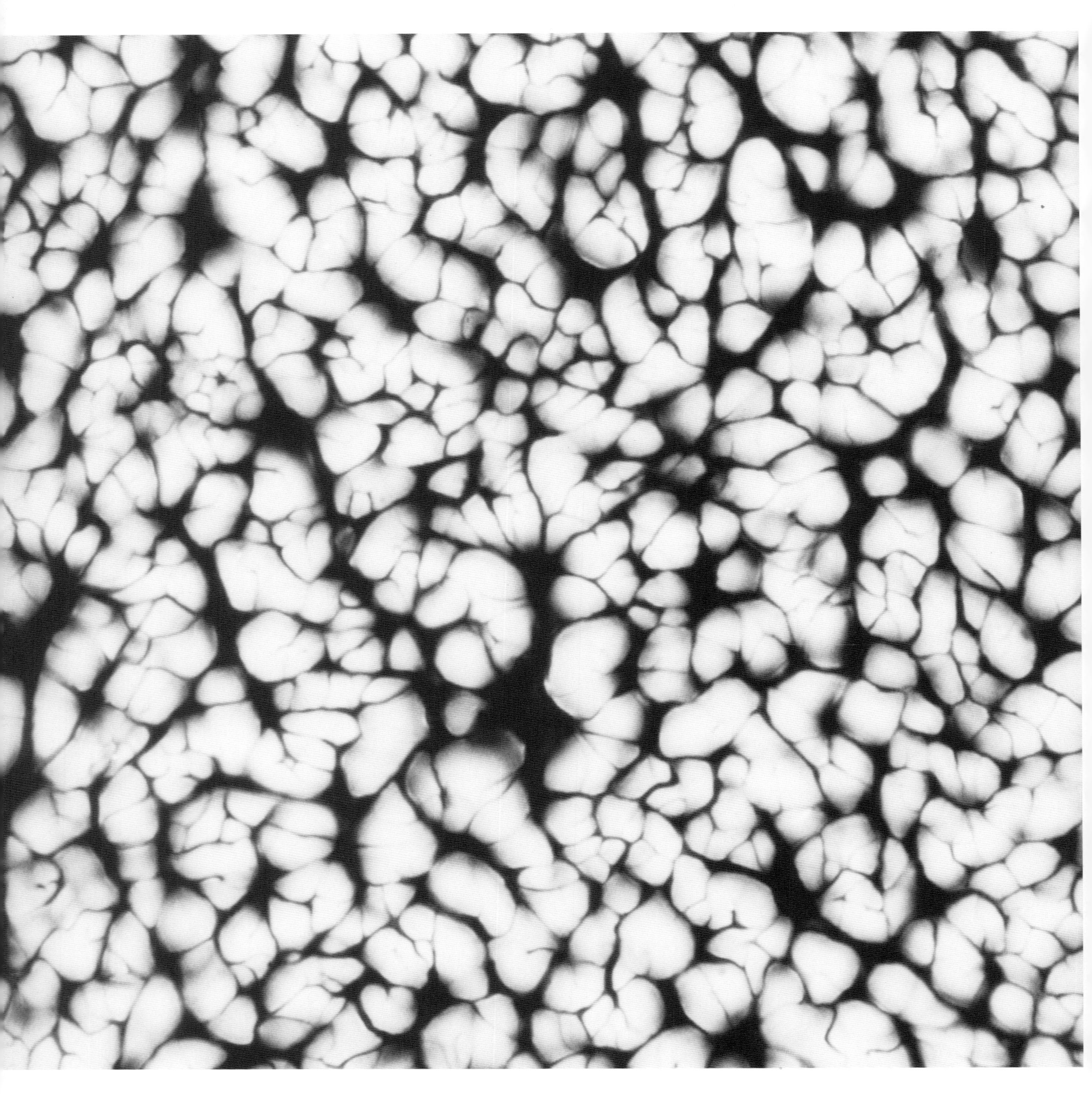

Tendon section [LM]

TENDONS ATTACH MUSCLE to bone and are inelastic. Seen in section, a tendon consists of bundles of closely packed collagen fibres (white) which give it great strength. The orange spaces between the fibres are loosely packed with a semi-fluid material known as areolar tissue, a type of connective tissue that fills the spaces between adjacent tissues and joins organs together.

Collagen fibres [TEM]

COLLAGEN FIBRES, seen here in cross section, strengthen tissues in much the same way as steel rods reinforce concrete. The major structural protein in the body, collagen reinforces skin, tendons, ligaments, bones and arteries. In some individuals, a genetic disorder of collagen production causes brittle bone disease.

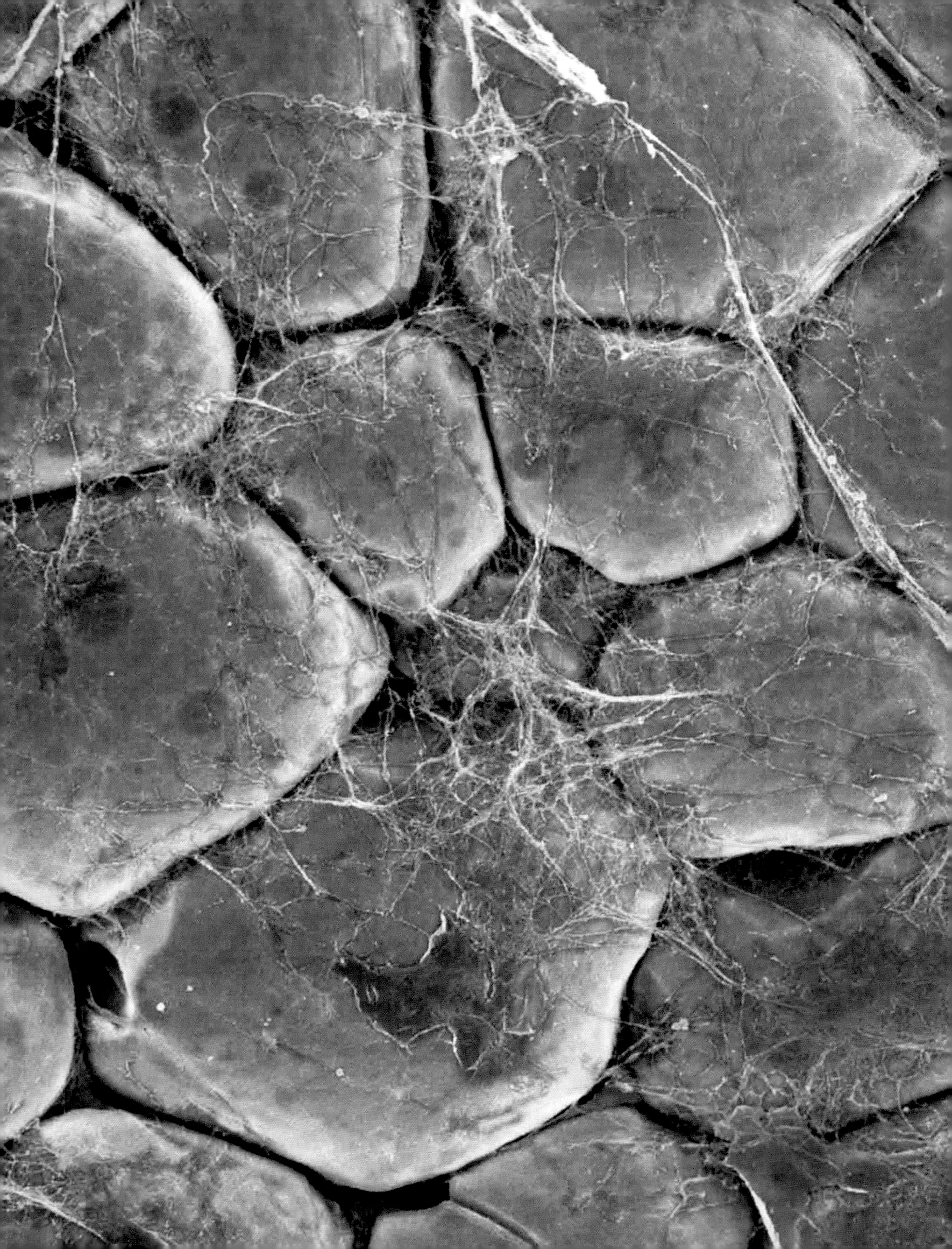

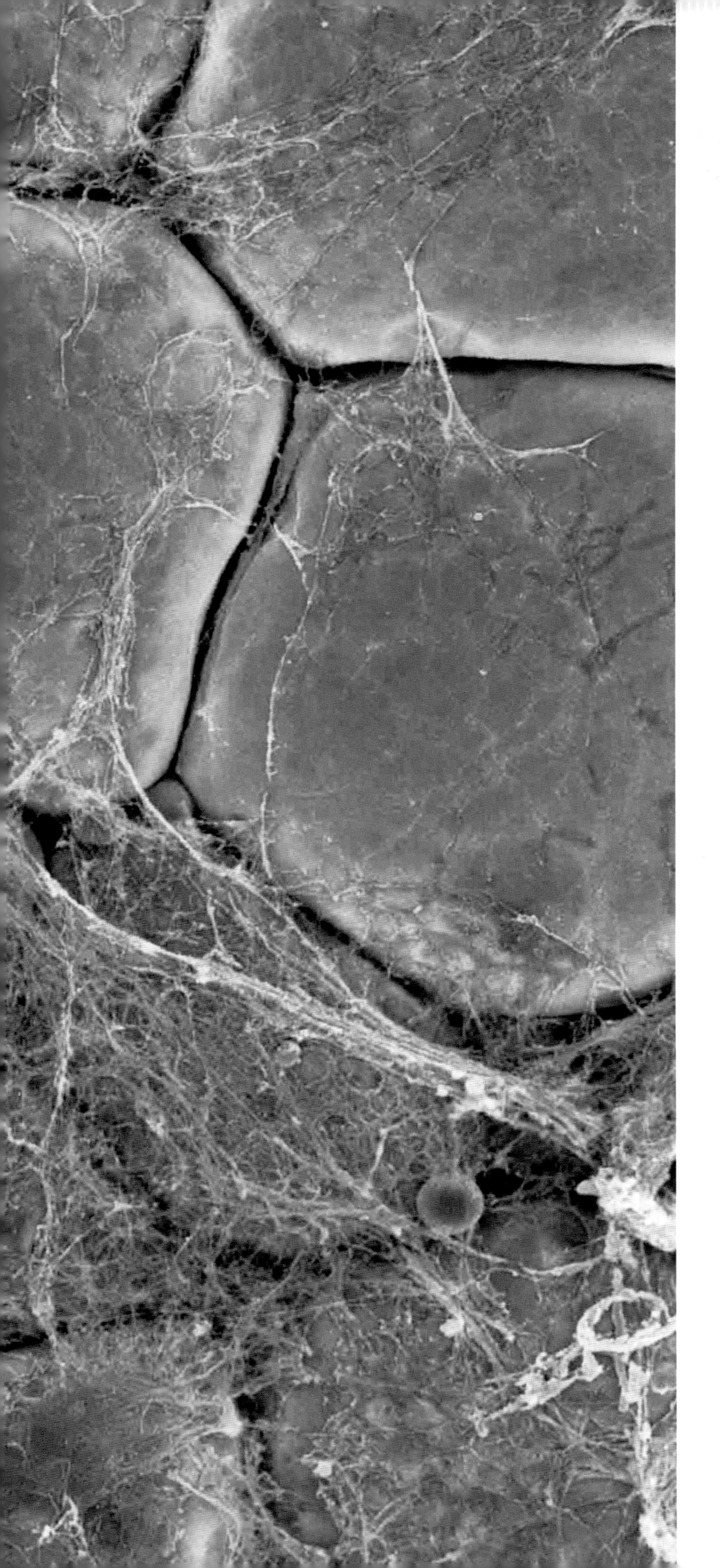

Fatty tissue [SEM]

ADIPOSE OR FATTY TISSUE has little structural strength, being composed of large cells filled with globules of oil or fat. Located beneath the skin and around internal organs such as the kidneys, fatty tissue provides insulation and is an important source of energy in concentrated form.

Heart tendons [Macrophoto]

POPULARLY KNOWN AS HEART STRINGS, these delicate but tough bands of connective tissue attach the muscles on the internal heart wall to the heart valves. The valves help control the flow of blood between the chambers of the heart.

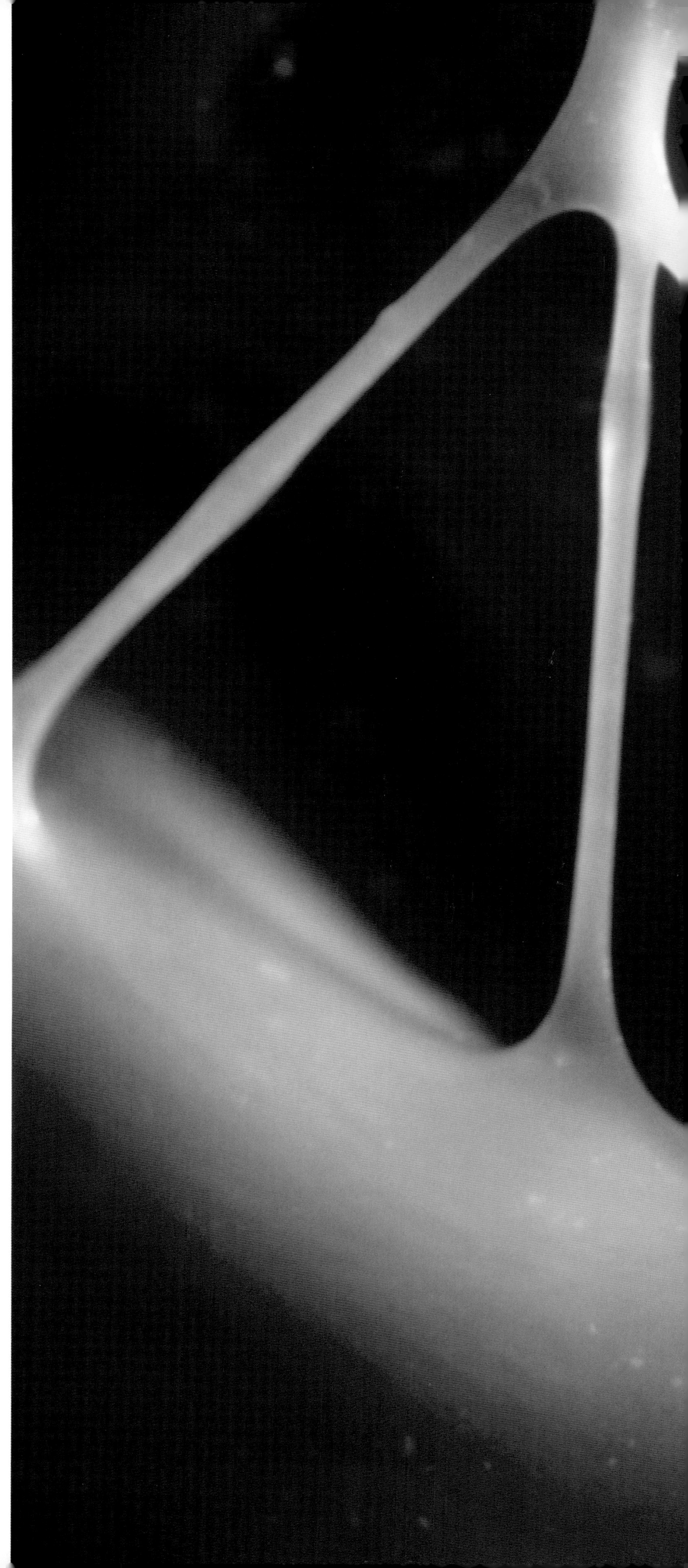

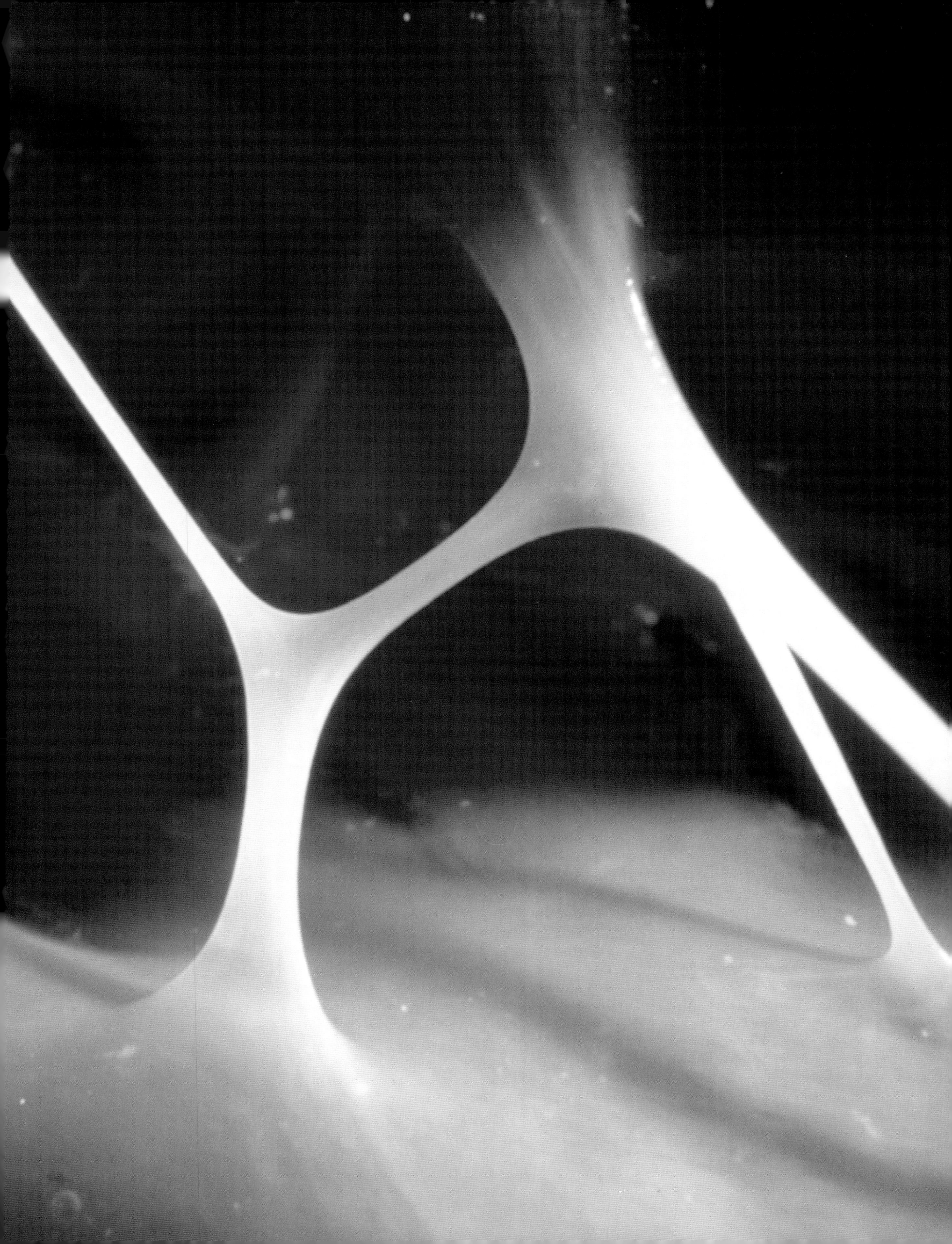

Collagen fibrils in bladder [SEM]

IN THE WALL OF AN EMPTY BLADDER, collagen fibrils are packed together in spirally twisted bundles. As the bladder fills, the bundles will untwist until they reach full extension. Because collagen is strong but relatively inelastic, the fibrils prevent overstretching of the bladder.

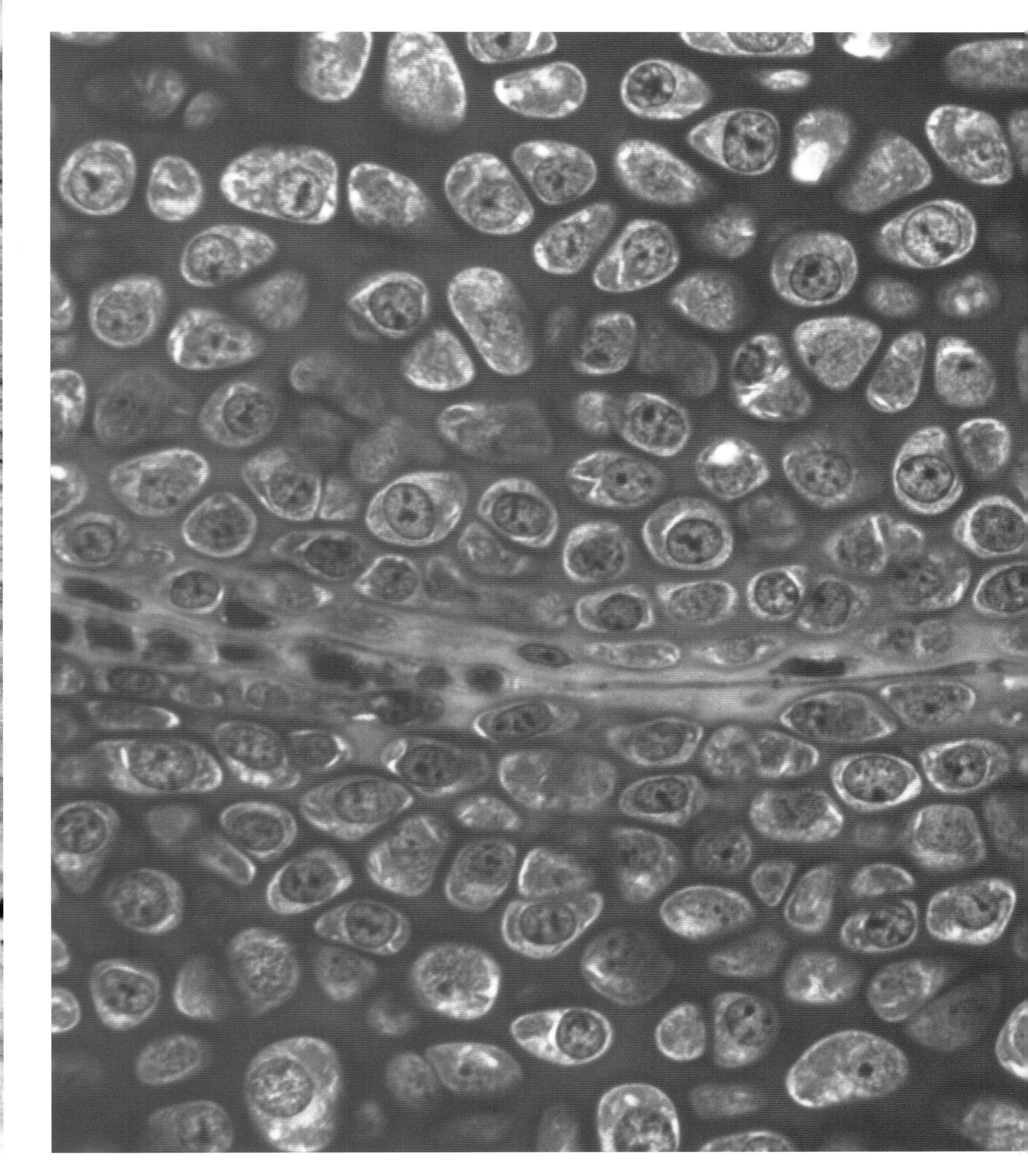

Hyaline cartilage [LM]

A SEMIRIGID CONNECTIVE TISSUE, hyaline cartilage has a slippery surface that reduces friction between joints at the end of bones. It is also found in the nose. This section is from the knee joint of a fetus. The pale purple cells are chondroblasts, which synthesize an extracellular matrix (dark purple) of collagen and other proteins in water. Cartilage is strengthened by collagen, while its high water content makes it compressible, protecting it from shocks.

Elastic cartilage in ear [SEM]

A SECTION THROUGH the outer ear shows the elastic cartilage (green) that maintains the ear's shape while making it flexible. Elastic cartilage contains the protein elastin. It is also found in the throat and epiglottis, the flap that closes over the airway during swallowing.

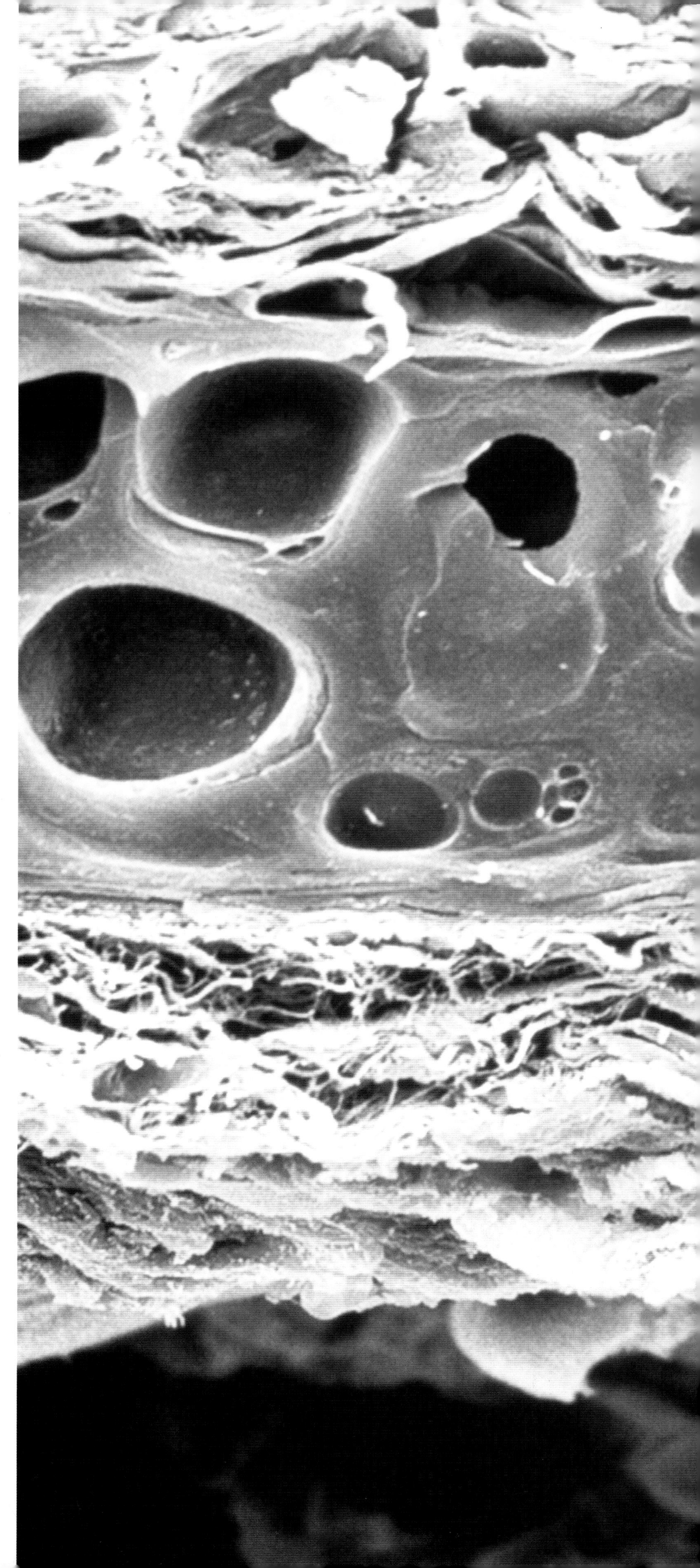

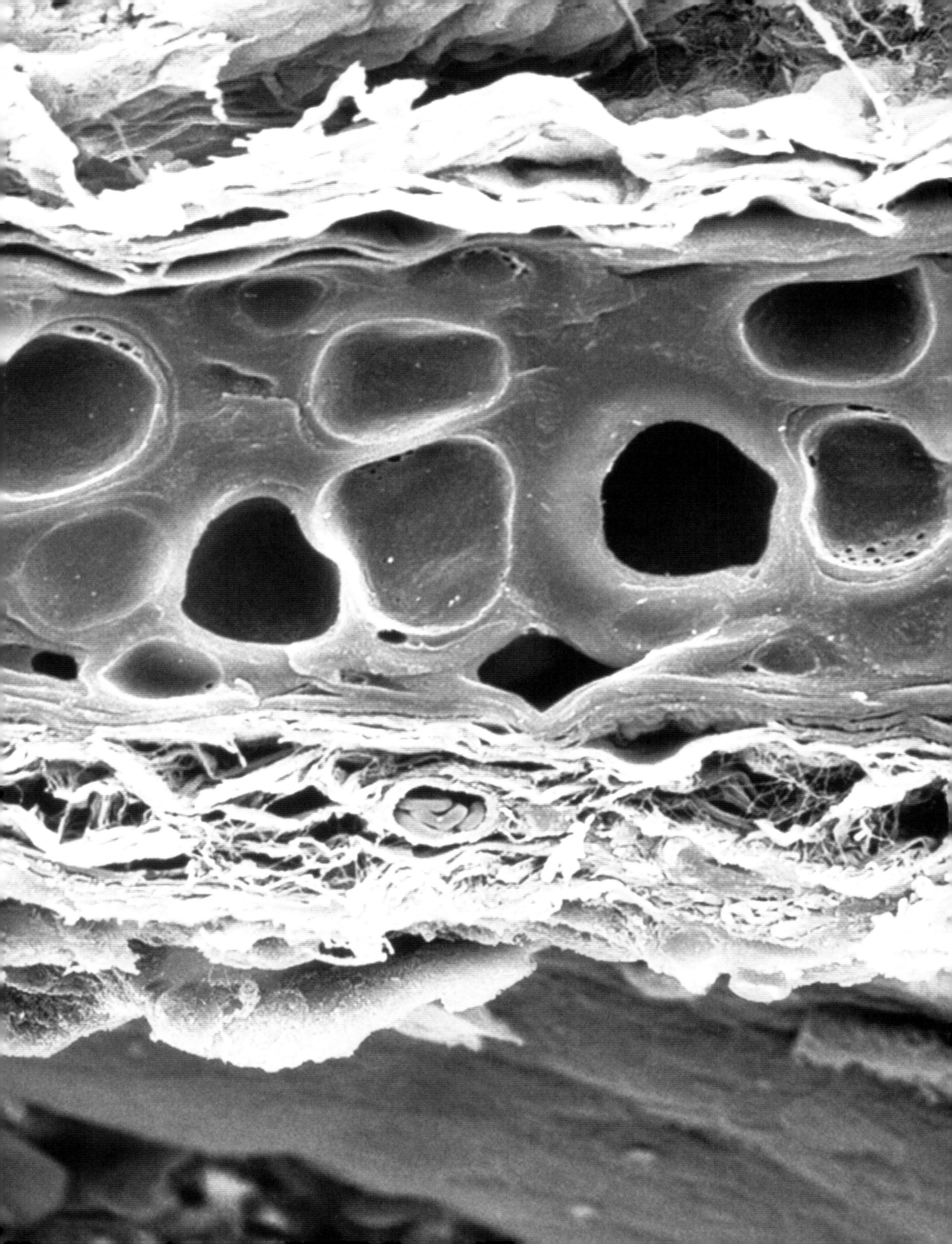

About two-fifths of our body weight is muscle. Some muscles are attached to bones and are used to move our bodies. Other muscles are found in the walls of internal organs and help transport fluids and other substances.

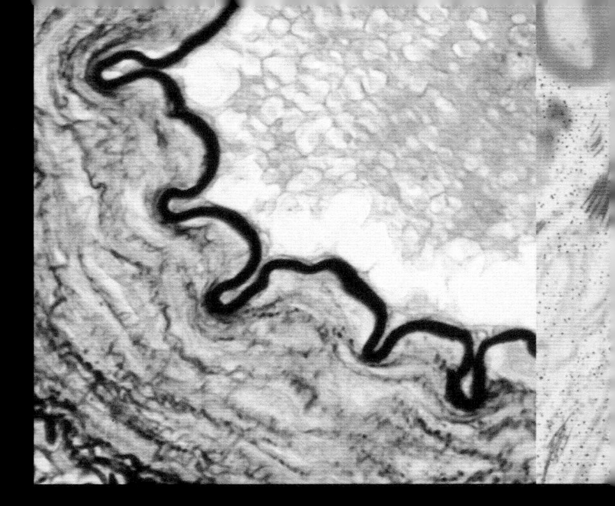

All muscles work by contracting. They cannot lengthen beyond their natural resting state, so to move a bone back and forth two sets of muscles – known as an antagonistic pair – are needed. A good example of an antagonistic pair is the biceps and triceps of the upper arm. By gently gripping your arm, you can feel the biceps contract as you bend the arm, then relax as you straighten it, when the triceps tighten.

These muscles and the other 638 muscles that move bones (and also control facial expression) are called skeletal muscles. They are called voluntary muscles as they can be contracted at will. Skeletal muscles, also known as striated muscles, are bundles of long fibres containing many pairs of protein microfilaments that work like chemical zips. When stimulated by a nerve signal, these microfilaments hook together and slide over each other, shortening the muscle by up to a third of its resting length.

Skeletal muscle fibres can be subdivided into two types known as slow-twitch fibres and fast-twitch fibres. Their different properties help explain why sprinters rarely become world-class marathon runners. Slow-twitch fibres contract slowly, use relatively little energy and are highly resistant to

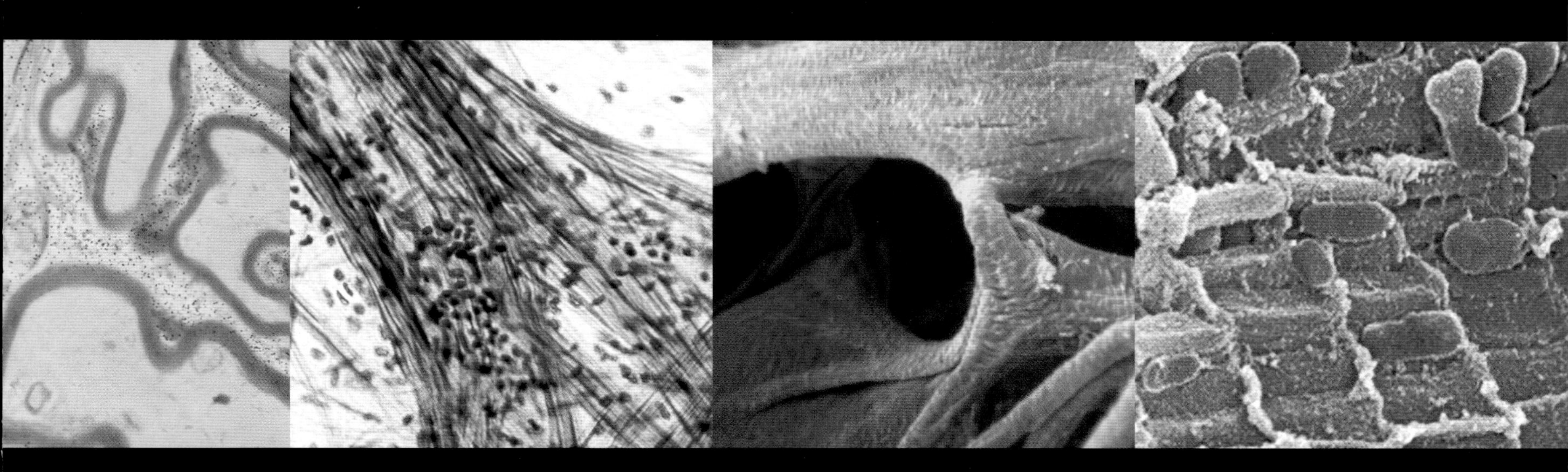

fatigue. The skeletal muscles of long-distance runners consist mostly of slow-twitch fibres. Sprinters and weightlifters, on the other hand, have a high proportion of fast-twitch fibres, which act in short, powerful bursts but soon tire. Aerobic exercises can improve the fatigue-resistance of fast-twitch fibres, but genetic heritage is the most important factor in determining which sports we excel at.

There are two other types of muscle tissue: cardiac muscle and smooth muscle. Both types are so-called 'involuntary' muscles which means they cannot be contracted at will. Cardiac muscle is a branching type of striated muscle that contracts automatically, powerfully and without tiring for decades on end. Smooth muscles are usually long, spindle-shaped fibres that contract slowly and rhythmically, moving food through the digestive system, controlling the flow of blood through blood vessels, and emptying the bladder of urine.

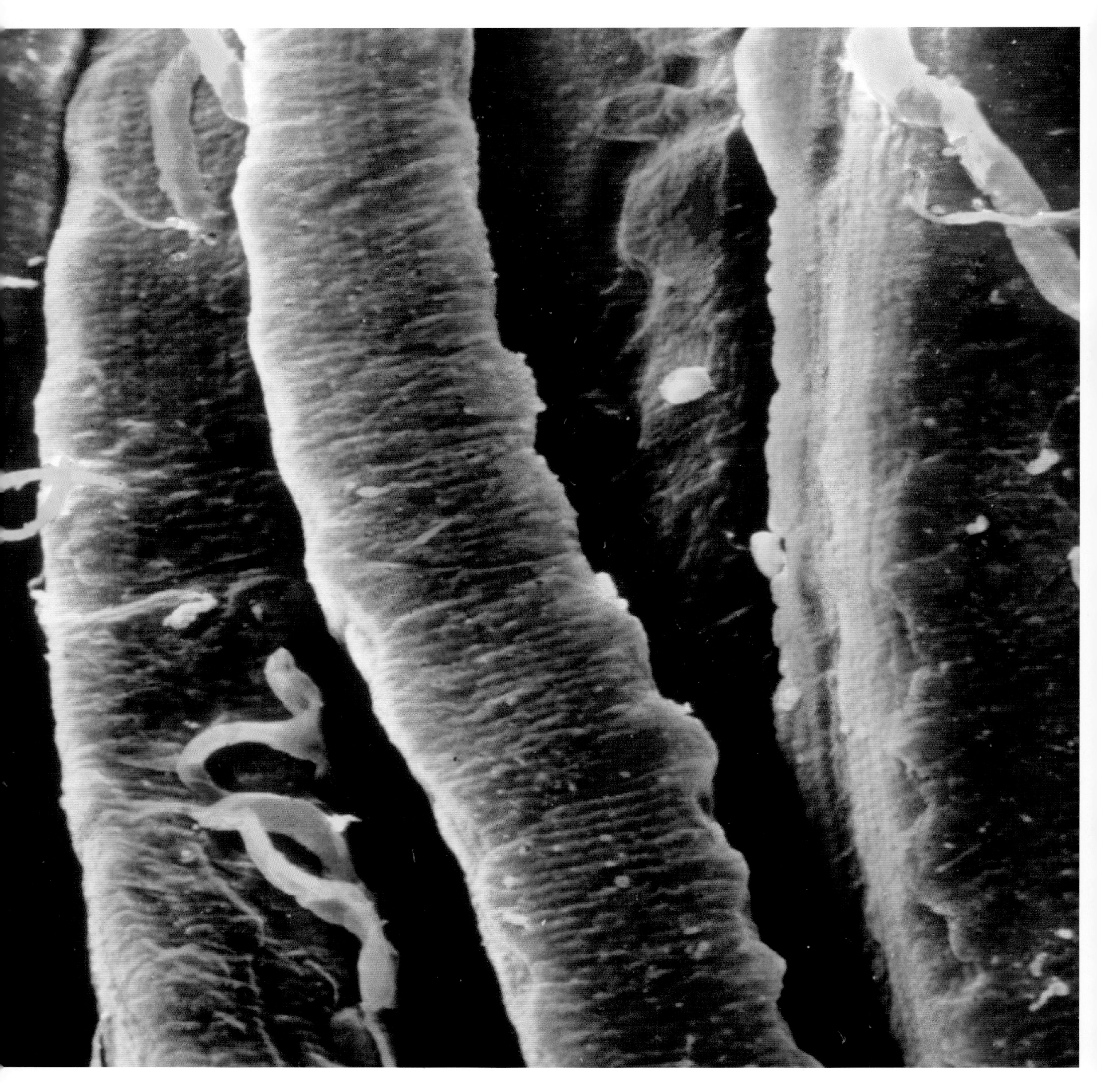

Skeletal muscle fibres [SEM]

A BUNDLE OF SKELETAL MUSCLE FIBRES (red) is shown with a few capillaries (blue). Skeletal muscles are under conscious control and are responsible for the movements of the skeleton and organs such as the tongue and eye. Each fibre is surrounded by a thin layer of connective tissue and is grouped into a bundle called a fasciculus. Most muscles are made up of many fasciculi.

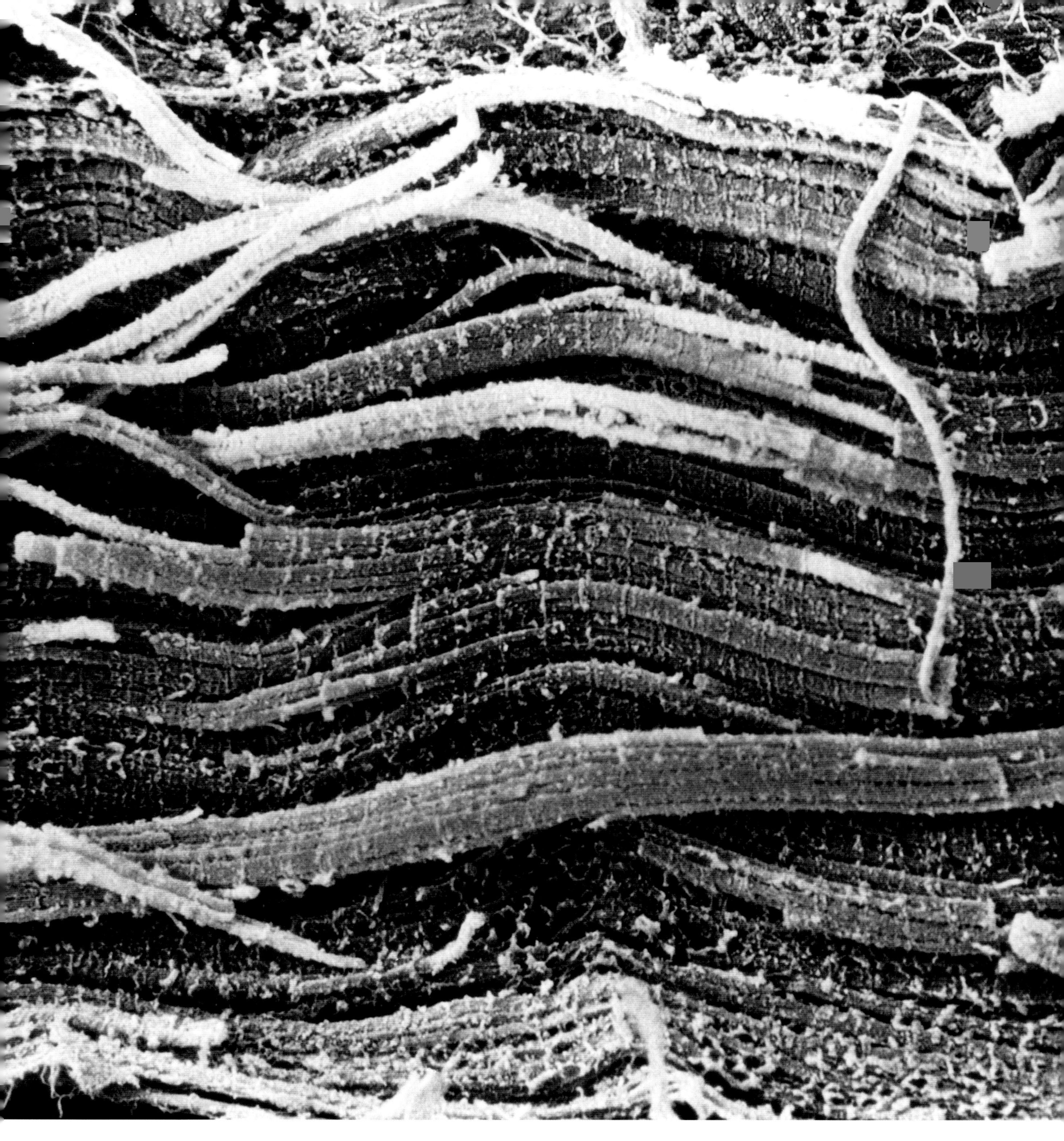

Mybrofils in skeletal muscle fibres [SEM]

SKELETAL MUSCLE FIBRES are thin cylindrical cells that can be up to 30cm/12in long. Running the length of each fibre are thin threads called mybrofils. Each mybrofil contains filaments of two types lying next to each other. One type is made of the protein myosin; the other is composed of the protein actin. When the muscle is stimulated by a nerve impulse, the actin filaments slide past the myosin filaments, causing the muscle to contract. Muscles can actively contract but not actively lengthen.

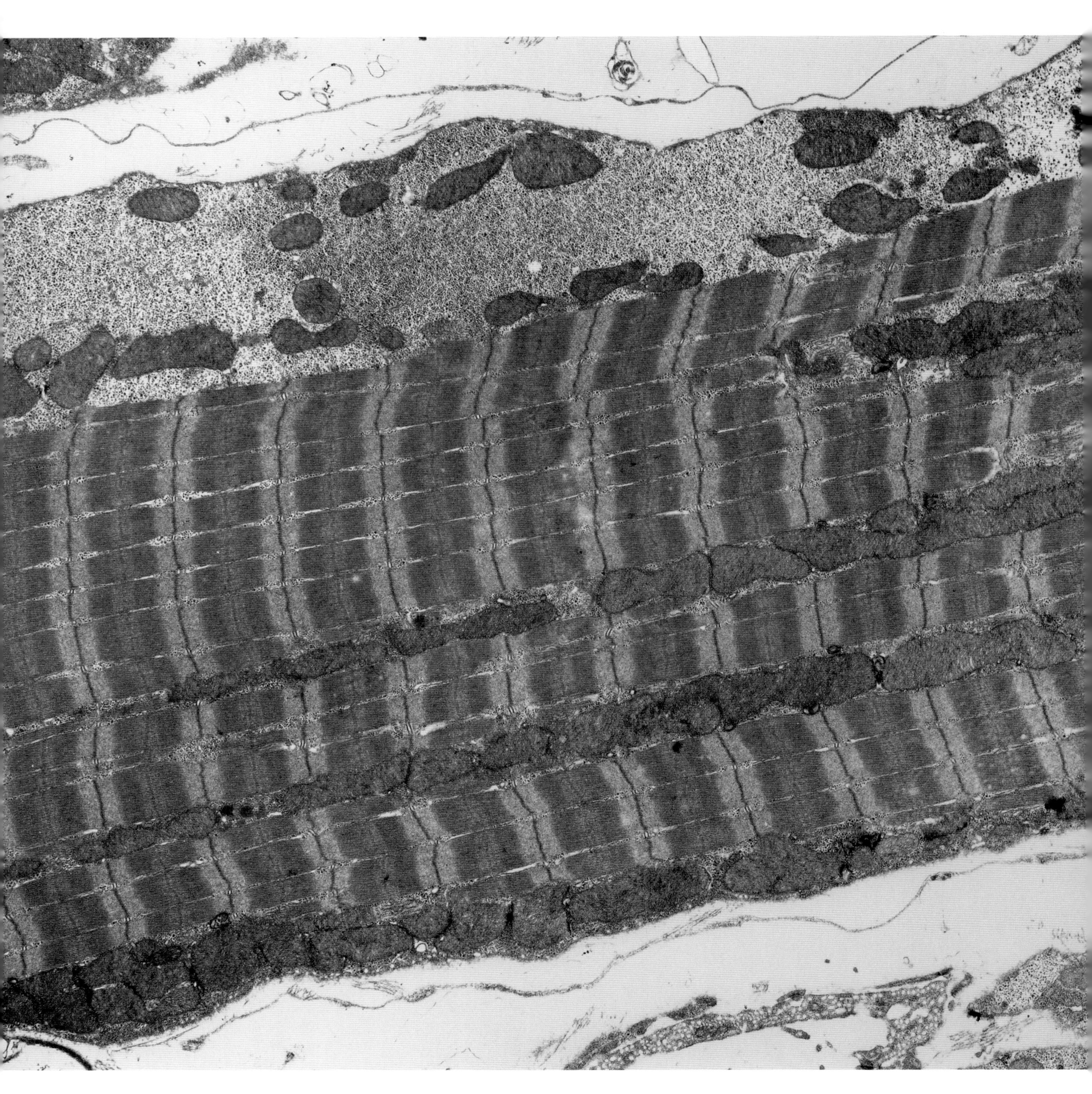

Striations in muscle fibre [TEM]

THE STRIPED PATTERN on this skeletal muscle fibre is produced by the regular arrangement of the two types of filaments that enable it to contract. The filaments, made of the proteins myosin and actin respectively, work a bit like a zip. When the muscle is stimulated, 'heads' on the myosin filament hook on to the actin filament, pulling it past them. Energy for this contraction is provided by mitochondria (small black dots).

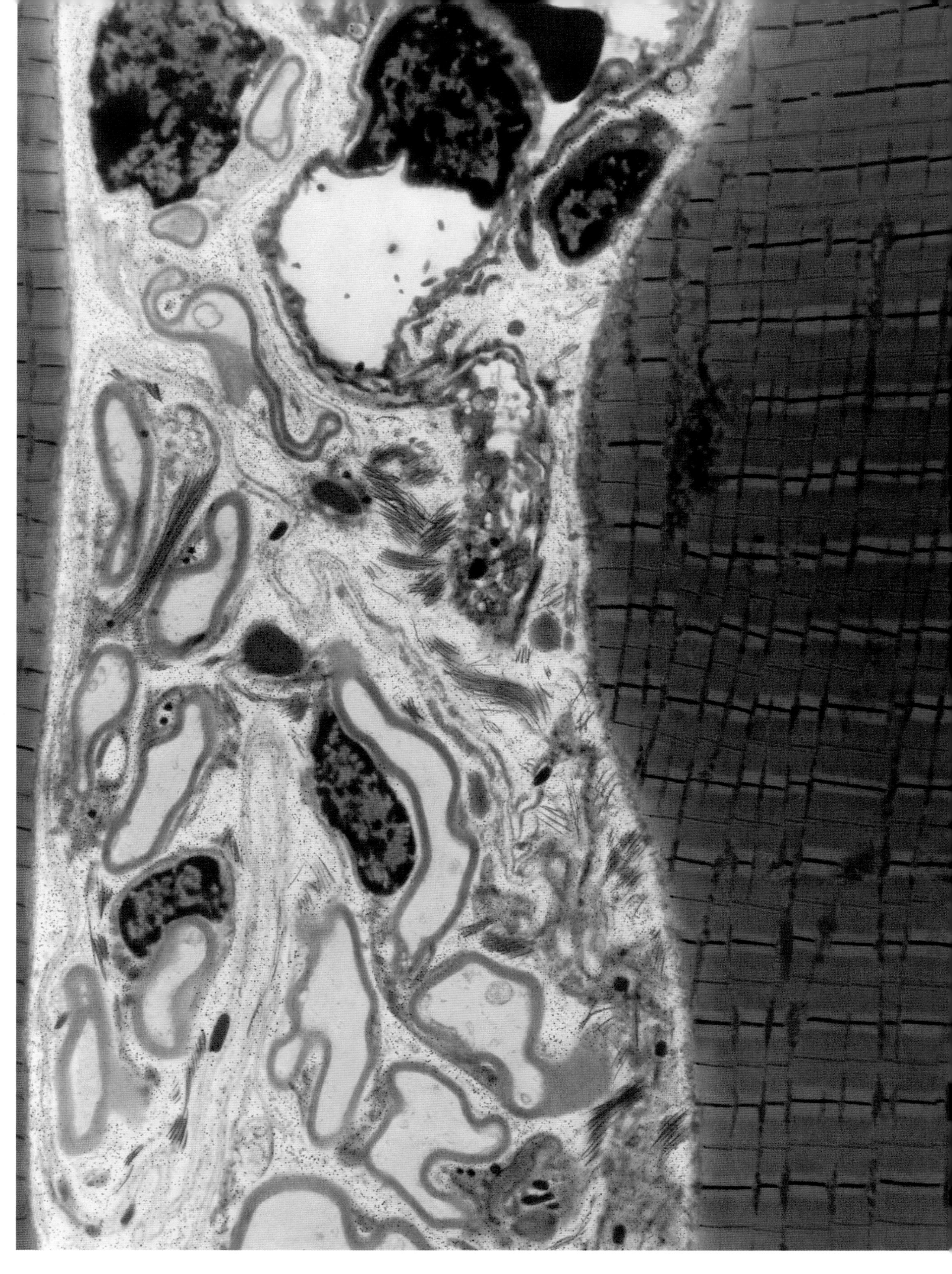

Skeletal muscle and nerve fibres [TEM]

SKELETAL MUSCLE (dark red) is stimulated to contract by signals carried by nerve fibres (pale green, inside dark green protective sheaths). In this image the muscle fibres run from top to bottom and are split into contractile units called sarcomeres by horizontal lines (black). The contraction of sarcomeres along the whole length of the muscle fibre causes it to shorten by up to 35 per cent.

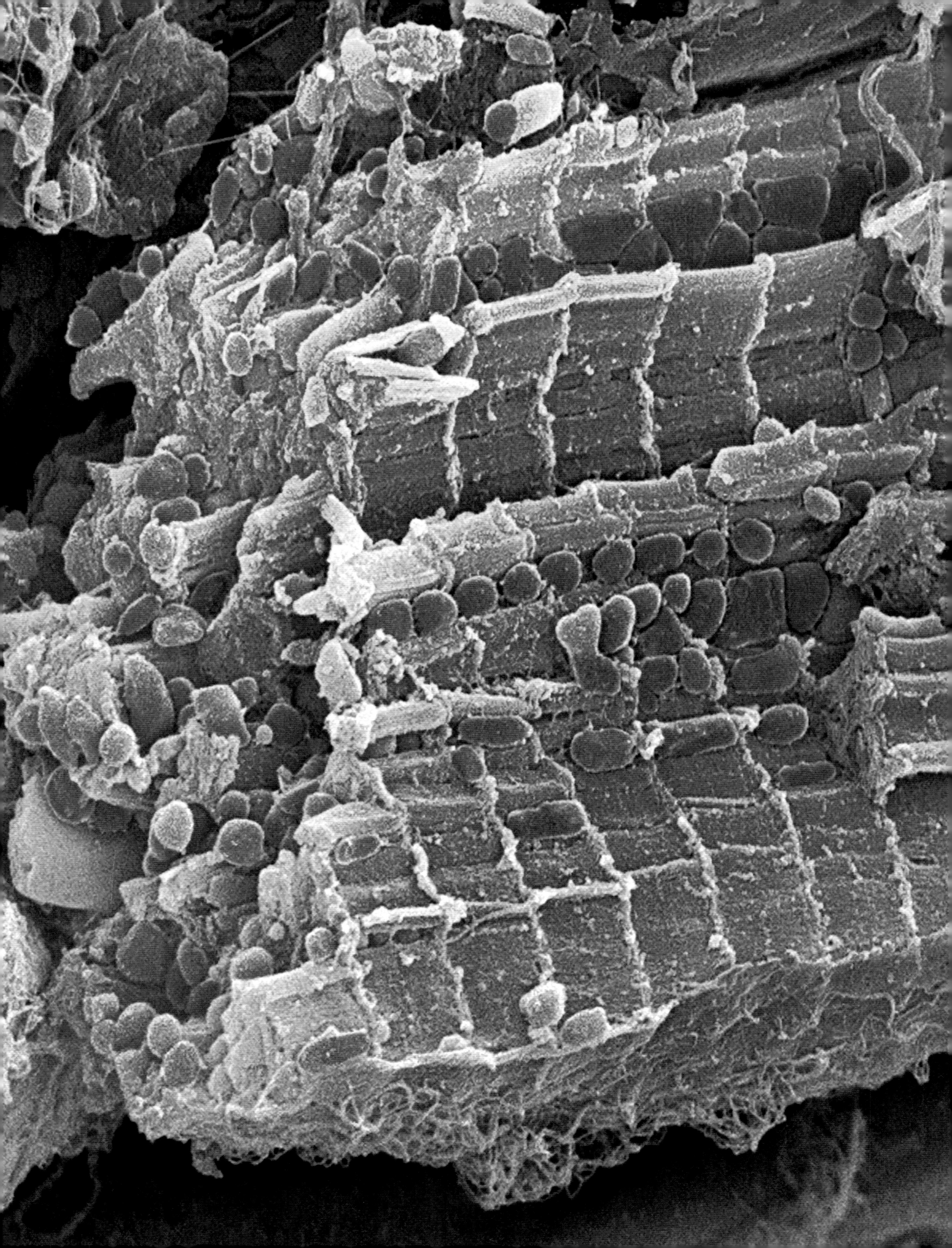

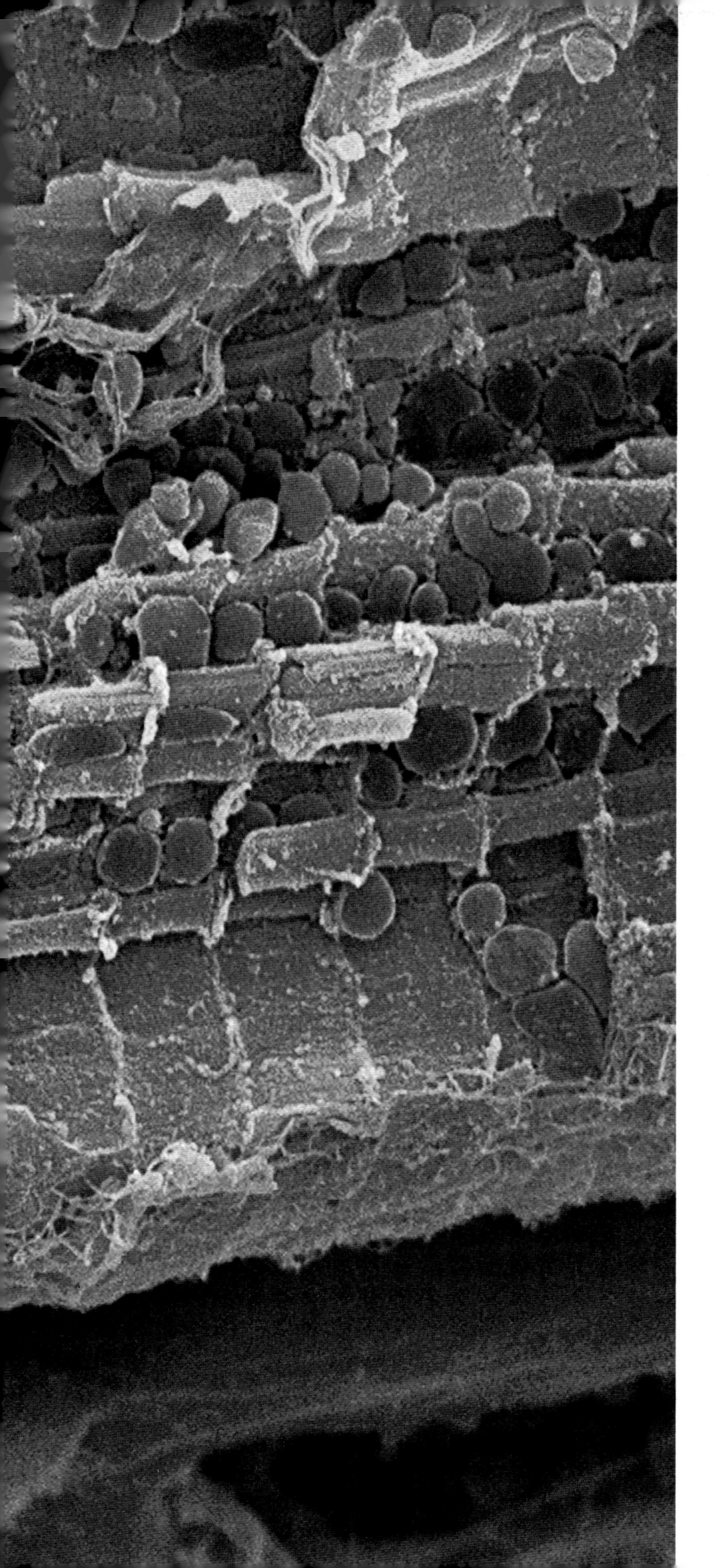

Cardiac muscles [SEM]

FOUND ONLY IN THE HEART, cardiac muscle under involuntary control pumps blood around the body without tiring. In order to beat, the heart produces regular electrical impulses, which cause the muscle fibres (blue) to contract. The large number of mitochondria (red) reflects the heart's high energy demands.

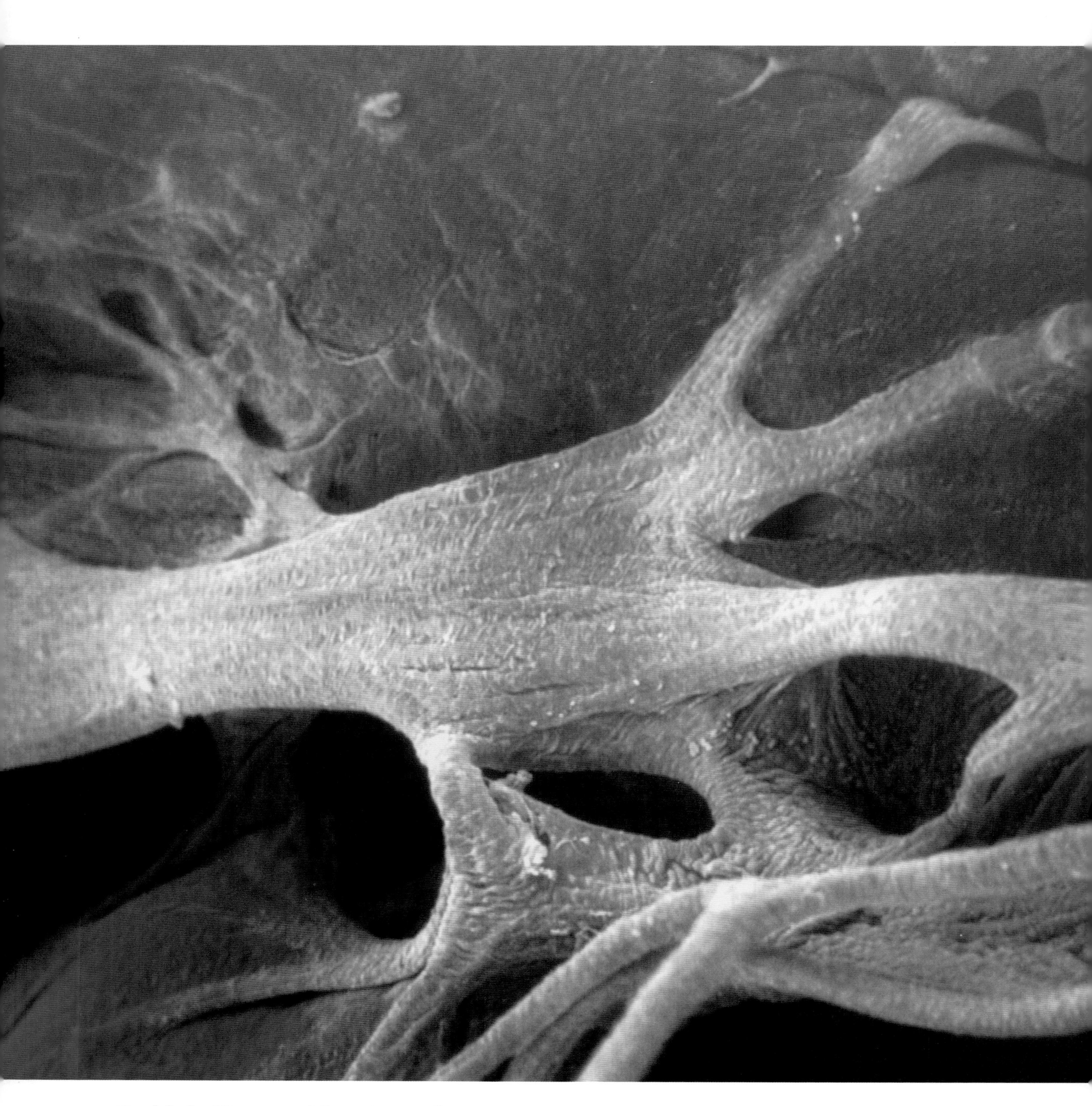

Purkinje fibres and heart muscles [SEM]

SOME PARTS OF THE HEART act as pacemakers, controlling the heartbeat. The green structures in this image are Purkinje fibres – modified cardiac muscle fibres that transmit an electrical impulse to the two ventricles, enabling their almost simultaneous contraction. The spread of excitation through the ventricles is very rapid, moving at up to 4m/13½ft per second.

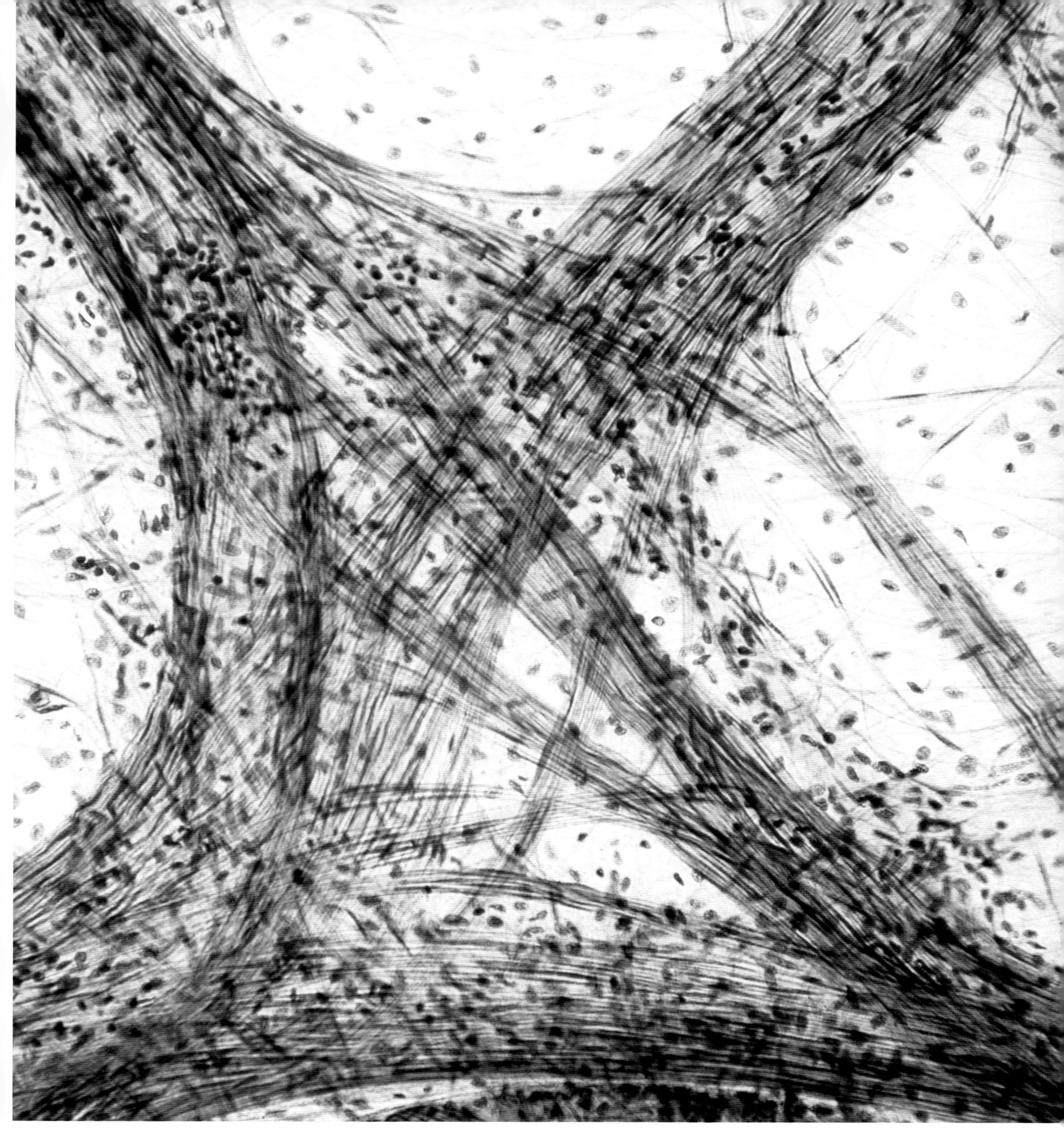

Smooth muscles of the bladder [LM]

THE SMOOTH MUSCLE FIBRES of the bladder can be seen as long, thin cells, each with a single, centrally located nucleus (purple dot). The bladder wall has three layers of smooth muscle fibre that intermingle and, unlike skeletal muscle, cannot be separated clearly. Smooth muscle is also called involuntary muscle, because it produces contractions which are beyond the conscious control of the individual.

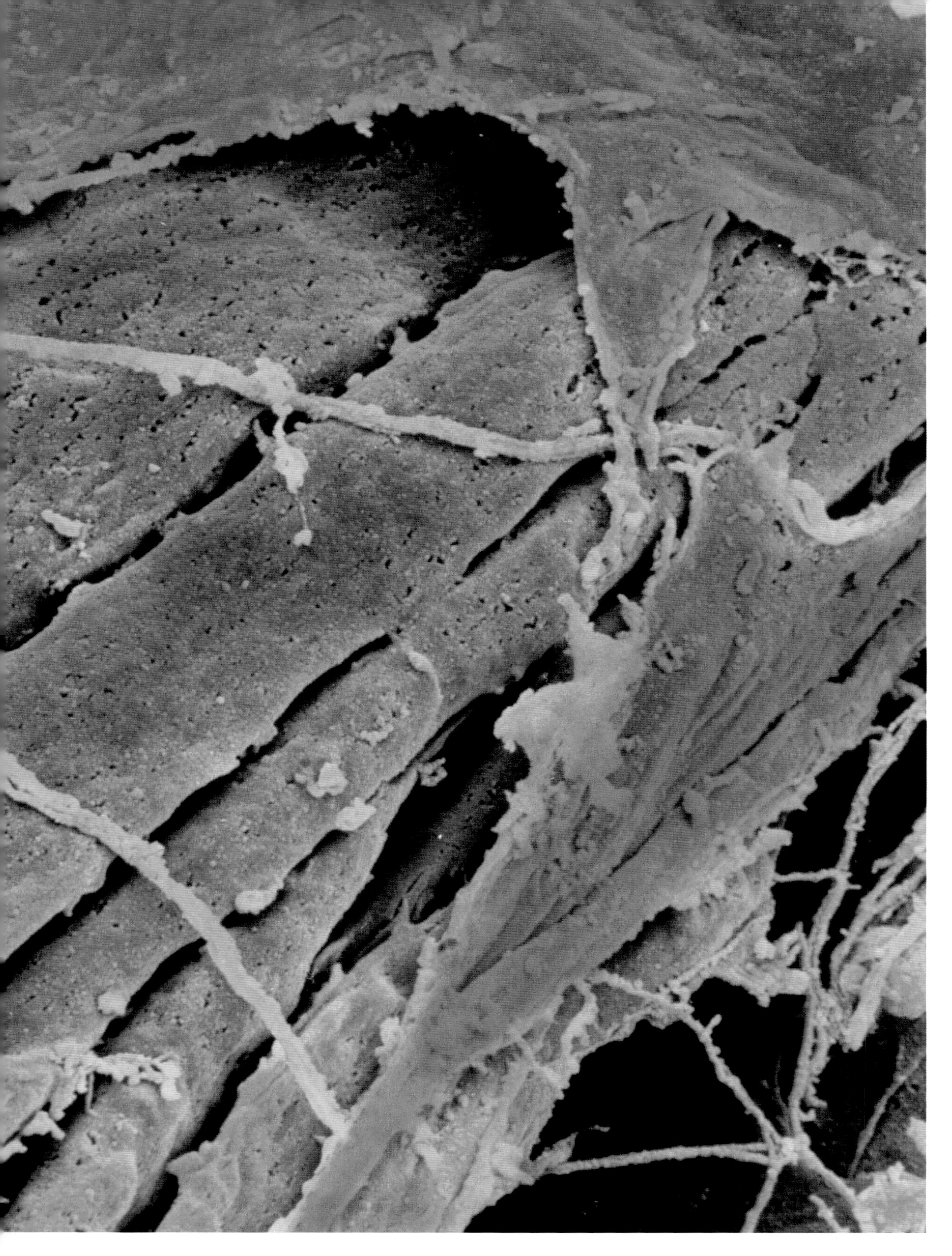

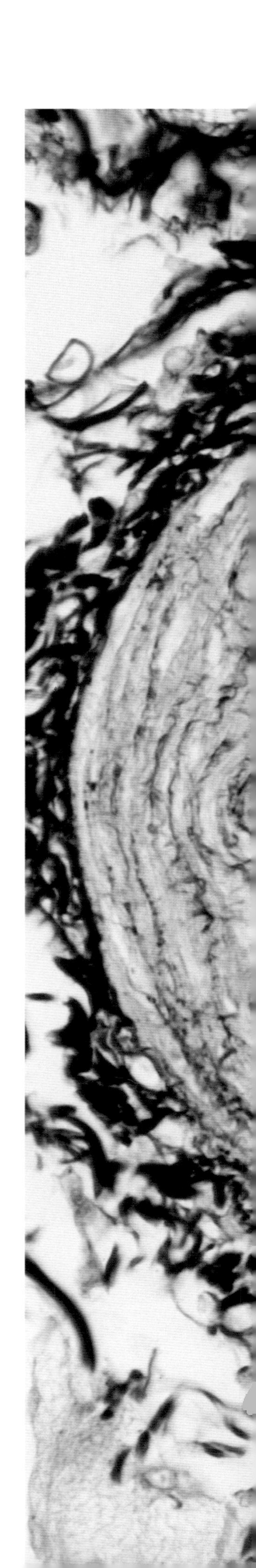

Smooth muscles in Fallopian tubes [SEM]

BANDS OF SMOOTH MUSCLE CELLS (pink) are seen in the wall of a Fallopian tube, the channel that conducts the female egg from the ovary to the uterus (womb), where the fertilized egg develops into an embryo. The bands of muscle help push the egg down the Fallopian tube by wave-like contractions.

Muscular renal arteries [LM]

ARTERIES HAVE THICK, MUSCULAR WALLS that enable them to carry blood away from the heart under high pressure. This artery, seen in section, is one of the pair of renal arteries that carries blood to each kidney. The muscle layer (blue) surrounds a thin, convoluted elastic lining known as the internal elastic lamina.

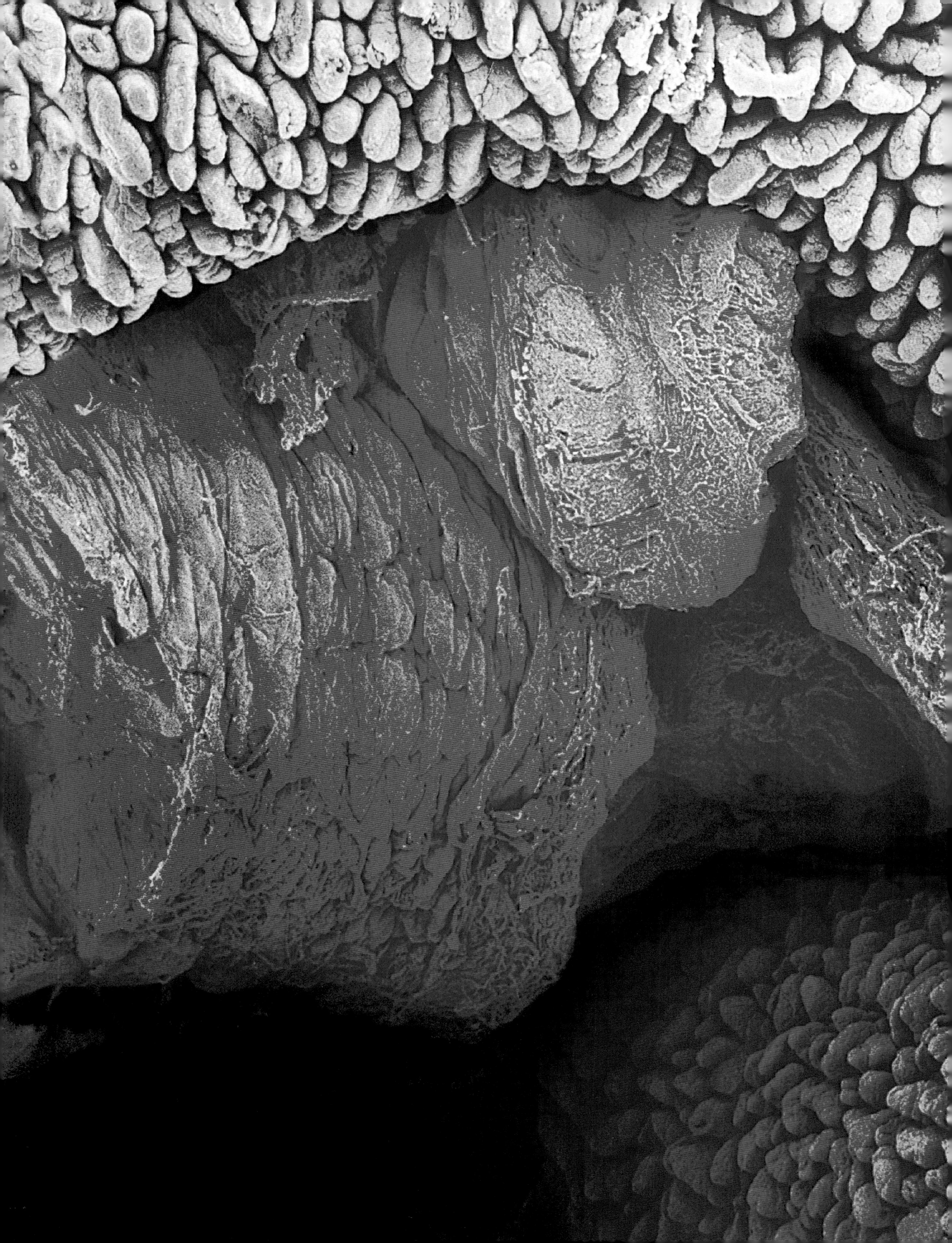

SYSTEMS

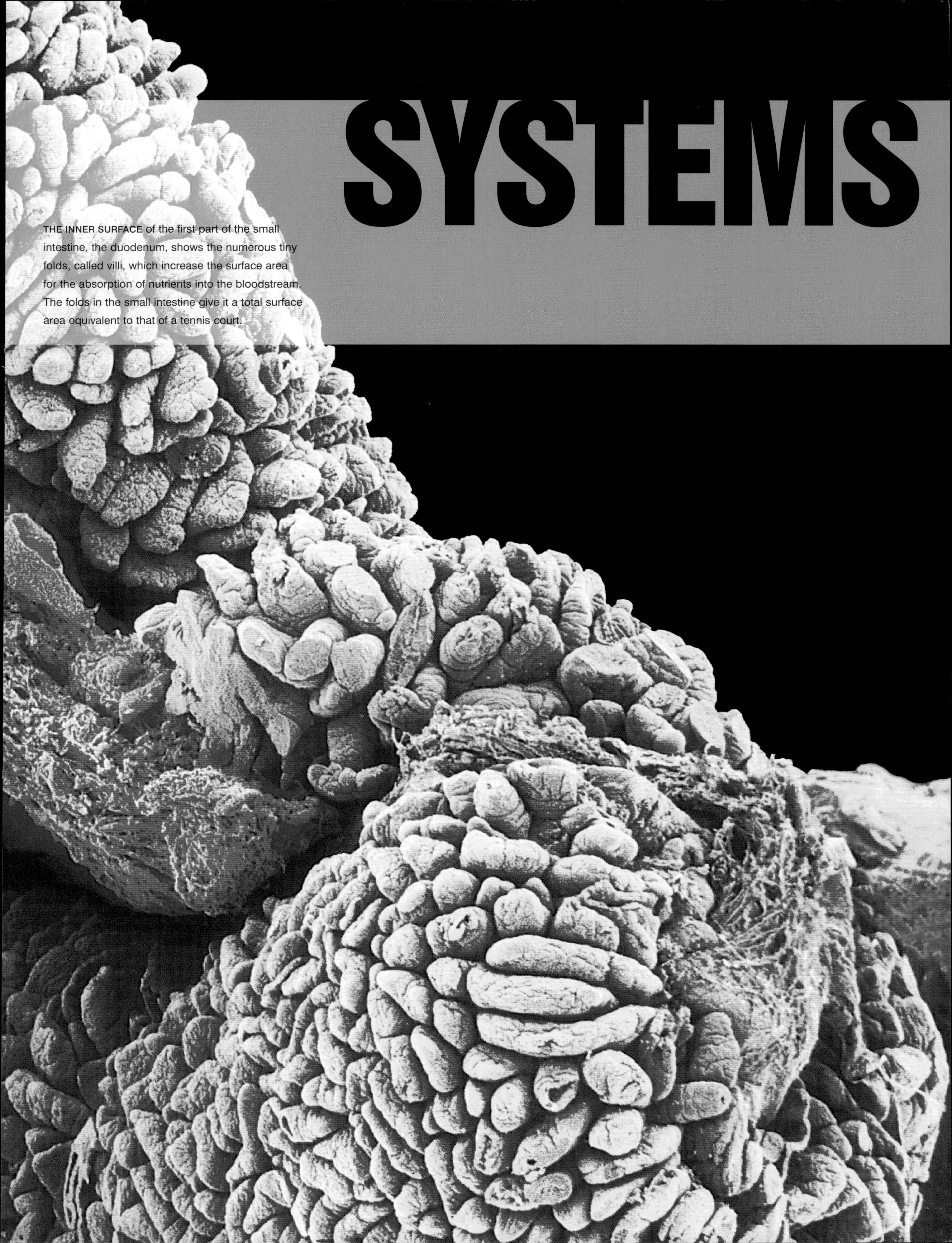

THE INNER SURFACE of the first part of the small intestine, the duodenum, shows the numerous tiny folds, called villi, which increase the surface area for the absorption of nutrients into the bloodstream. The folds in the small intestine give it a total surface area equivalent to that of a tennis court.

Without skeletons, we would not be able to move and would collapse under our own weight. The bones in the body provide a structural framework and act as levers operated by the muscles. Bones also protect delicate organs such as the brain, heart and lungs, and produce the body's red and white blood cells in the bone marrow.

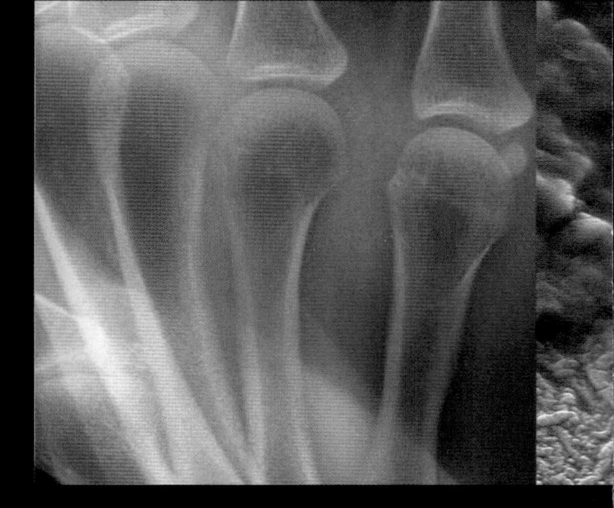

The skeleton of a baby has about 350 bones. They are mostly made of cartilage, which is softer and more flexible than bone, and therefore less likely to be broken in the rough and tumble of childhood. As a child grows, the bones harden and some of them fuse together, so that by the age of twenty-five the skeleton has 206 mature bones. Adults retain tough, smooth cartilage on the surfaces at the ends of bones, where, lubricated by a liquid called synovial fluid, it allows friction-free movement around joints. Cartilage is also found in the external ears, nose and windpipe.

Bone is incredibly tough. Weight for weight, it is stronger than wood, concrete or steel. A block of bone the size of a matchbox can support nine tonnes – four times as much as concrete. A bridge made out of bone material would weigh only a fifth as much as one made of steel.

A bone's strength comes from its composition and structure. A limb bone such as a femur is made of two types of bone: compact bone in the shaft, and spongy bone at the ends. Compact bone is deposited in concentric tubes arranged around nerves and blood vessels. This tubular arrangement,

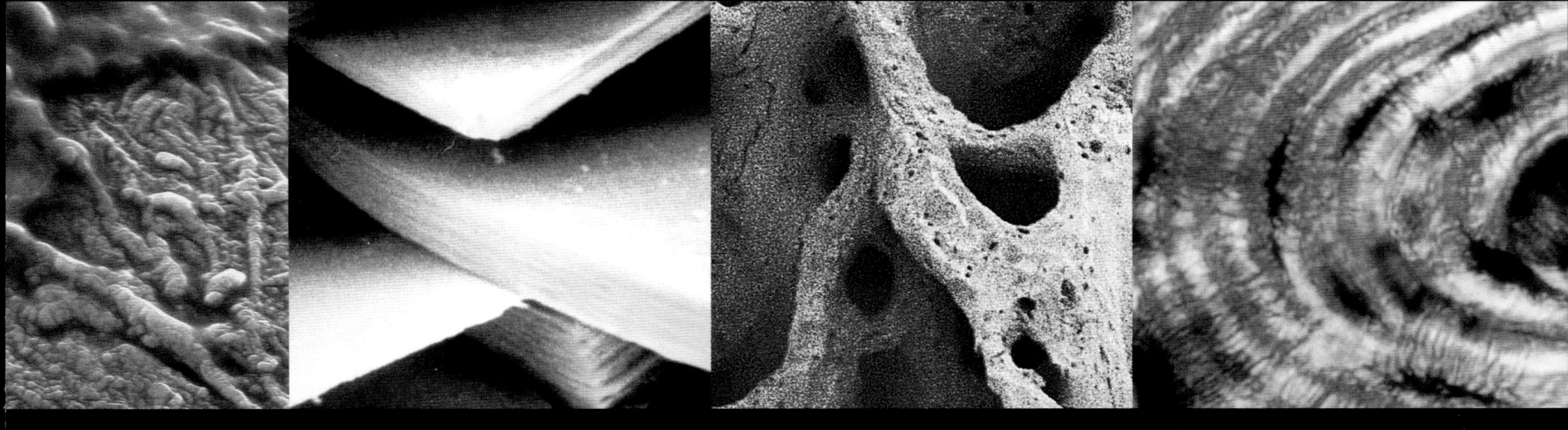

copied from nature by engineers and architects, keeps weight down and makes the bone resistant to fracturing, because cracks tend to stop at the boundaries between tubes. Spongy bone is deposited in a criss-cross pattern, forming a honeycomb structure that has excellent shock-absorbing properties.

Bone is a living, plastic material – properties not immediately obvious to anyone who is only familiar with the rigid scaffolding of skeletons displayed in a museum. Bones are constantly being reshaped, with new material being added and old material being removed. This remodelling does not take place properly in zero-gravity. Without the forces imposed by gravity, bone begins to break down, losing mass and strength – a serious problem for astronauts who must spend long periods living in conditions of weightlessness.

Human skeletons [Computer artwork based on X-ray]

THE 206 BONES OF AN ADULT human provide protection and structural support, while the joints between them allow locomotion. There are several different types of joint, each with a different range of movement. For example, ball-and-socket joints at the shoulders and hips enable movement in almost any direction, while simple hinge joints at the knees allow movement in only one plane.

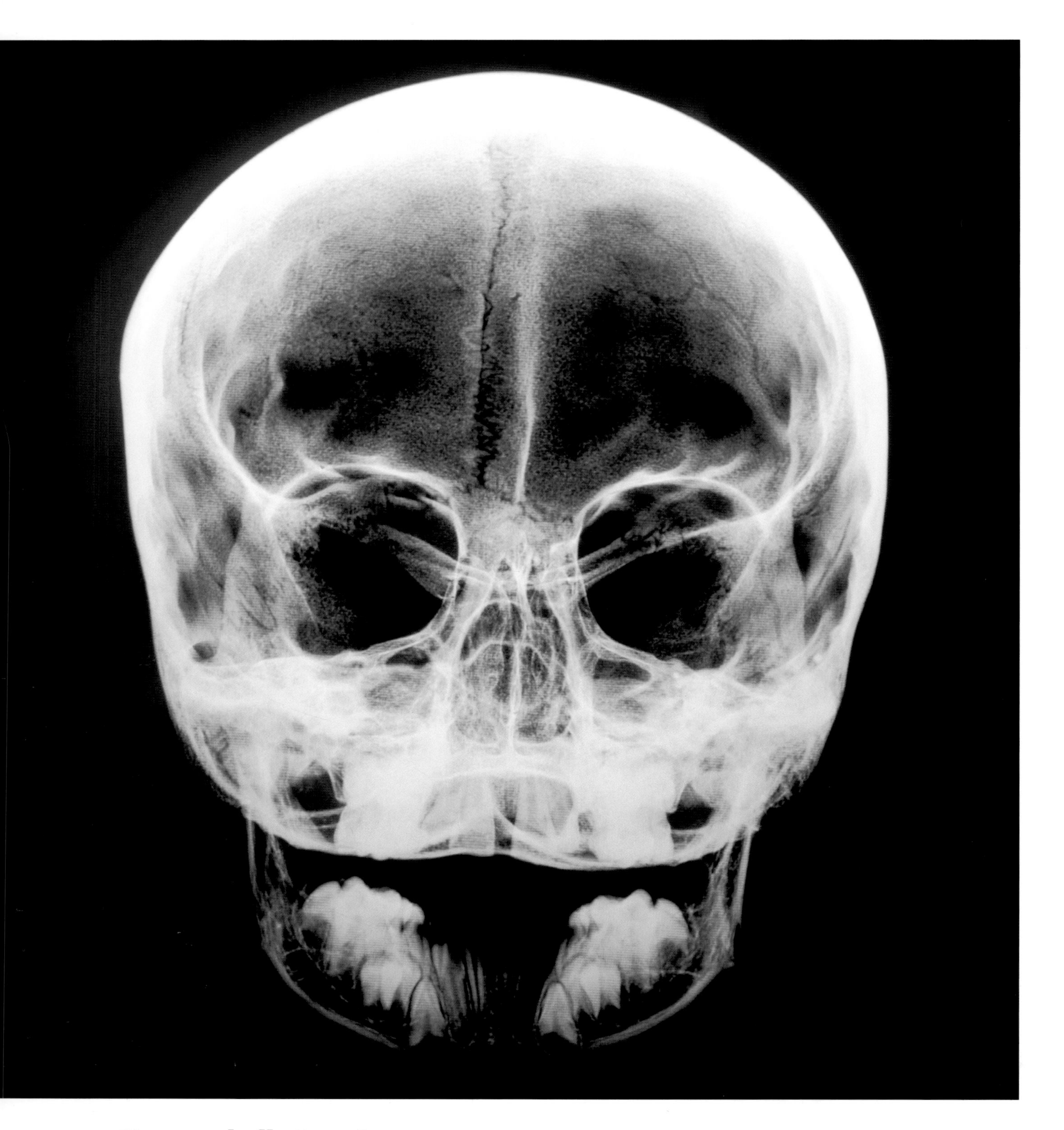

Human skulls [X-ray]

THE SKULL CONSISTS OF TWENTY-TWO BONES fixed together by rigid joints called sutures (seen here, in white, running up from between the eye sockets). The lower jaw is an exception as it hinges around its anchor points. Together these bones provide protection for the brain and support the organs of sight, smell and hearing. Many of the bones are hollow, thereby reducing the weight of the skull.

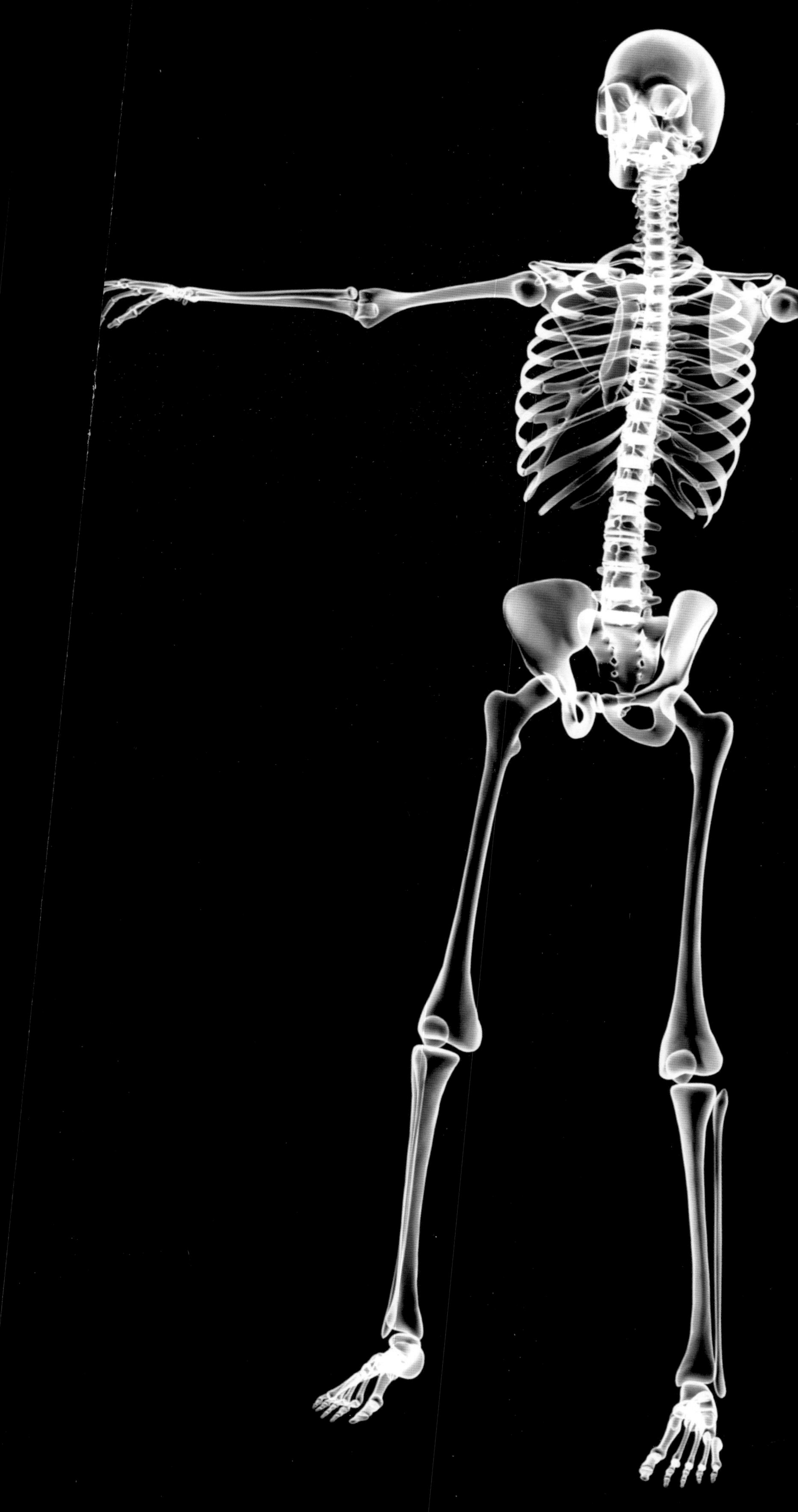

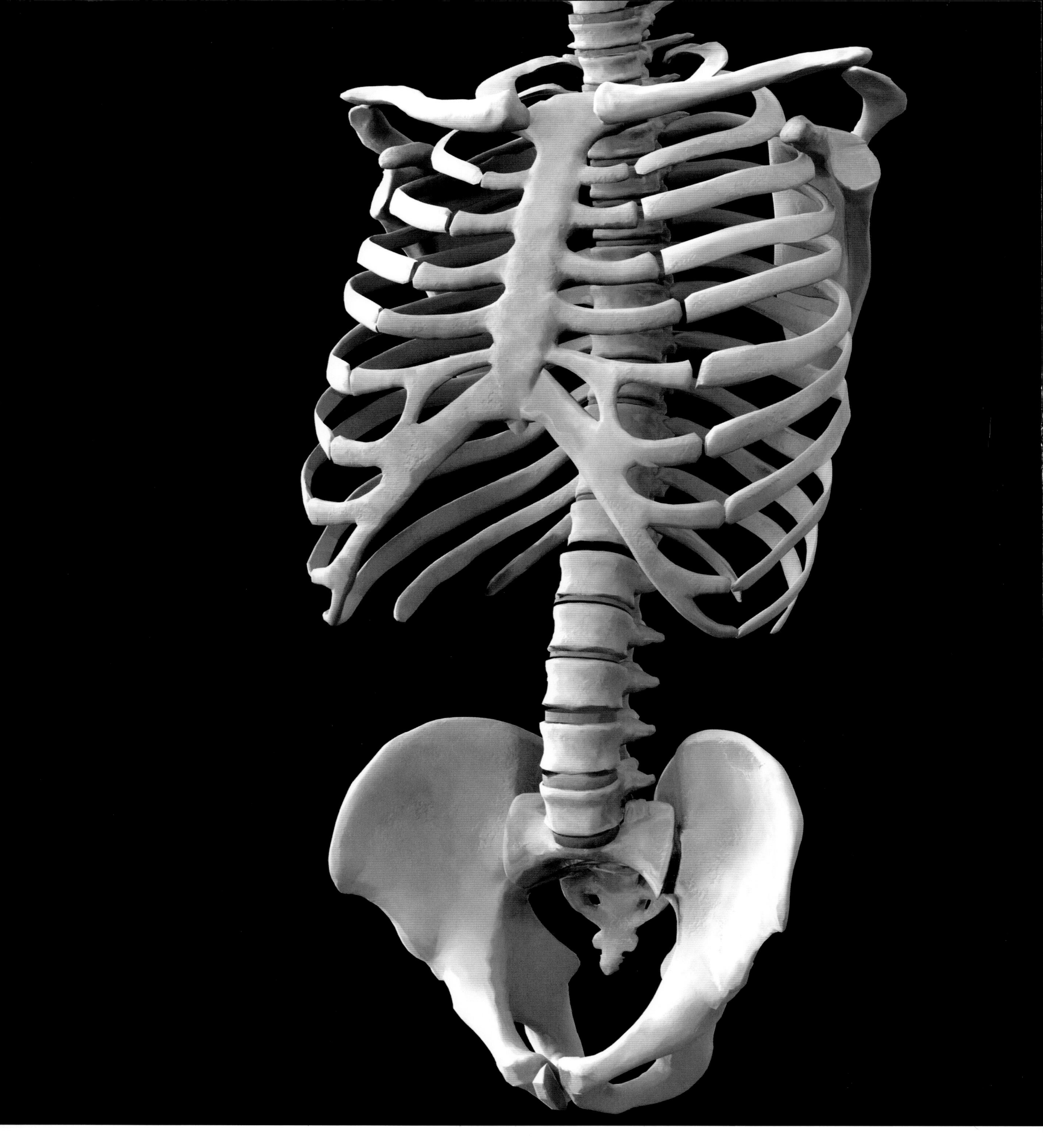

Bones of the human torso [Computer artwork]

THE BACKBONE RUNNING VERTICALLY down the centre supports the weight of the torso and encloses the delicate spinal cord. It is connected to the pelvis (lower centre), whose bones connect with the leg bones (not seen) and also protect the lower abdominal organs. The twelve pairs of ribs connected to the backbone and the sternum (front, upper centre) form a cage around the heart and lungs. The scapula (shoulder blade) bones articulate with the arm bones, and the clavicles (collarbones) are attached to the top of the ribcage.

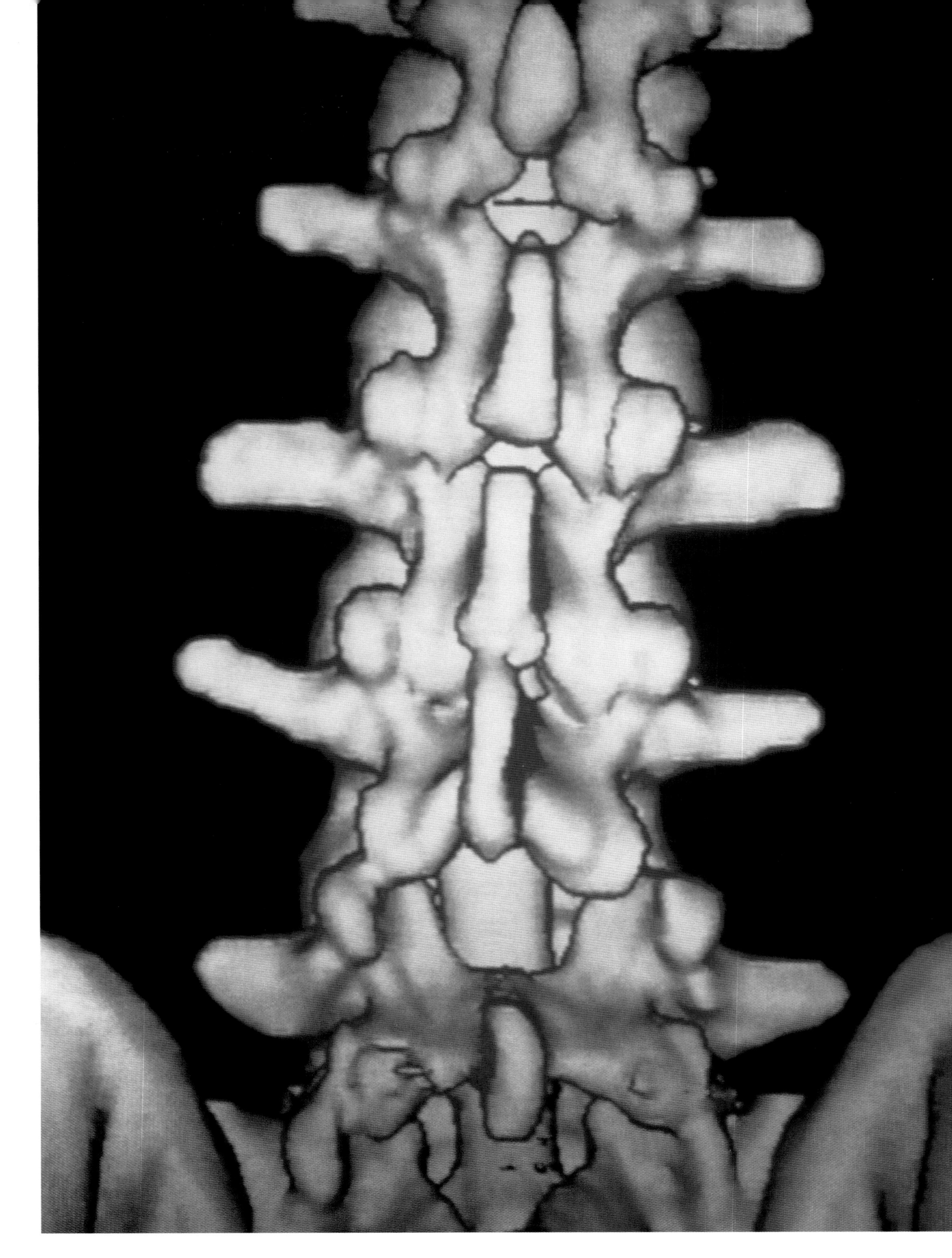

Lower spines [CT]

THE SPINE IS MADE UP OF TWENTY-NINE cylindrical bones called vertebrae. Shown here from the rear of the body are five vertebrae from the lower spine. Between the vertebrae are flexible joints that both stabilize the spine and allow movement. The bony projections extending behind the cylindrical part of each vertebra provide attachment sites for the muscles and ligaments that hold the spine together. The pelvis can be seen at lower left and right.

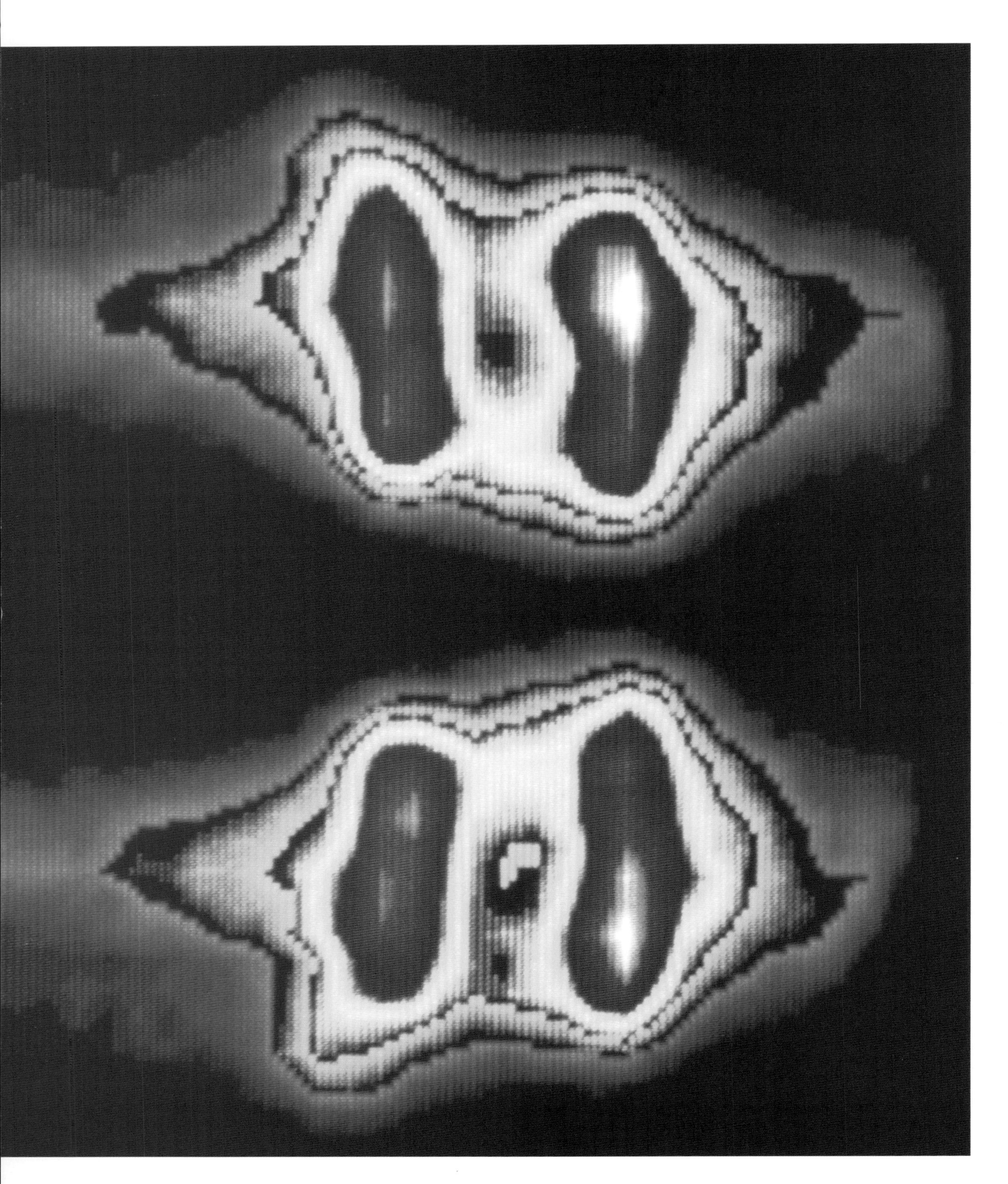

Human knees [Gamma camera scan]

A GAMMA SCAN OF NORMAL HUMAN KNEES reveals the joint cartilages in red and white. A gamma scan is a map of radioactivity emitted by technetium-99m (Tc-99m), an isotope injected intravenously into the subject. Tc-99m is readily taken up by bone and cartilage. Gamma scans can be used in the diagnosis of bone cancer. Cancerous bone concentrates the isotope more strongly, and this shows up as 'hot spots' on the scan.

Bones of the hands [X-ray]

HUMANS OWE THEIR MANUAL DEXTERITY to the complex arrangement of their hand bones (yellow, orange and pink). Each finger has three phalanges, while the thumb has only two. The palm comprises five metacarpal bones, which articulate with the eight small carpals of the wrist. The two bones of the lower arm are the radius and the smaller ulna.

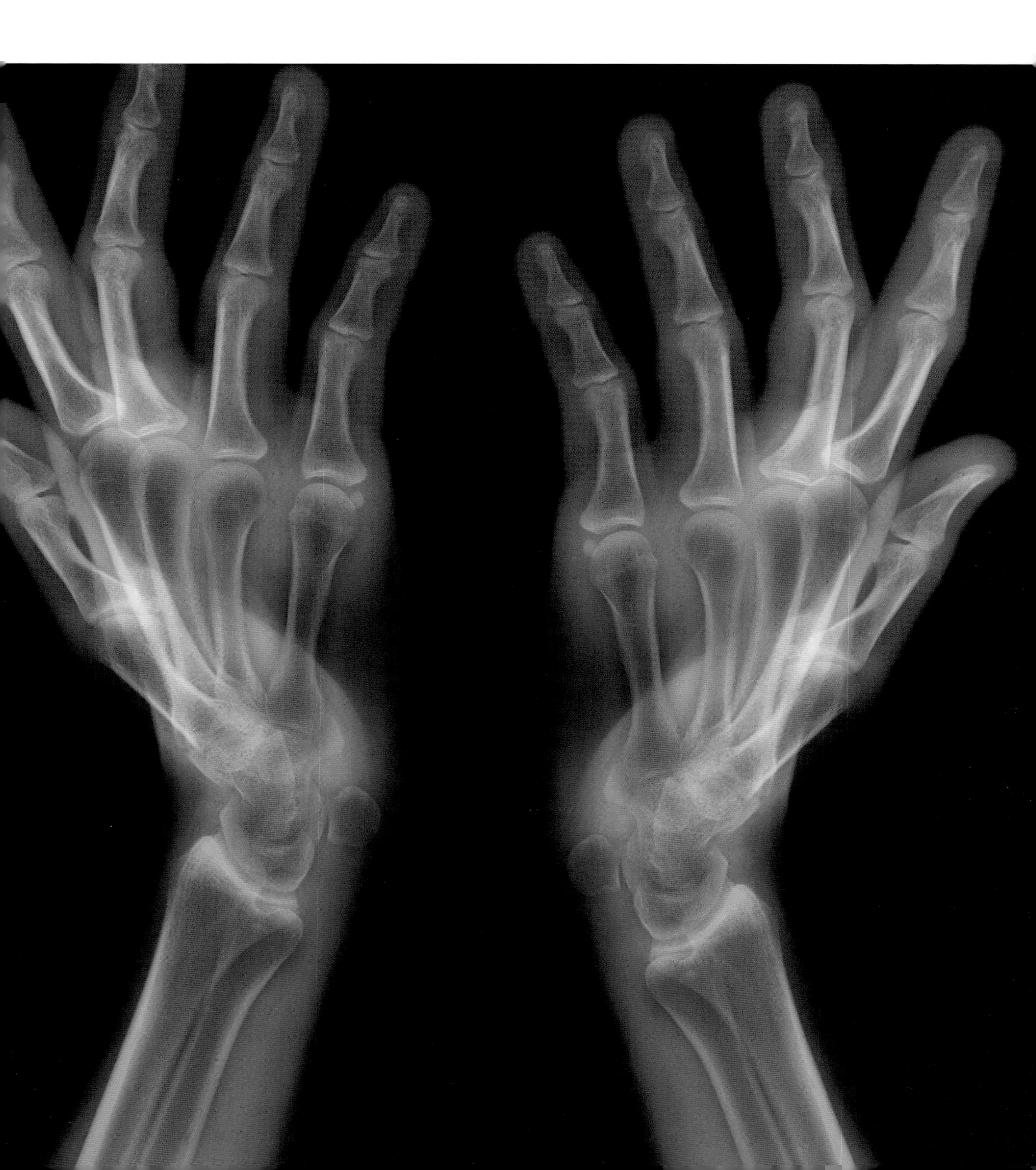

Bones of the foot [X-ray]

SEEN FROM THE SIDE, A FOOT on tiptoe shows the bones (pale green) that support the weight of the body and act as levers during movement. The heel bone, the largest in the foot, is one of seven tarsal bones that form the ankle and rear of the foot. Immediately above the heel bone is the talus, which articulates with the tibia (shin bone). The smaller leg bone to the left is the fibula. The elongated metatarsals join the tarsal bones to the toes (phalanges).

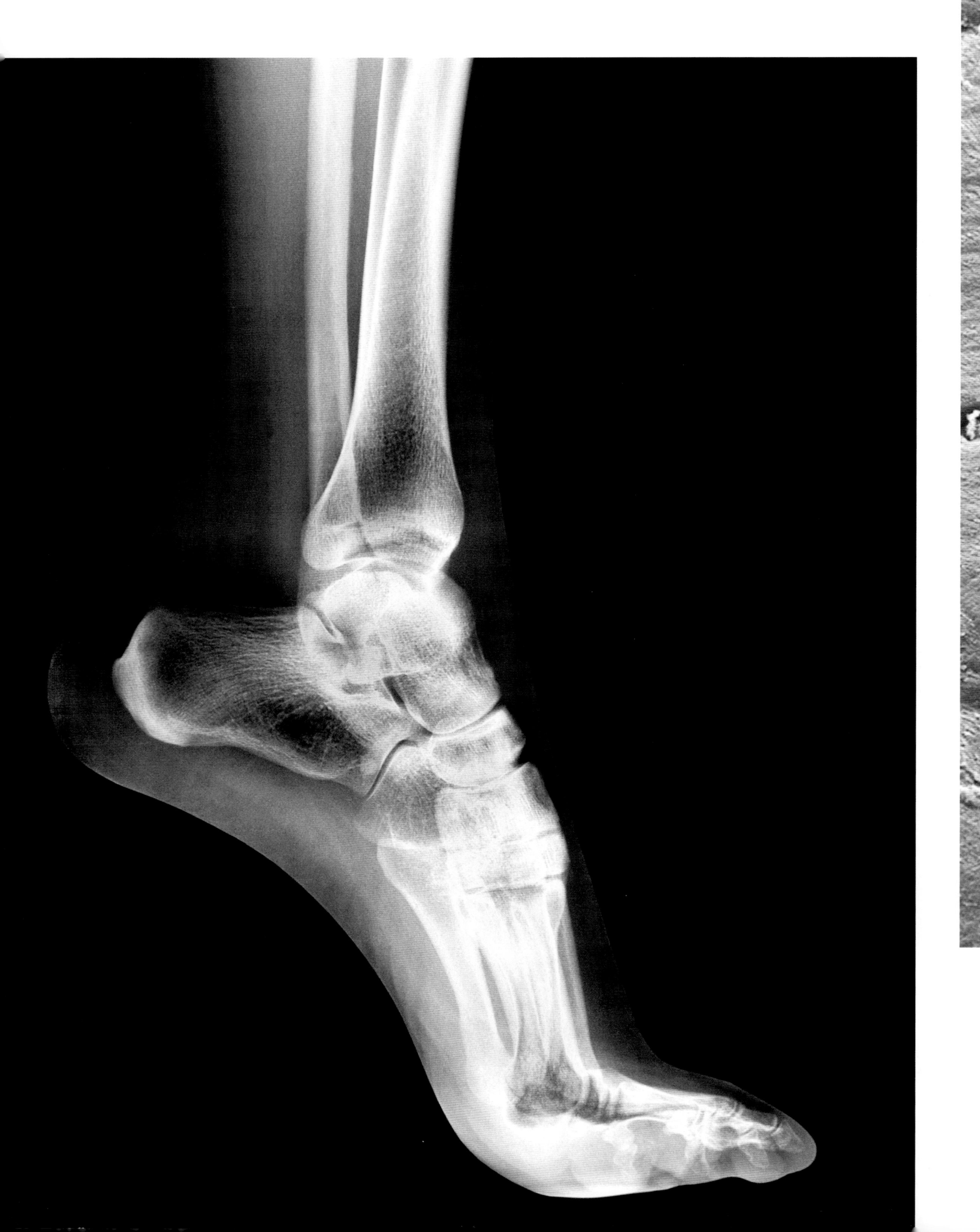

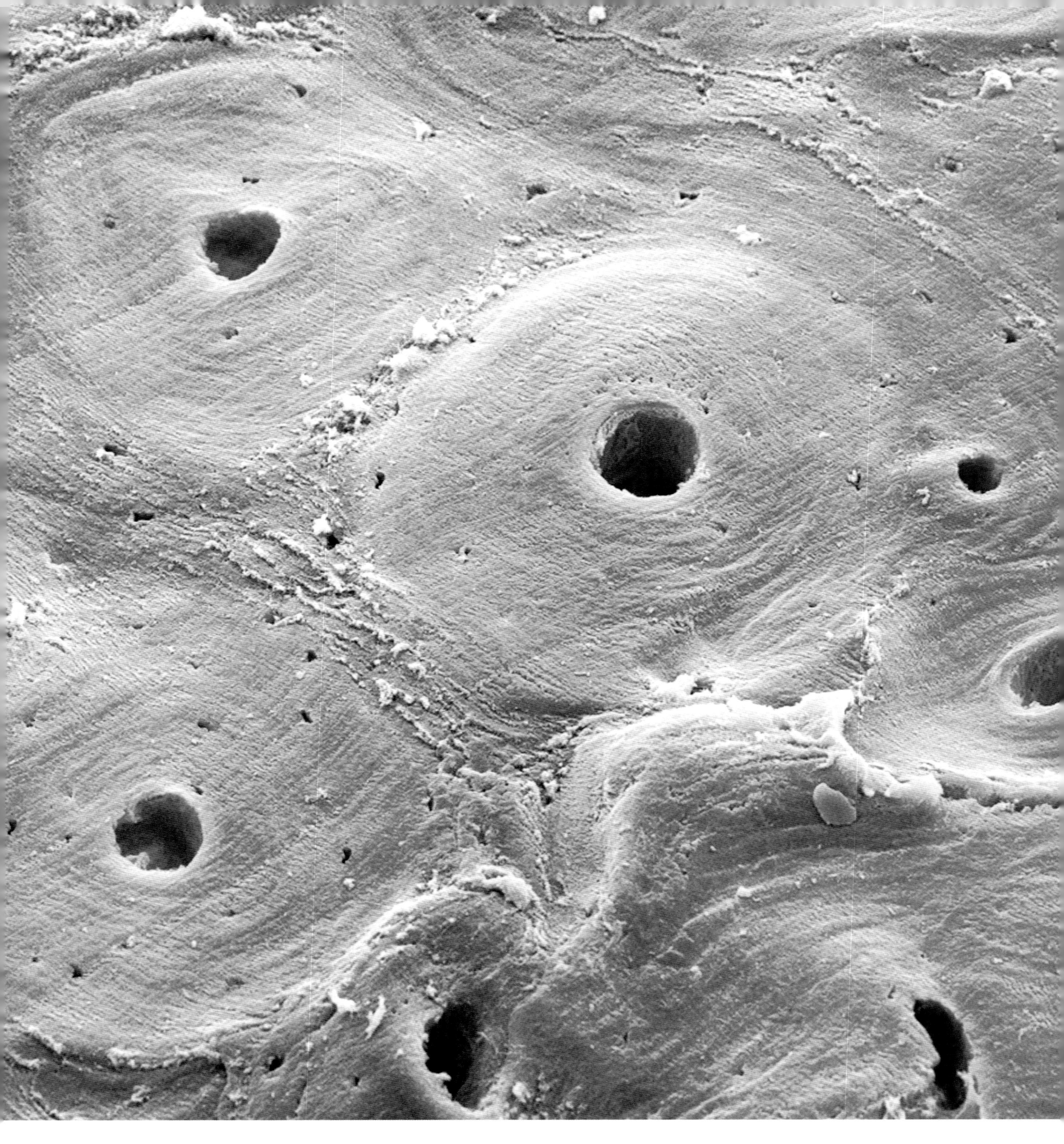

Compact bone [SEM]

A SECTION THROUGH COMPACT BONE shows the typical structure of concentric bony layers surrounding the channels known as Haversian canals, which formerly housed blood vessels and nerves. The small elliptical pits in the bony matter are the locations of osteoblasts, the bone-forming cells. Although bones may appear rigid and unchanging in adults, they are made of plastic material that is constantly being reabsorbed and refashioned in response to mechanical stresses.

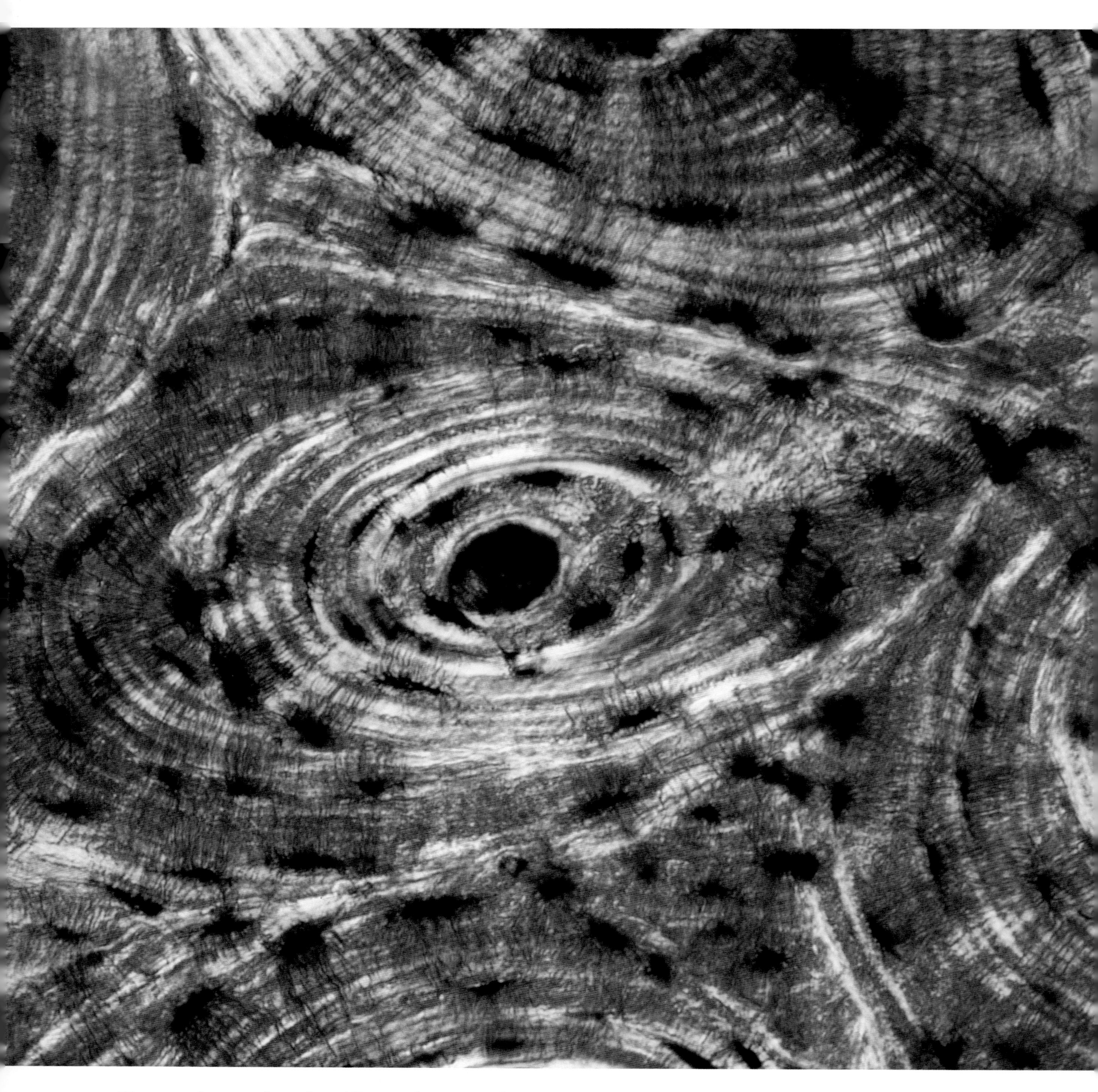

Human bone showing Haversian canals [LM]

THE LAMELLAE (BONY LAYERS) of a compact bone are arranged around channels called Haversian canals, which contain blood and lymph vessels, and nerves. A bone consists of a great many canals and their surrounding lamellae, known collectively as Haversian systems. Bone decomposition occurs concentrically from the edges of the Haversian canals.

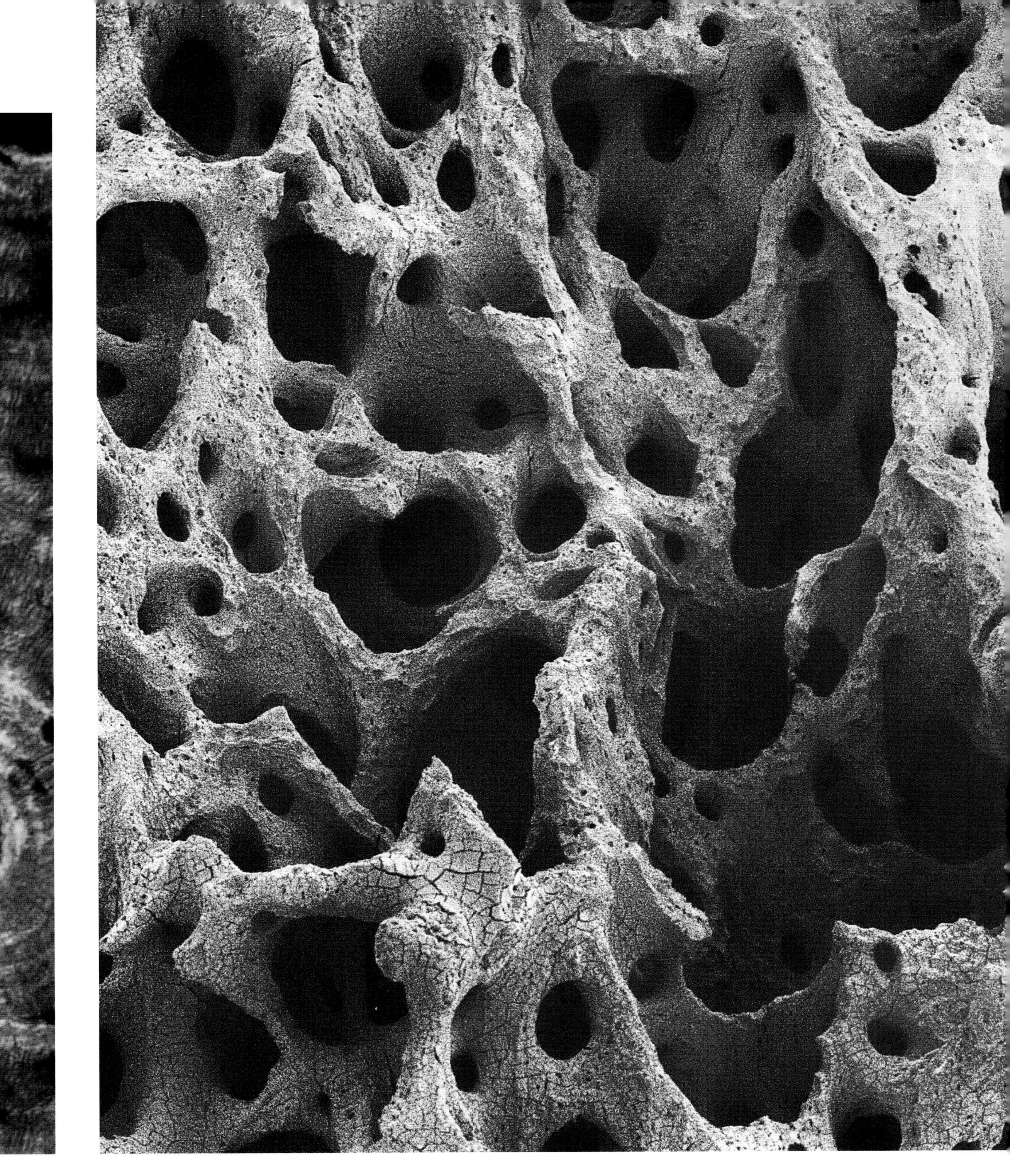

Spongy bone [SEM]

A BONE'S ABILITY TO ABSORB STRESS is due largely to the presence of spongy bone, a honeycomb meshwork found beneath the dense compact bone layer and also at the ends of the bone. The spaces between the framework are normally filled with bone marrow, which manufactures blood cells.

Lamellae in compact bone [SEM]

BONE CONSISTS OF TWO TYPES of bony tissue: compact bone for strength, and spongy bone for resilience. In this sample of compact bone from a femur (thigh bone), the strength is supplied by lamellae (layers) of compacted collagen and minerals running along the axis of the bone. In spongy bone the lamellae are arranged in a honeycomb pattern, which is not so strong but gives good shock-absorbing properties.

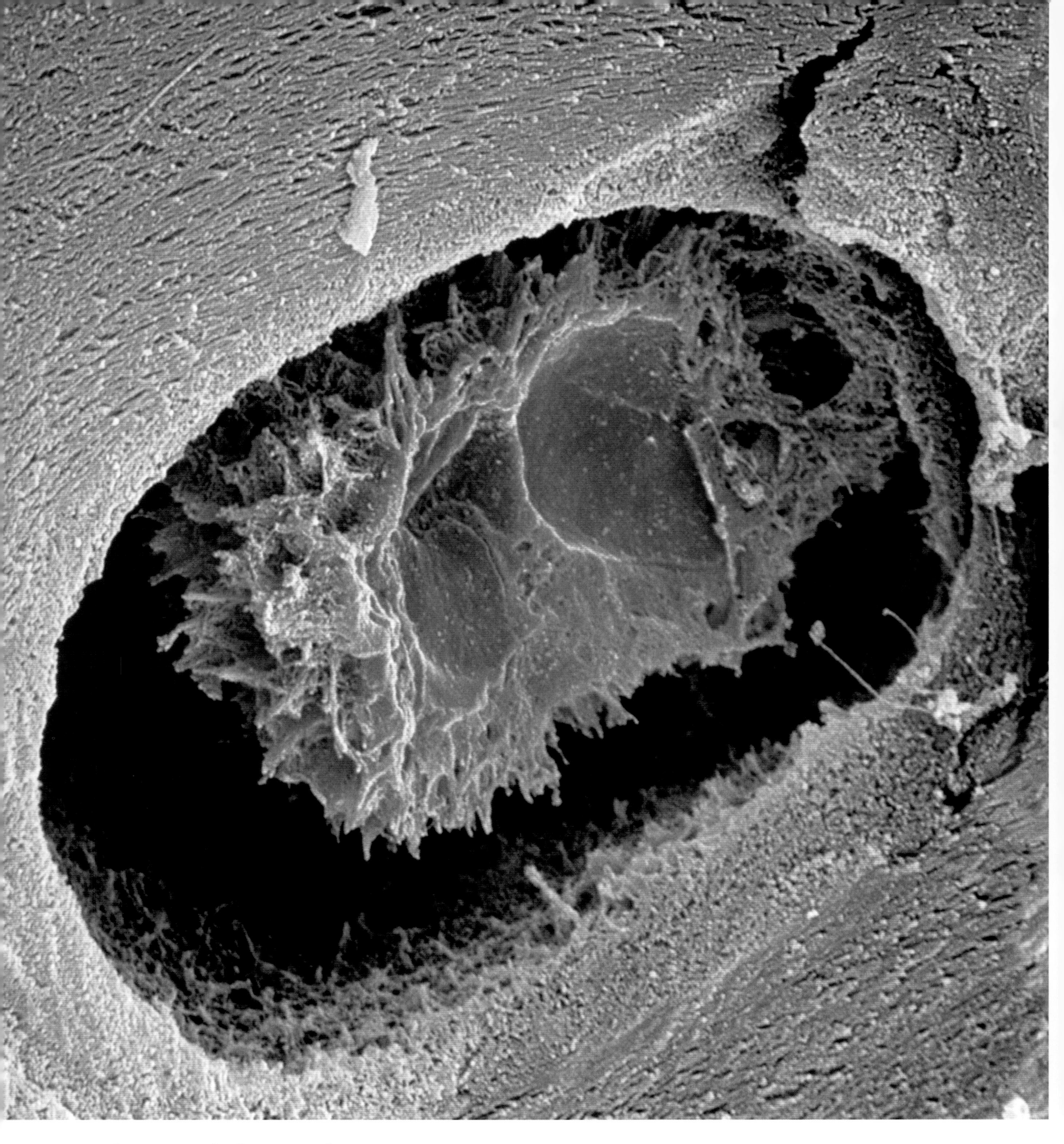

Osteocyte in bone cavity [SEM]

BONE IS PRODUCED BY THE MINERALIZATION of proteins synthesized by cells called osteoblasts. The cells gradually become enclosed by a matrix of bone, and in this state they are called osteocytes. This image shows an osteoblast that has become trapped in bone cavity. The large dark concave region within the osteoblast was the site of the cell nucleus.

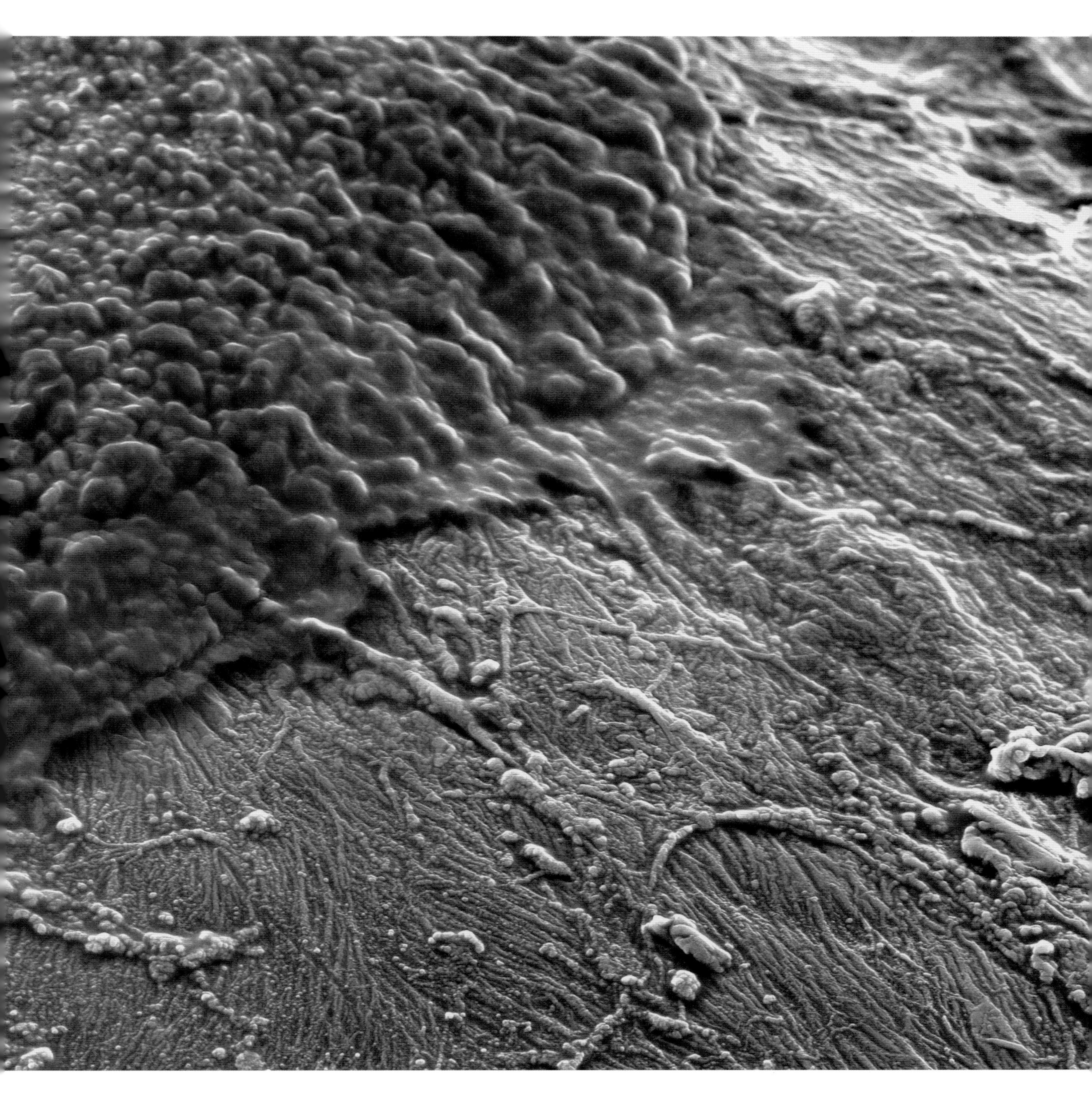

Bone reabsorption by osteoclasts [SEM]

PART OF AN OSTEOCLAST, a giant multinucleate cell (upper left), is shown reabsorbing bone substance (lower right). Osteoclasts are a type of bone cell that is involved in the breakdown and shaping of the bone matrix during bone repair, restructuring and growth. After osteoclasts have burrowed into bone, osteoblasts enter the cavities and deposit new bone.

The job of the digestive system is to break down food for absorption into the body. Food is medicine as well as sustenance, and it is the digestive system that releases the energy, vitamins, minerals and water from the bulk of the resource that is eaten. The digestive system enables the body to utilize all the nutrients contained in our food. It extracts the protein, energy, vitamins, minerals and water on which we depend from all that we eat.

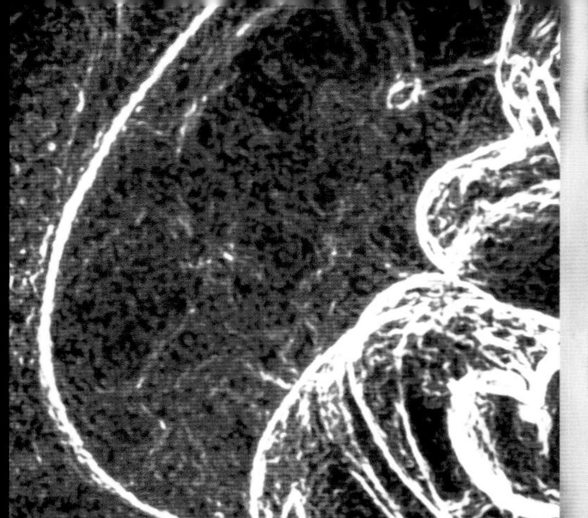

The process occurs in five well-defined phases: ingestion, fragmentation, digestion, absorption and egestion. The system begins at the mouth and ends several hours later and 8m/28ft or 9m/30ft farther on, at the anus. In between lies a passageway lined with special cells that secrete chemicals to break down the food into smaller and smaller units until it can be absorbed. A combination of voluntary and involuntary muscular movements passes the food from region to region.

When food enters the mouth, enzymes in the saliva begin to break down starch into a sugar called maltose. Enzymes are amazing proteins that accelerate chemical reactions without being changed themselves. To convert starch into sugar by boiling would take several hours; enzymes do the job in a minute.

In the stomach the food is churned up and broken down further by more enzymes, aided by a strong solution of hydrochloric acid. Food is not absorbed in the stomach, but alcohol can cross the stomach

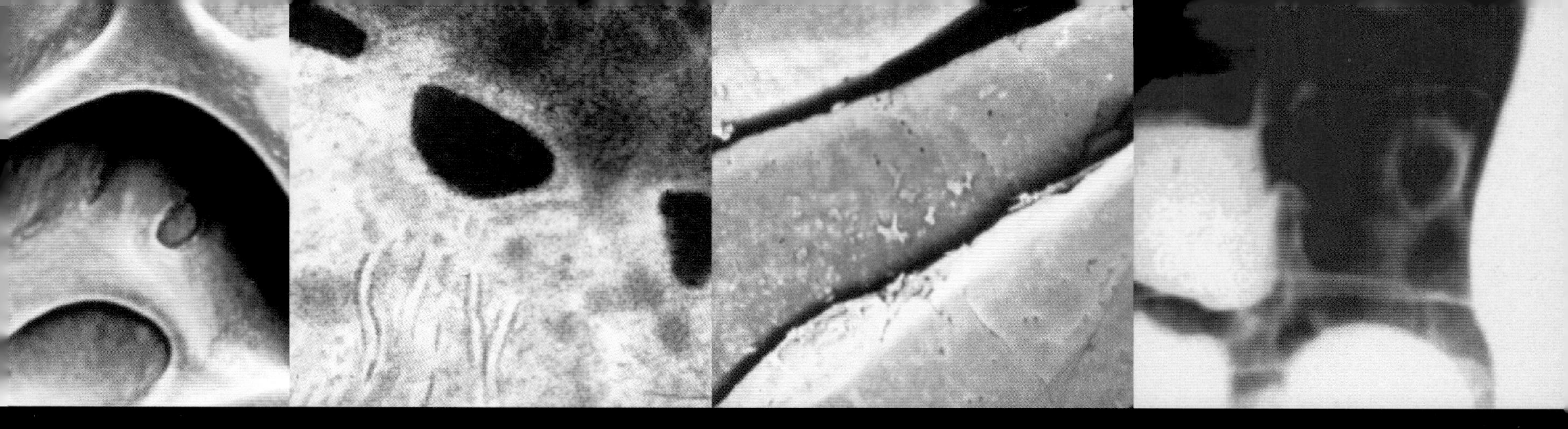

Most of the nutrients in food are absorbed in the small intestine, which despite its name is more than 5m/16½ft long, with a surface area of about 550sq m/660sq yd. The large intestine acts like a recycling plant, absorbing what nutrients remain, together with some of the 8 litres/14 pints of water that are poured over food during the digestive processes. The colon, the last section of the large intestine, is home to huge populations of bacteria that break down certain carbohydrates – those found in beans, for example – that humans cannot digest. The by-products are gases such as methane and hydrogen sulphide.

While most of the body's systems are faithful servants that work silently without complaint, we are aware – sometimes embarrassingly so – of the digestive processes. From the mouth to the stomach, then through the twisting tunnels of the intestines, before the unwanted parts are ejected, the digestive system keeps us informed of what is going on.

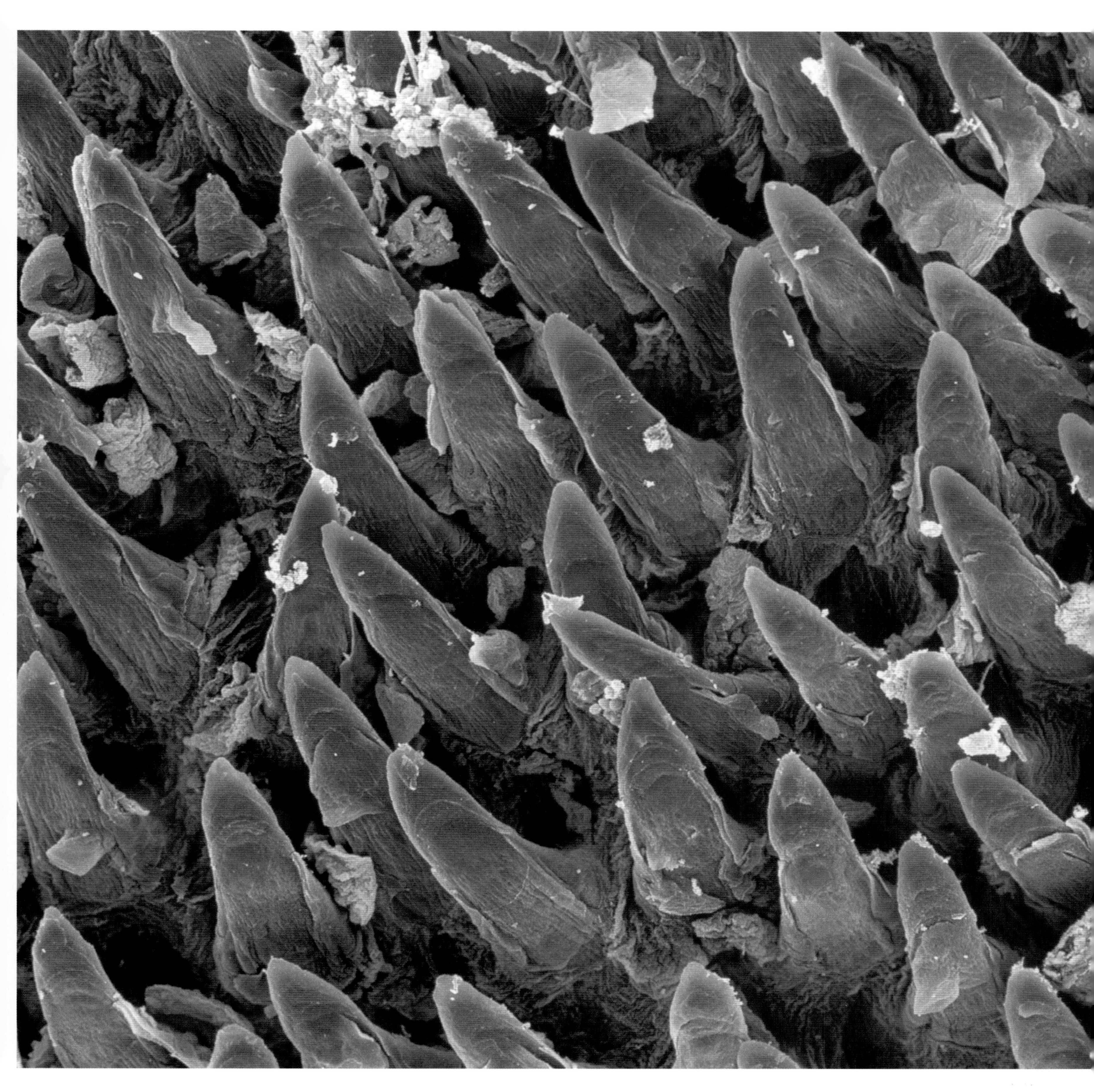

Papillae on the tongue [SEM]

THE ROUGH TEXTURE of the upper surface of the tongue is due to these small projections, called filiform papillae. Filiform help in the mechanical processing of food and also transmit tactile information to the brain.

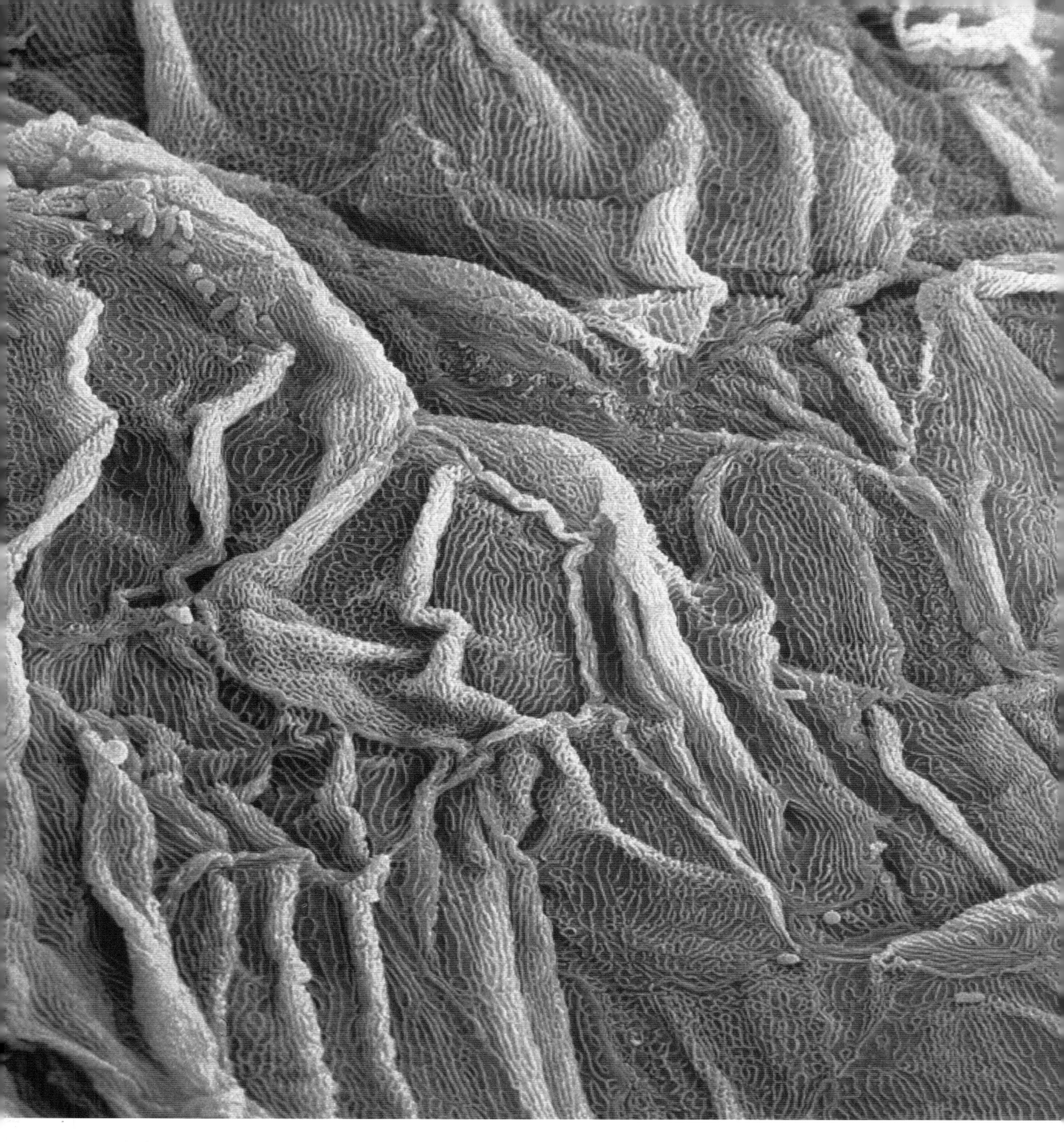

Lining of oesophagus [SEM]

THE OESOPHAGUS IS A MUSCULAR TUBE, about 25cm/10in long in adults, that carries food from the back of the throat to the stomach. Food is propelled down the oesophagus by reflex muscular contractions called peristalsis. It takes about two seconds for fluids to reach the stomach, and between four and eight seconds for solids.

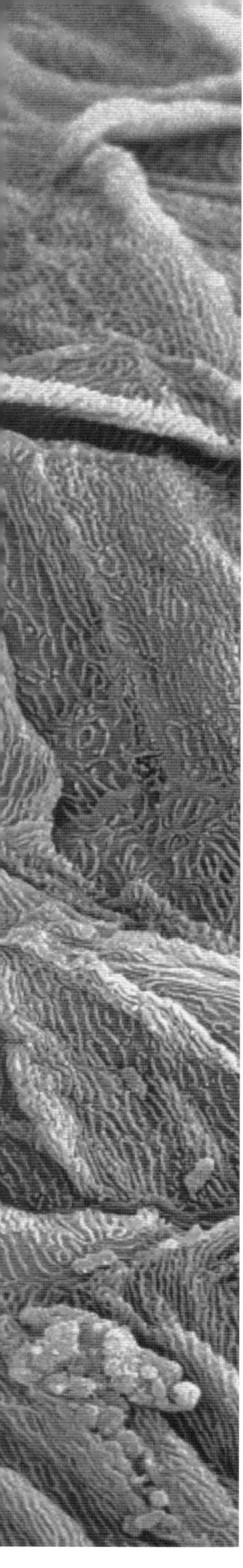

The stomach [X-ray]

THE YELLOW MASS IN THE PICTURE is the stomach, a muscular, sac-like organ that receives food from the oesophagus, at top right. In the stomach, food is prepared for digestion by the action of gastric juices, mainly hydrochloric acid and enzymes. Partly digested food leaves the stomach in stages – fluids first, then carbohydrates and finally fats. Depending on the type of meal, the process can take up to four hours.

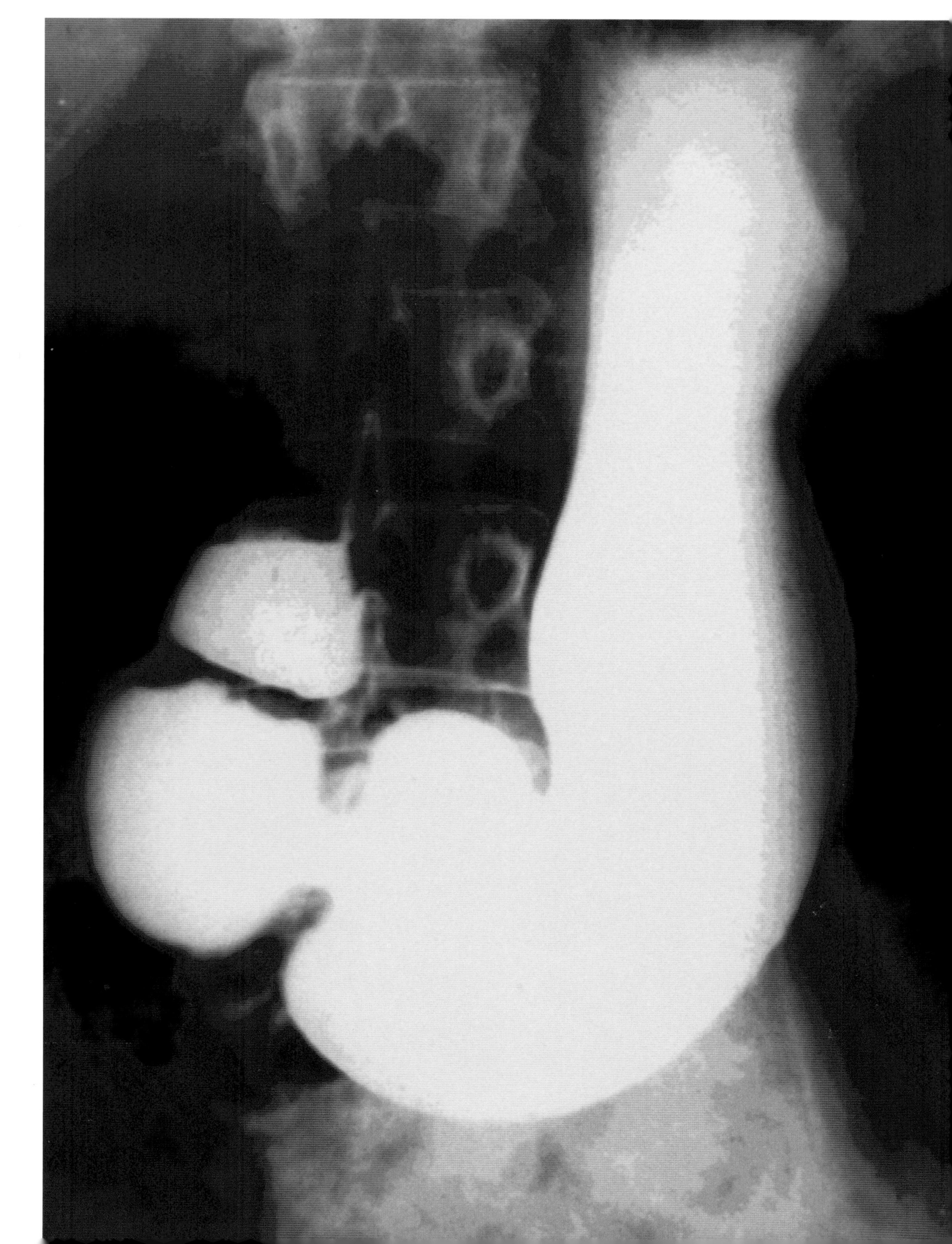

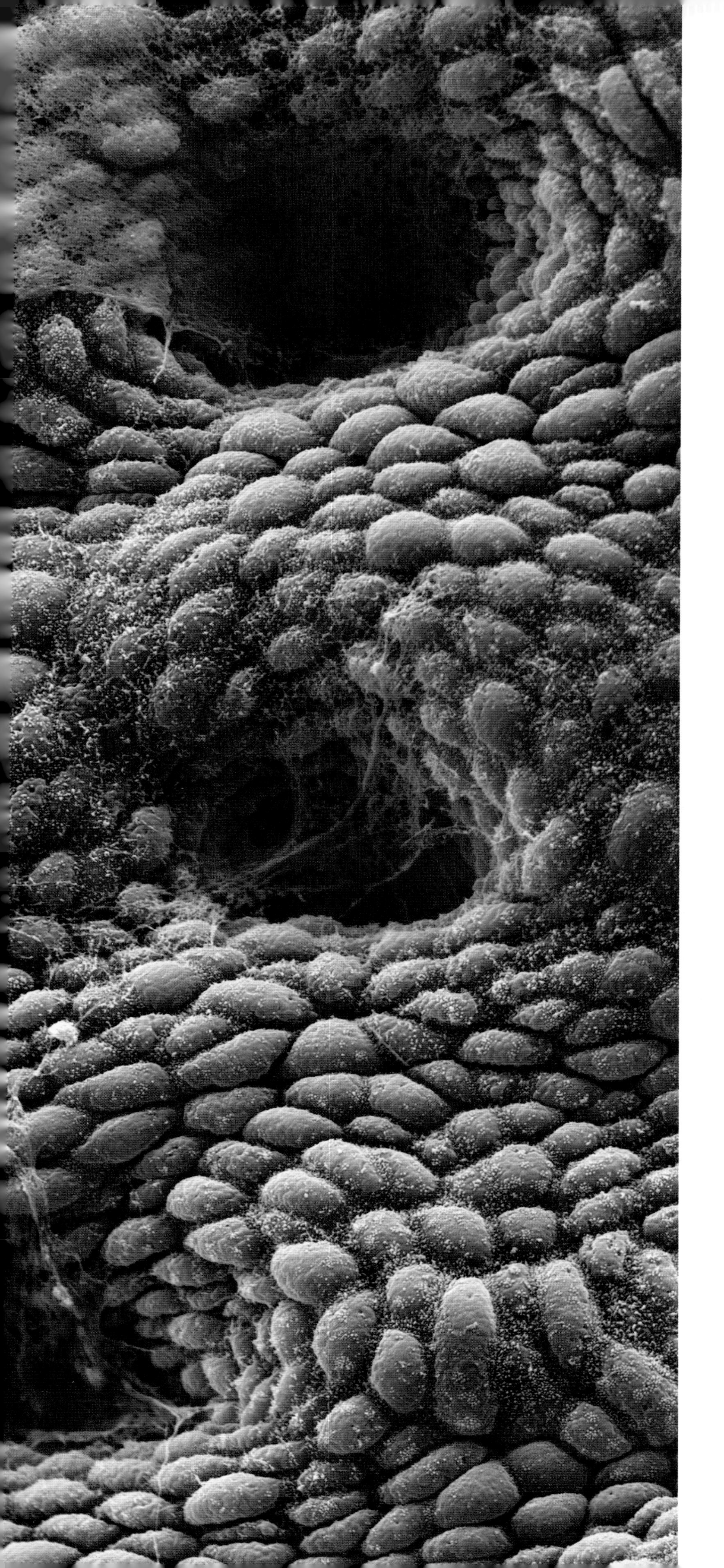

Stomach lining [SEM]

THE STOMACH IS LINED with simple columnar cells that secrete mucus. The mucus protects the stomach from gastric acid, which makes the enzymes work efficiently, helps break down tough tissues in meat, and also kills some of the harmful bacteria that enter the body in food.

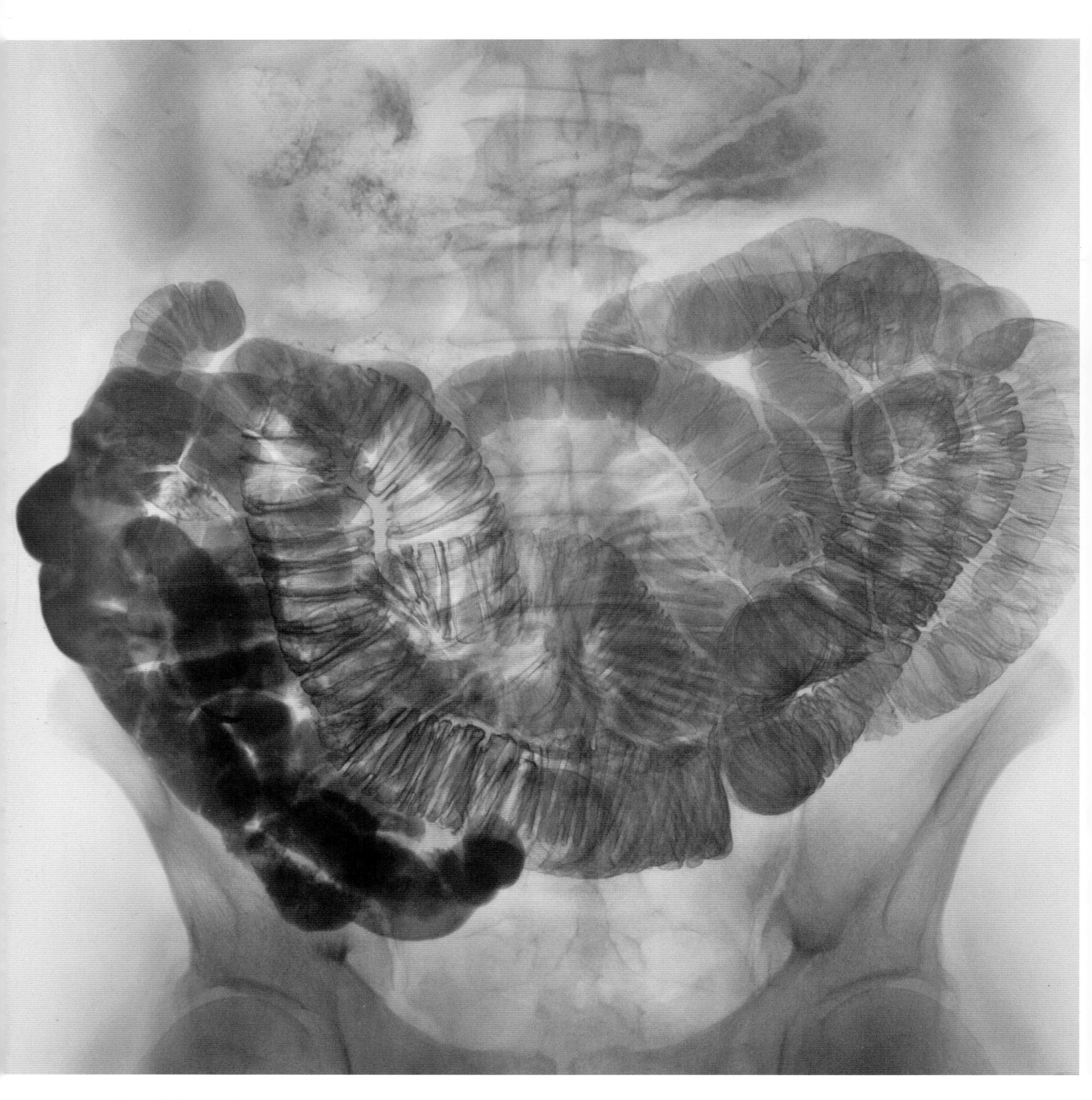

Small intestine [X-ray]

FOOD IS DIGESTED MAINLY IN THE SMALL INTESTINE, a coiled tube about 5m/16½ft long in adults. The banding within the intestine is caused by circular folds in the mucosal lining, which increase the area through which nutrients can be absorbed. Food passes down the small intestine at a rate of about 1cm/½in a minute.

Duodenum wall [SEM]

MOST DIGESTION TAKES PLACE IN THE DUODENUM, a tube about 25cm/10in long that forms the first part of the small intestine. The inside walls of the duodenum are covered in folds called villi (blue), which greatly increase the absorptive and secretory surface. Villi are more numerous in the duodenum than anywhere else in the intestine. Food is absorbed into the blood through capillaries in each villus. The brown crescent in this cross section is the muscular wall that keeps the duodenum in shape.

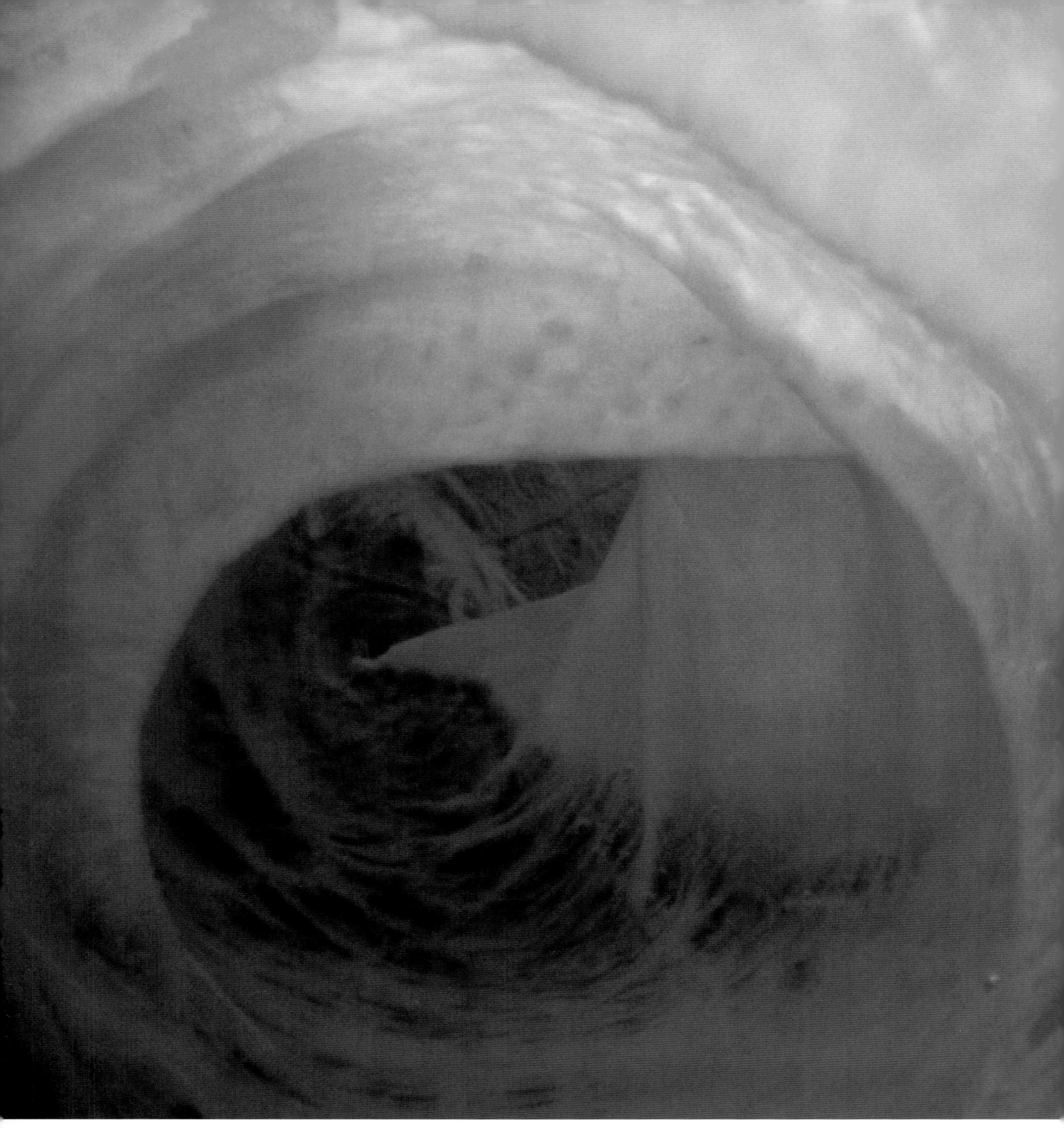

Inside the small intestine [Endoscope]

IT IS HERE THAT THE BULK OF FOOD is digested, aided by secretions from the intestinal walls, and from the liver and pancreas. Pancreatic juices help neutralize the acid from the stomach and contain digestive enzymes. Bile secreted by the liver also neutralizes acid and helps digest fats. The intestinal juices contribute water to aid absorption of digested food.

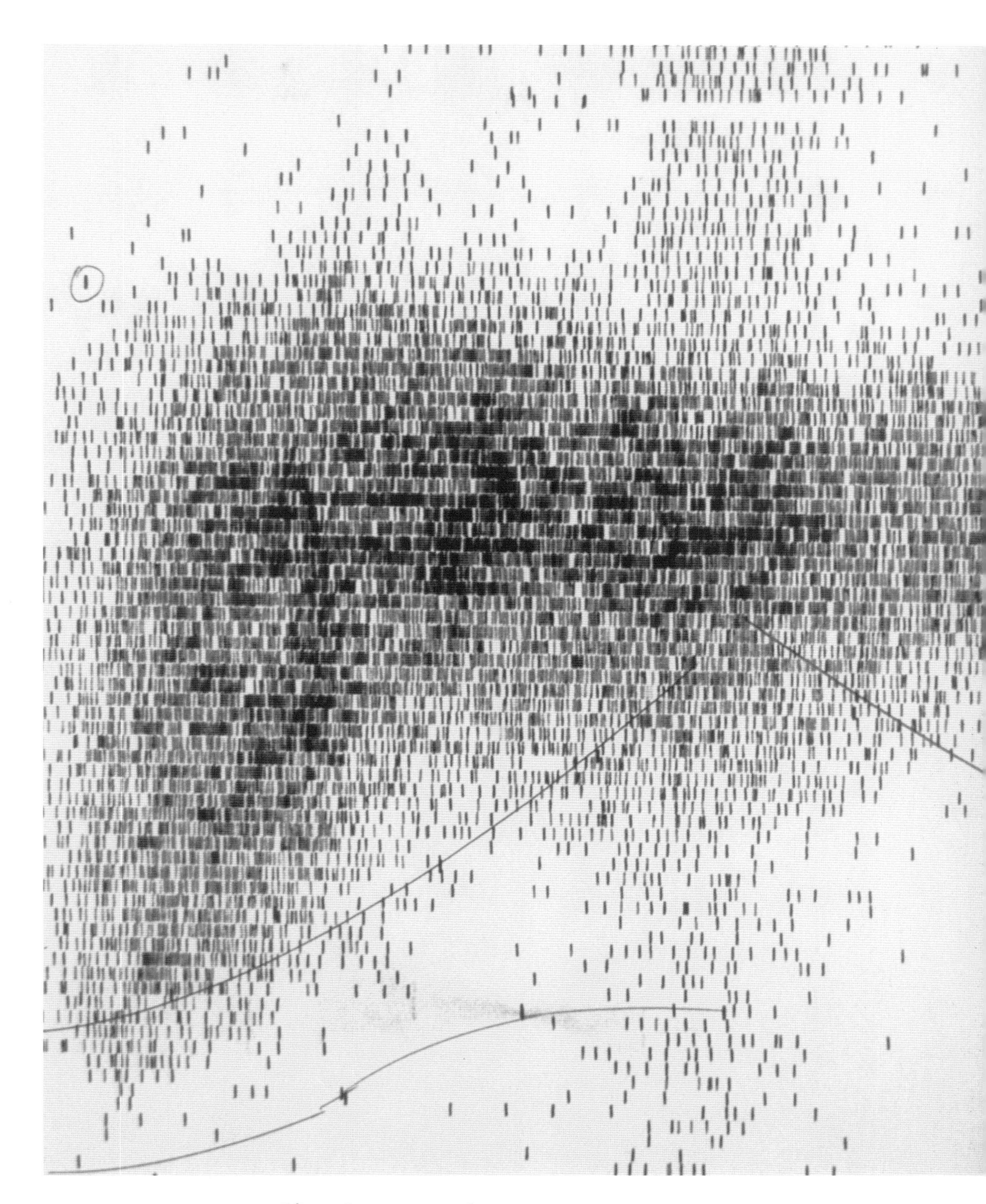

Liver [Gamma scan]

EACH DAY THE LIVER produces up to 1 litre/2 pints of bile, which emulsifies fats, making it easier for the digestive enzymes to work on them. As well as producing bile to help digest food in the small intestine, the liver processes absorbed food, converting it into storage products such as glycogen. This image was produced by a gamma scan, which detects radioactive substances introduced into the body. Although gamma scans do not produce precise anatomical details, they can provide a clearer picture of an organ's functioning than other imaging techniques.

Intestinal microvilli [TEM]

THIS INTESTINAL CELL ABSORBS nutrients from digested food. To increase the surface area available for absorption, each cell is covered in thousands of microvilli (pink). These long, finger-like projections have a core of microfilaments (dark pink) containing proteins that can contract, giving primitive motility to the microvilli.

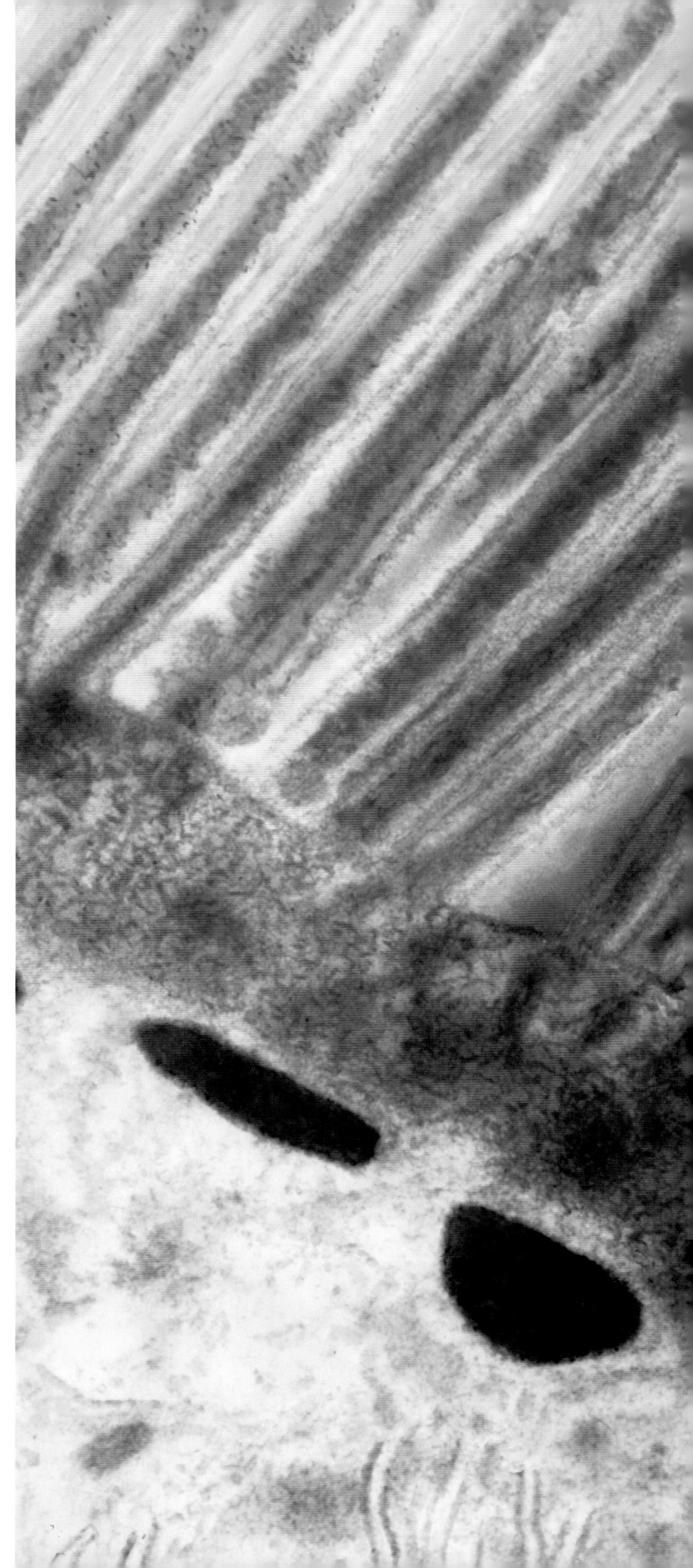

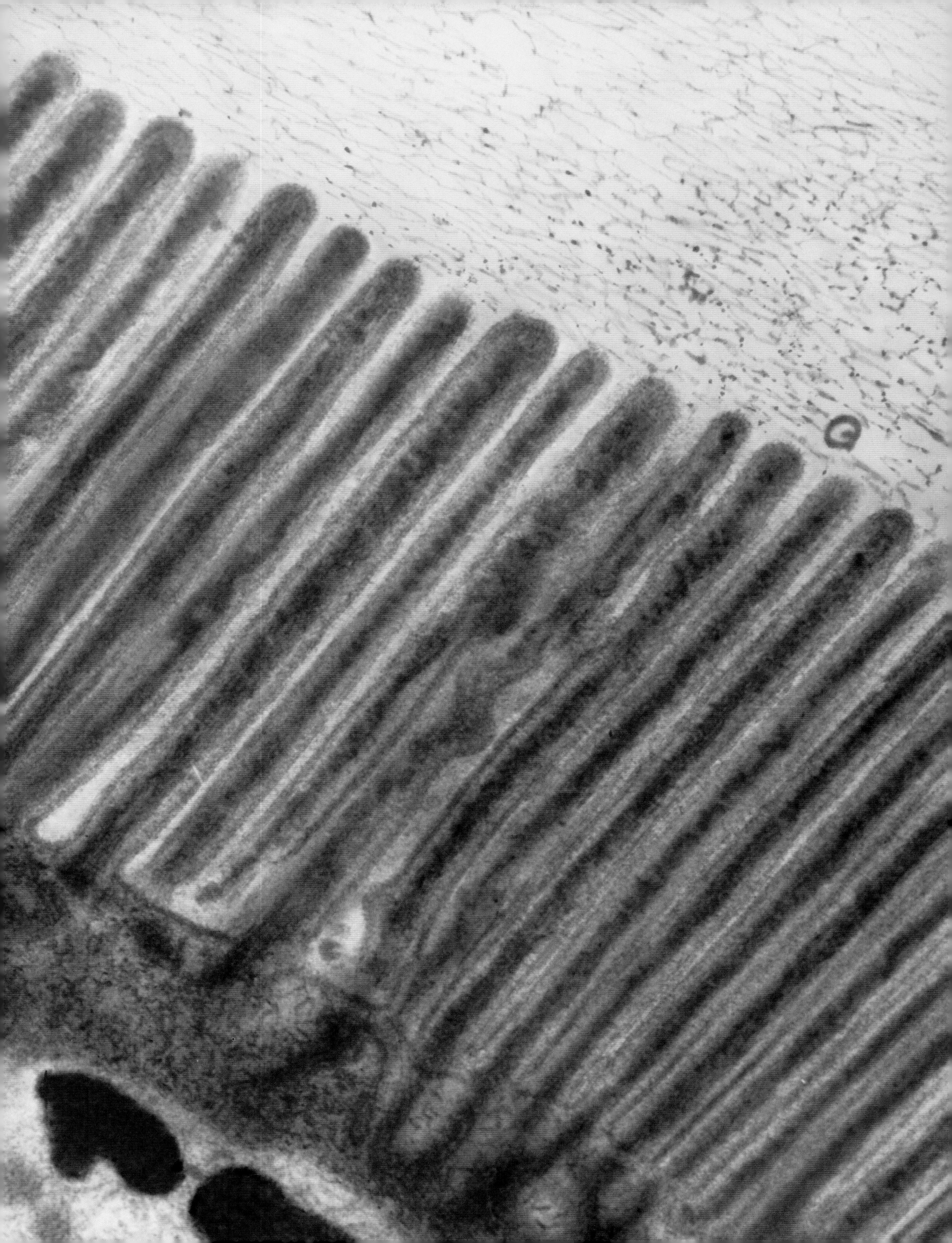

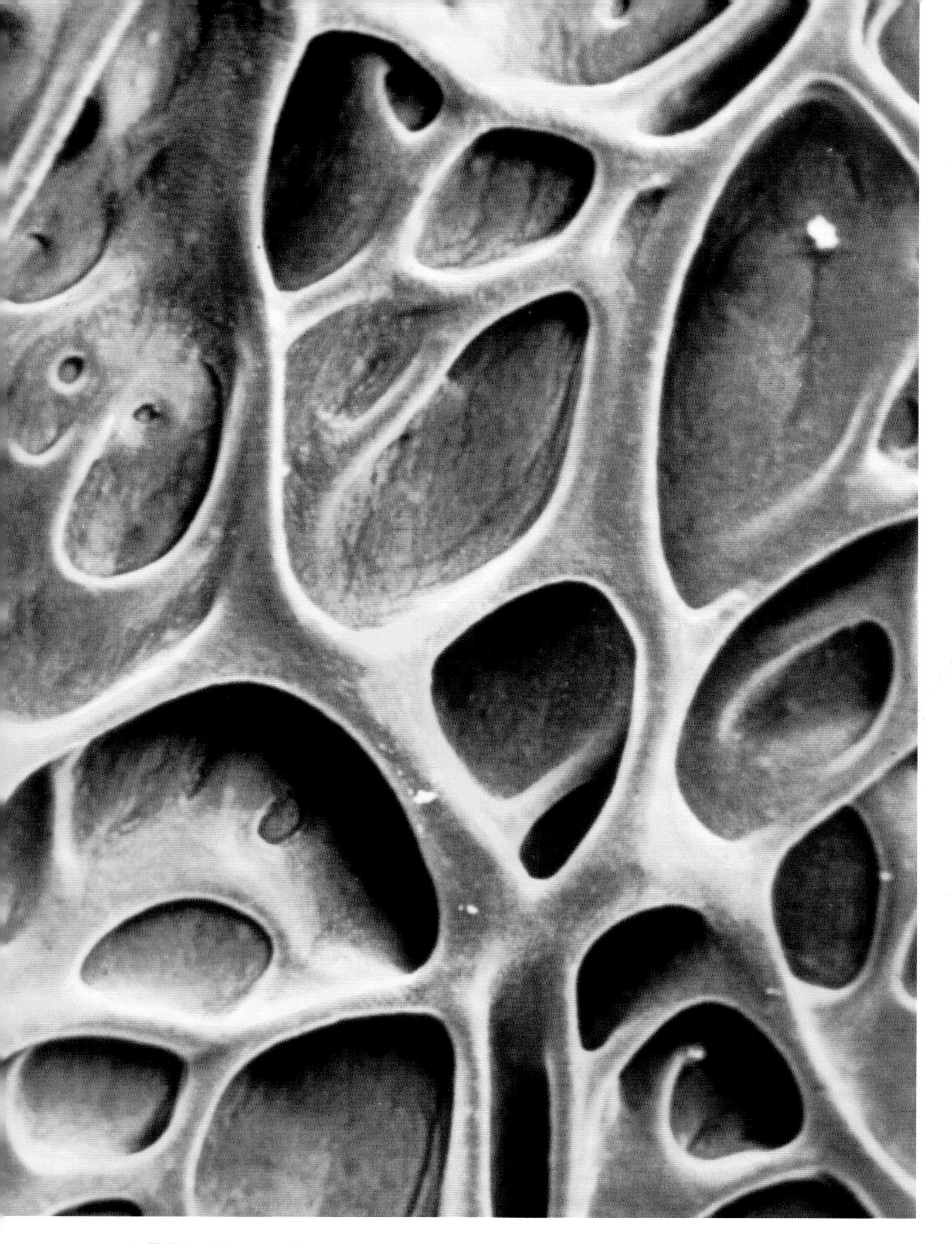

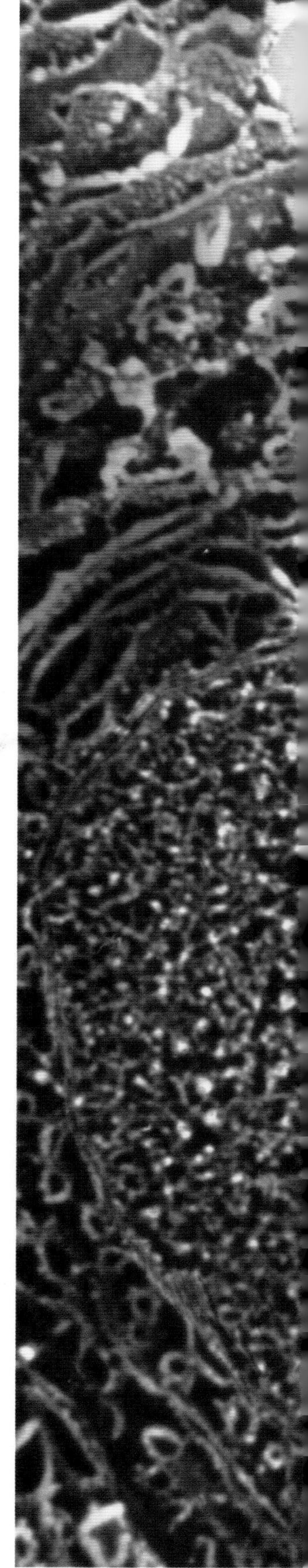

Gall bladder wall [SEM]

THE GALL BLADDER is a small storage vessel for bile, an alkaline digestive fluid produced by the liver. The gall bladder releases the bile into the duodenum which is part of the small intestine leading out of the stomach. Bile is secreted when required, particularly to aid fat digestion. The folds in the image are called rugae, and give mechanical strength to the gall bladder wall.

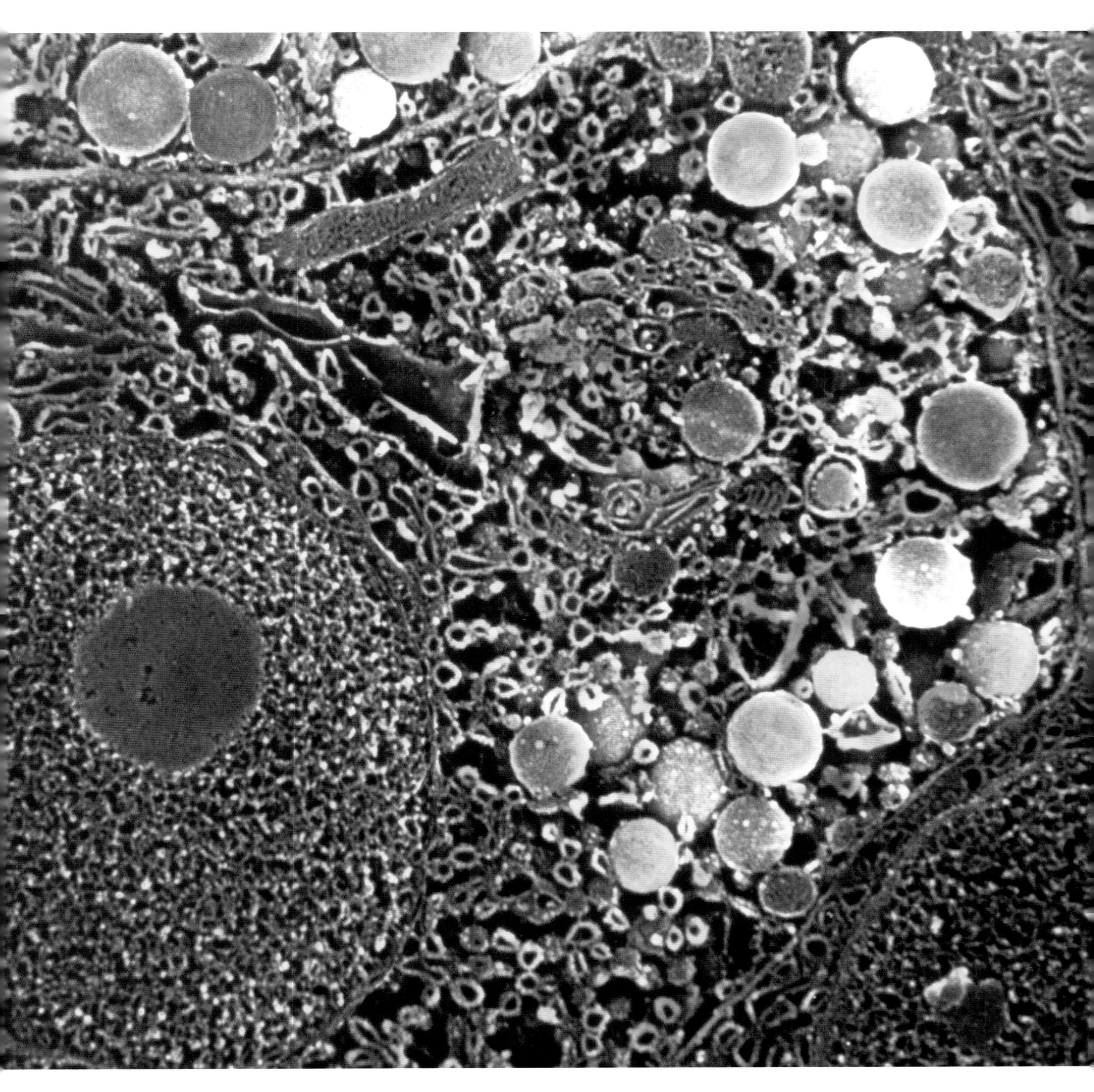

Enzyme-producing cell in pancreas [SEM]

MANY OF THE DIGESTIVE ENZYMES that break down food in the small intestine are produced by cells in the pancreas, a large gland located just below the stomach. This pancreatic cell, with blue nucleus and a red nucleolus, contains granules of enzymes (brown, white and orange circles) in an inactive form. When the enzymes are secreted, they reach the gut through the pancreatic duct and are activated by chemicals produced by cells lining the small intestine.

Large intestine [X-ray]

AN X-RAY REVEALS the three regions of the large intestine (purple). On the left is the caecum and the ascending colon, which is relatively immobile. The first part of the colon contains the caecum and the appendix. In the centre is the transverse colon, a mobile region that here forms a U-shaped loop. It leads to the descending colon (right), which descends to the rectum. The main function of the colon is to absorb water and minerals from digested food.

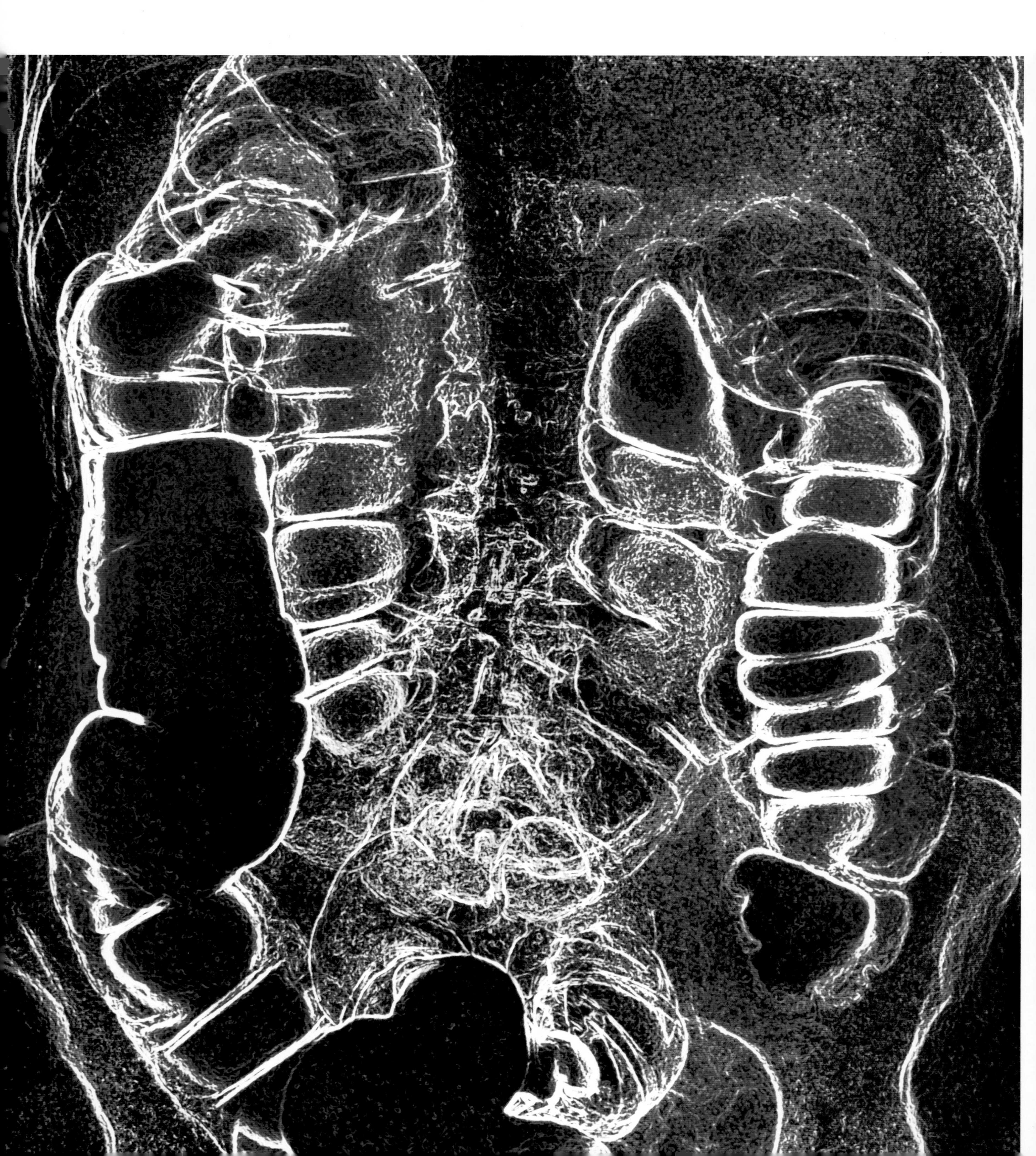

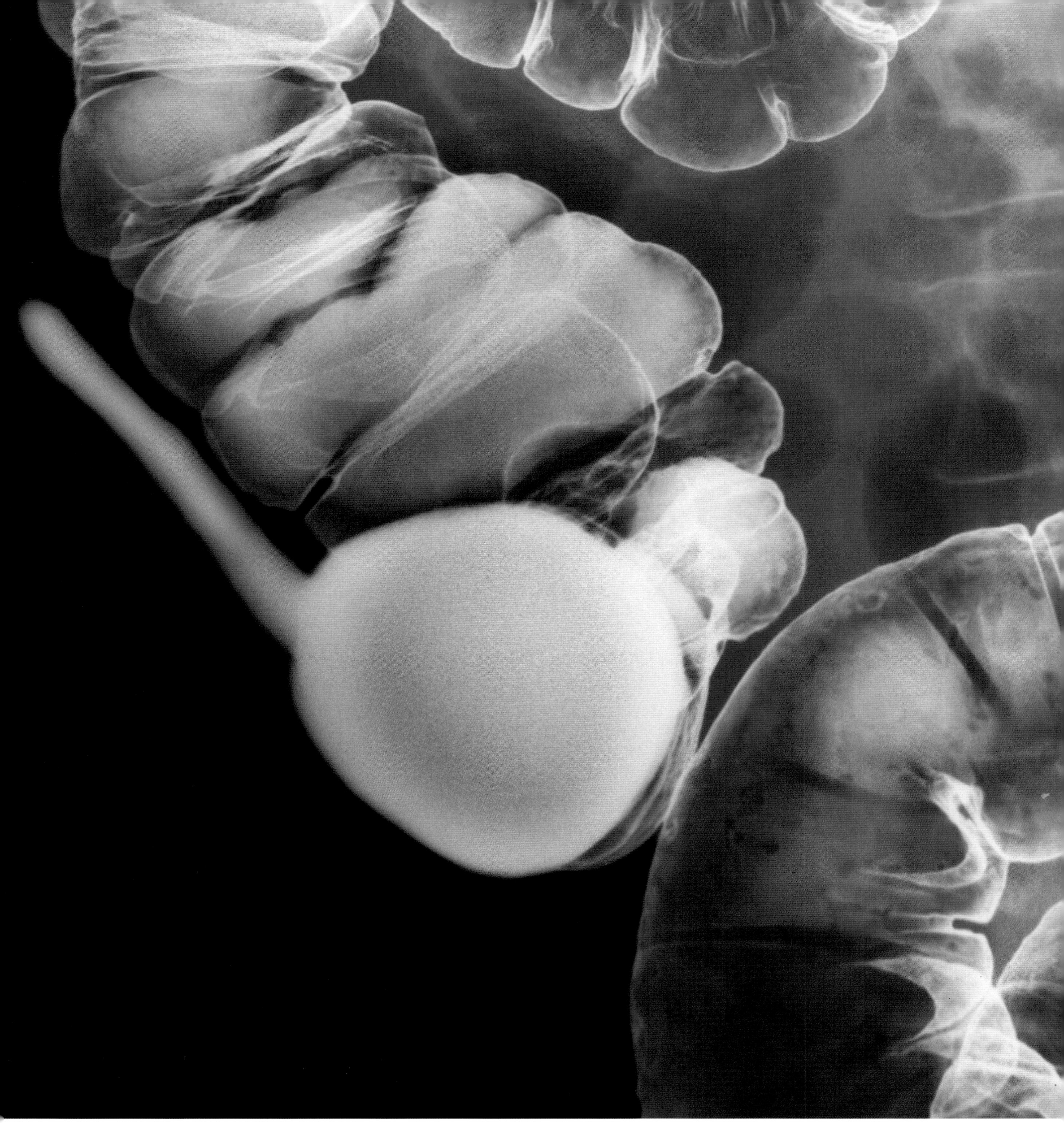

Appendix [X-ray]

THE APPENDIX (lower left) is a worm-like, blind tube, about 9cm/3½in long in adults, leading off the caecum, the first section of the large intestine. Although it is associated with the digestion of cellulose in herbivores, it has no known function in humans. Appendicitis – inflammation of the appendix – is a common and potentially serious condition.

Gland in the colon [SEM]

THIS IS ONE OF MANY TUBULAR GLANDS that secrete mucus in the colon. Underneath the mucus layer is a layer of smooth muscle that contracts and relaxes to propel the contents of the colon towards the rectum. Water, sodium, vitamins and other minerals are absorbed in the colon. The watery residue is reduced to semisolid faeces.

Lining of the rectum [SEM]

THE RECTUM, the last section of the large intestine, has a mucus lining made up of polygonal cells (red) covered in tiny, hair-like microvilli that greatly increase the absorptive area of the rectum. The upper mucosal layer is continually being shed and replaced. Debris (pink) can be seen.

Water makes up 65 per cent by weight of the human body. All of the chemical reactions essential for life take place in a watery environment. One role of the urinary system, consisting of kidney, bladder and associated ducts, is to keep the volume and composition of body fluids more or less constant. This is part of the process known as homeostasis.

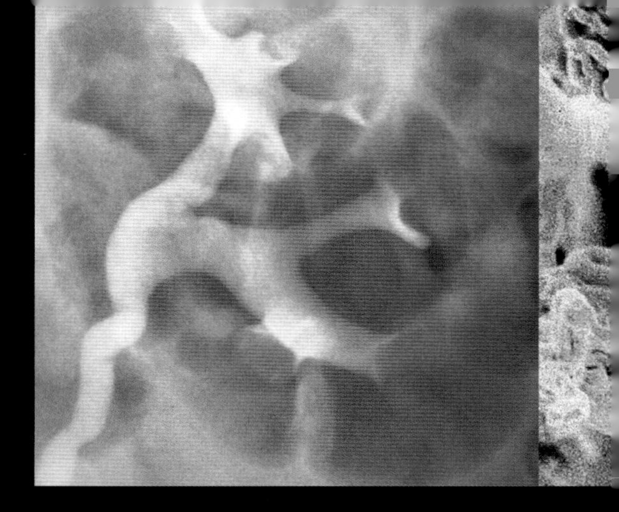

The other job of the kidneys is to excrete waste products, especially urea, a compound formed in the liver by the breakdown of proteins. The importance of the kidneys can be gauged from the fact that people with kidney failure may be restricted to a daily intake of 0.5 litres/1 pint of fluid and about 40g/1½oz of protein.

There are two kidneys, one on each side of the rear abdomen, just below waist level. About 160km/100 miles of blood vessels inside the kidneys deliver up to 2,000 litres/440 gallons a day. Each kidney has a million tiny filters, which extract wastes from the blood and produce about 7.5 litres/13 pints of filtrate an hour. If all this was expelled as urine, the body would quickly become completely dehydrated. In fact, over 99 per cent of the fluid is reabsorbed in the kidney tubules

Urine output varies with fluid intake, exercise and outside temperature, but averages about 1.5 litres/ 3 pints a day.

Each kidney releases urine into a ureter, a duct that leads to the bladder, where the urine is stored until it is expelled. The bladder can stretch and holds 0.5 litre/1 pint or more. Urine is passed by the opening of two sphincter muscles – one under involuntary control, the other under voluntary control. Young children pass urine whenever the bladder fills until they develop control over the voluntary sphincter. However, the desire to urinate begins when the bladder is about half full, and when it is very full it takes serious willpower to prevent urination.

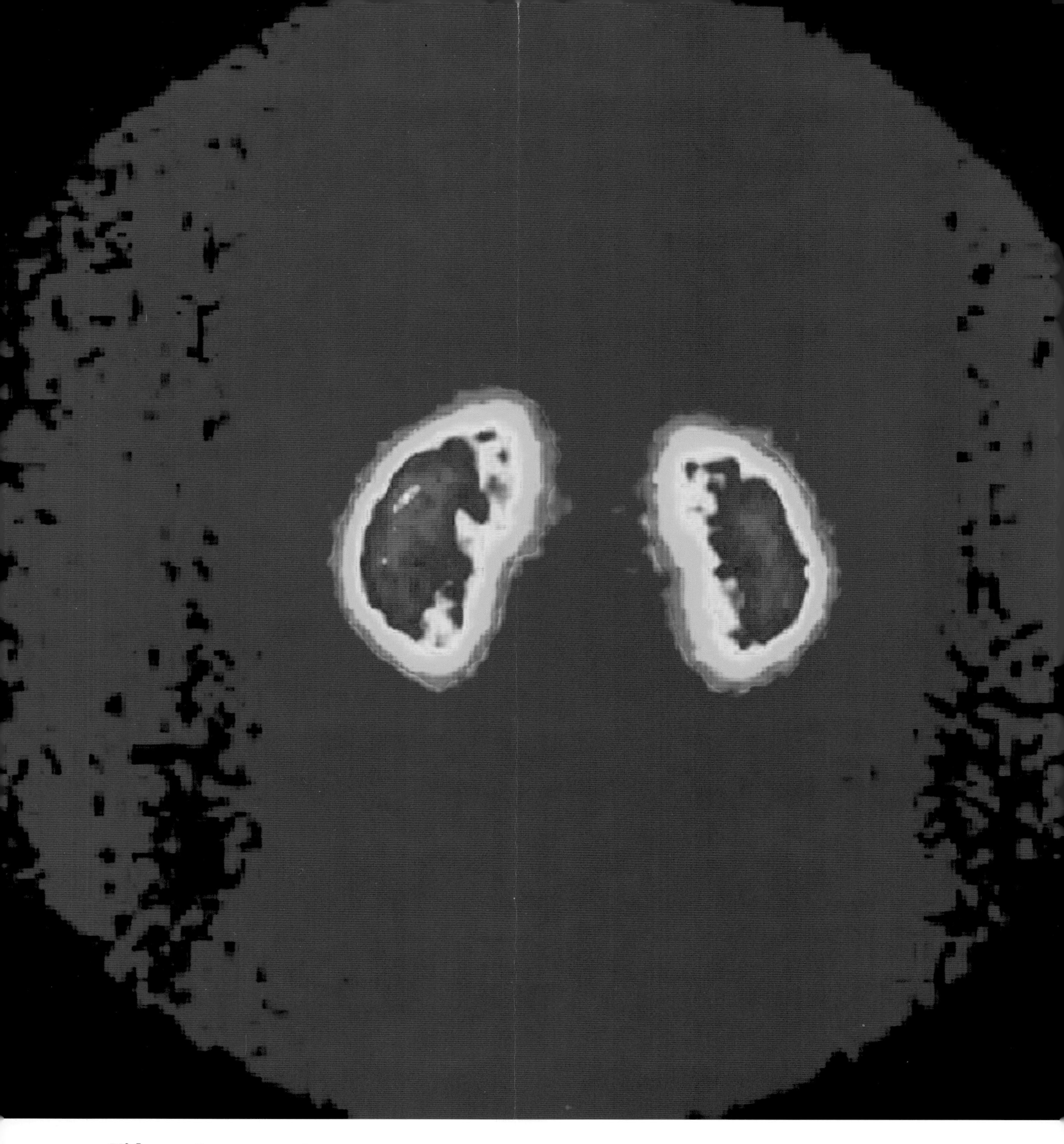

Kidneys [Gamma scan]

THE UPTAKE OF BLOOD by a pair of kidneys seen from behind is shown in red. The kidneys filter waste products from the blood, passing them in the form of urine to the bladder (not seen). Gamma scans are used to assess renal blood supply and diagnose kidney disorders.

Blood supply to a kidney [Resin cast]

THE KIDNEYS RECEIVE THE LARGEST BLOOD supply of any organ, as indicated by this resin cast showing the dense network of capillaries arising from divisions of one of the kidney's renal arteries. Every minute about 1 litre/2 pints of blood under high pressure flows to each kidney, where it is filtered. During a single day, all of the blood in the body passes through the kidney about 400 times.

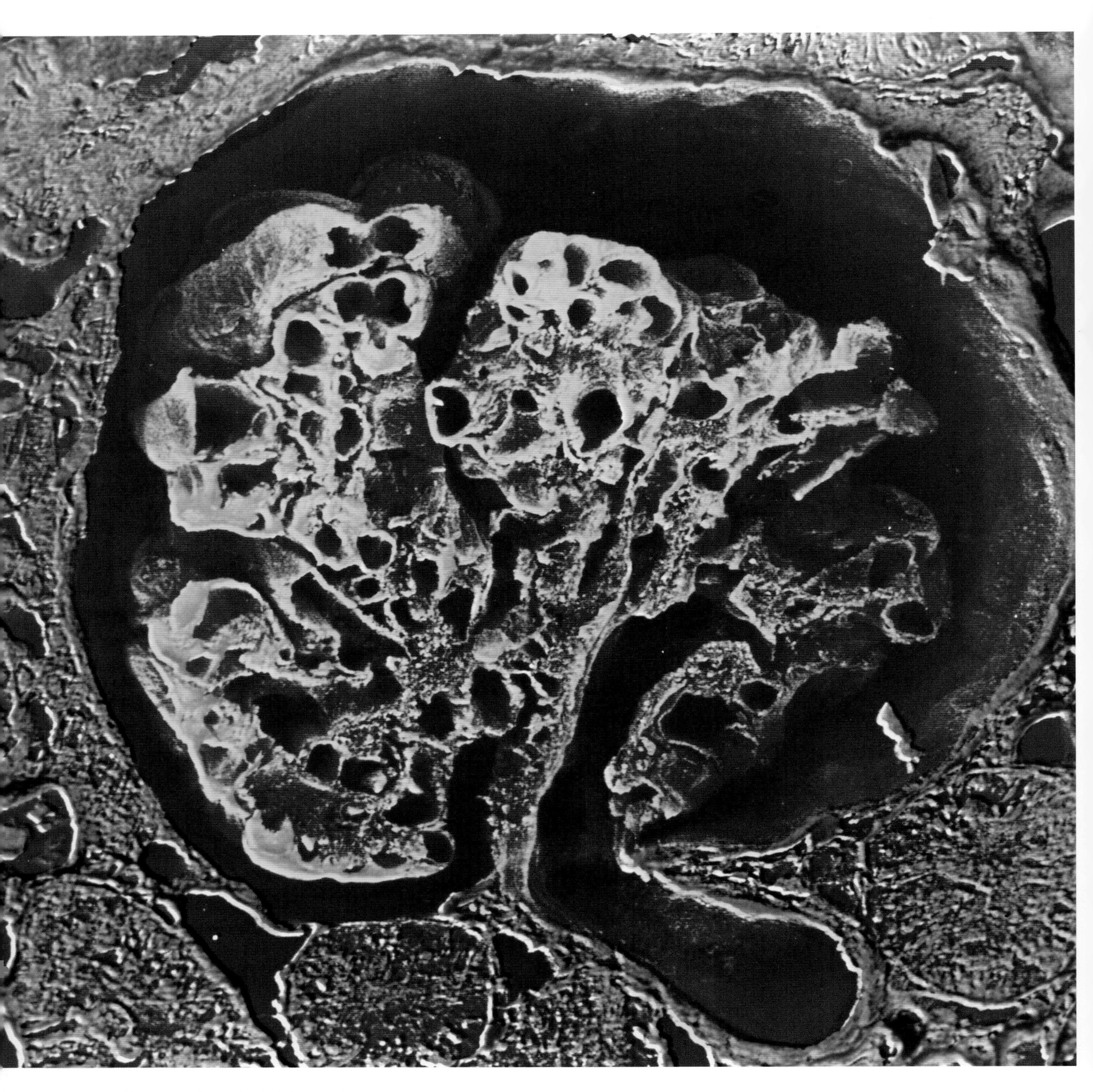

Glomerulus in kidney 1 [SEM]

BLOOD PASSING THROUGH THE KIDNEYS is filtered by glomeruli, tight balls of capillaries, one of which is shown here (centre). Fluid passes out of the capillaries and drains into a long tube (right). Here, essential substances and some water are reabsorbed. The remaining unwanted fluid, containing toxins from the blood, drains to the bladder as urine.

Glomerulus in kidney 2 [SEM]

THE KIDNEY'S FILTRATION MECHANISM, including the glomerulus shown here, is so efficient that it can produce about 7.5 litres/13 pints of filtrate an hour. Of this amount, only about 0.125 litre/¼ pint becomes urine; the rest is reabsorbed. Urine output varies with fluid intake and outside temperature, but averages about 1.5 litres/3 pints a day, most of which is produced during waking hours.

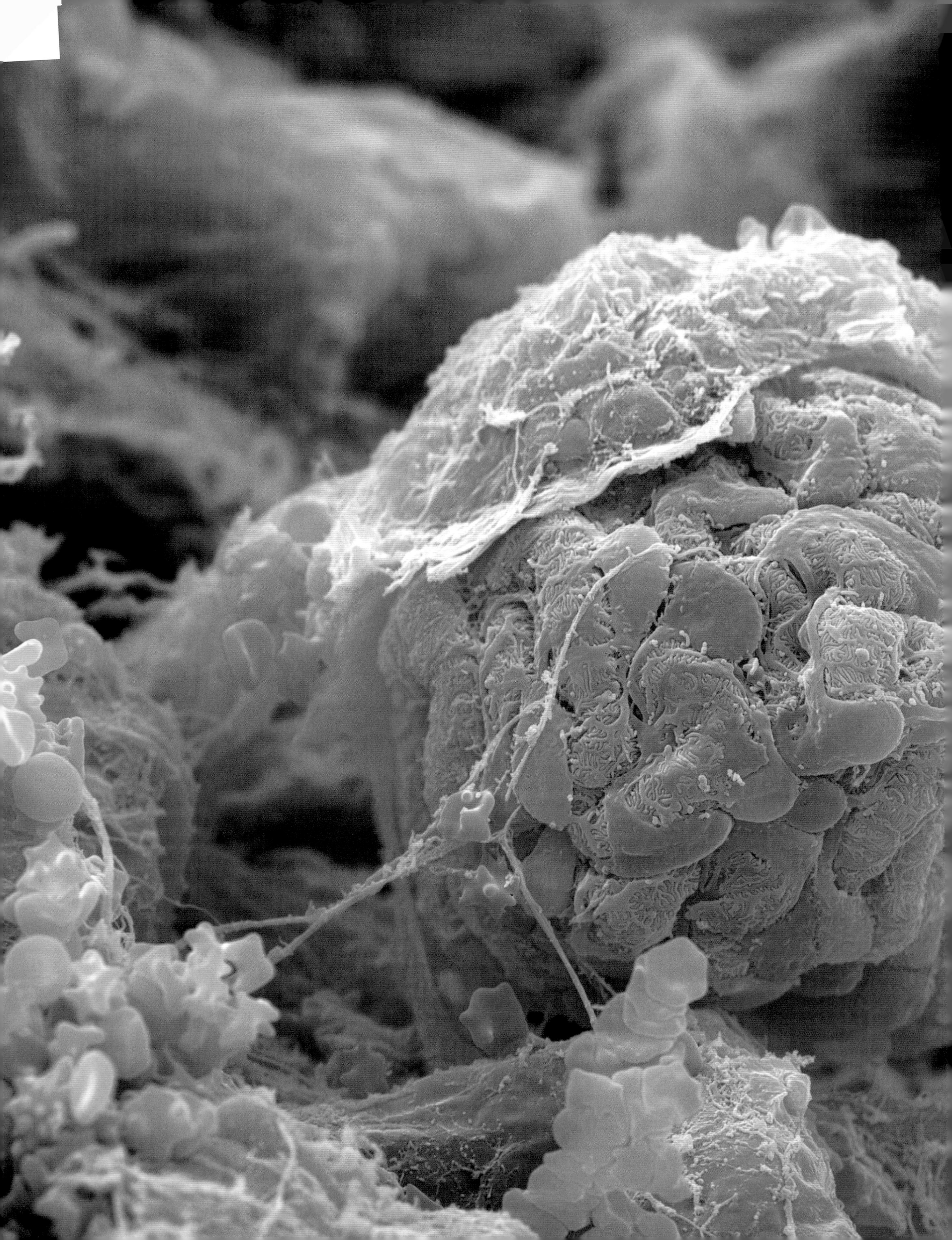

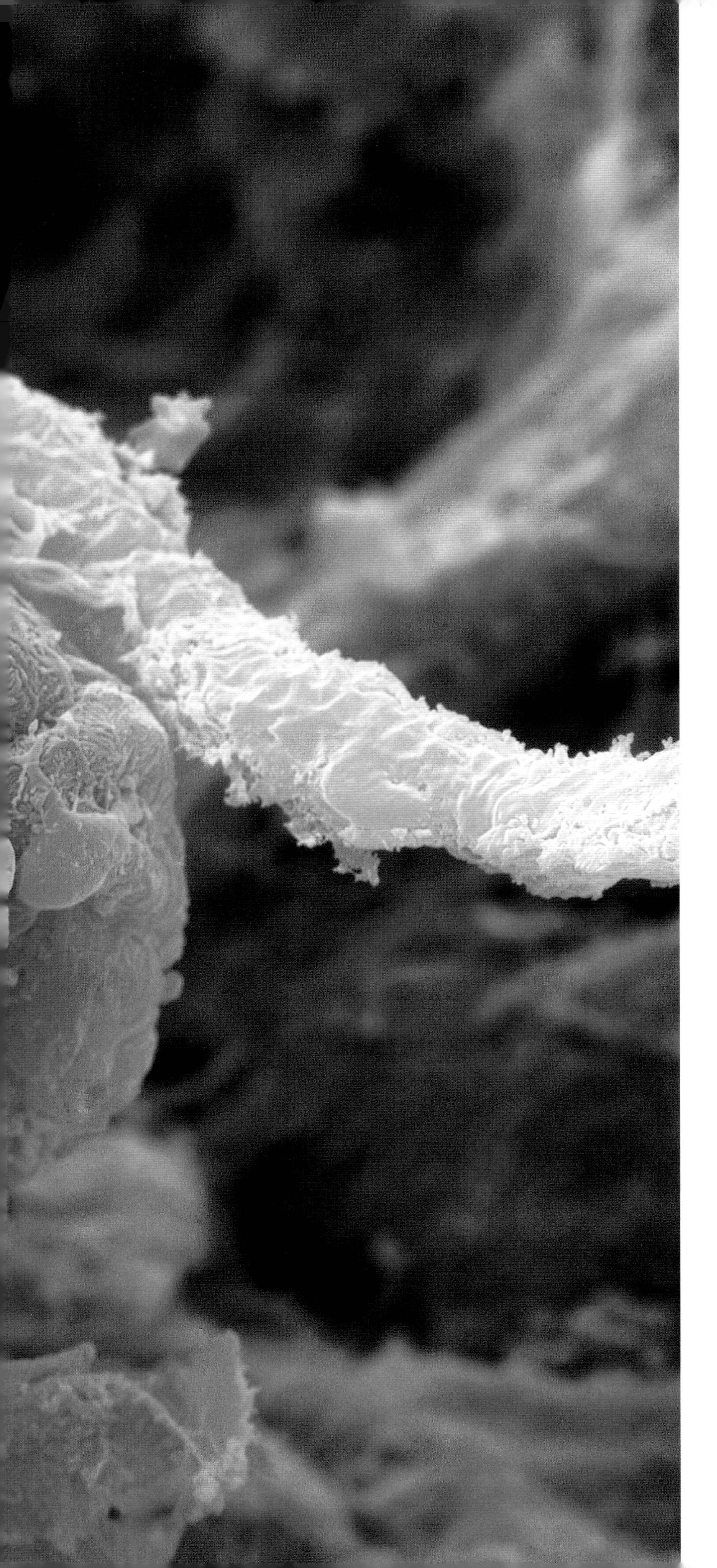

Glomerulus in kidney 3 [SEM]

ANOTHER IMAGE OF A GLOMERULUS shows the knot of capillaries enclosed in a renal capsule, where blood passing through the kidney is filtered. The walls of the capillaries and the lining of the capsule are composed of cells with thin slits in between. Most of the constituents of the plasma pass freely through these slits, but large molecules – mainly proteins – are filtered out by a membrane between the capillaries and lining.

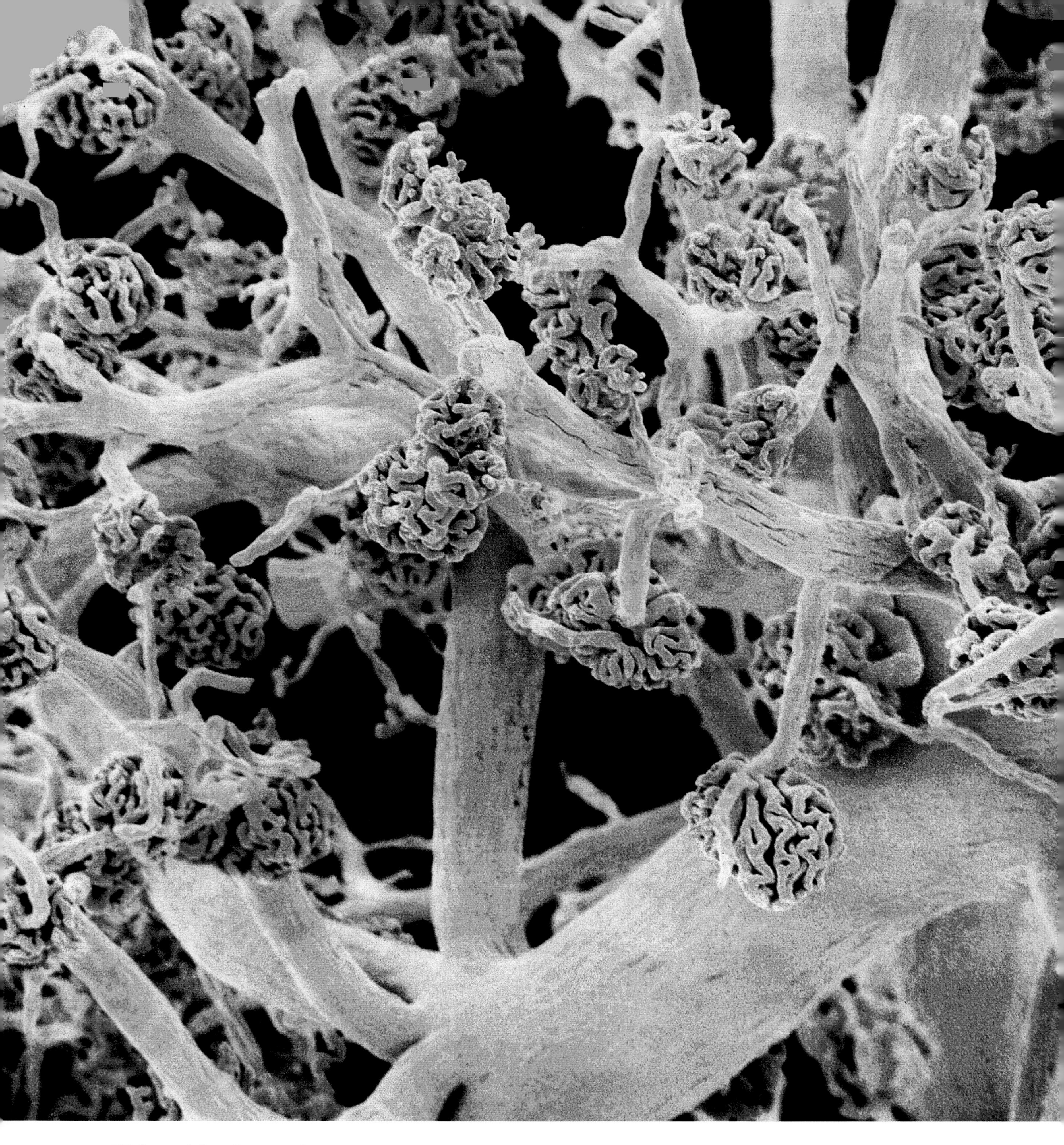

Kidney blood vessels [Resin cast/SEM]

THE INTRICATE NETWORK of blood supply is shown in this resin cast, made by injecting resin into a kidney and then dissolving away the surrounding tissues. Vessels (green) carry blood to glomeruli (red), where toxic waste is filtered from the blood. Each kidney contains about 1 million glomeruli, and such is the functional reserve that the essential excretory function can be carried out by two-thirds of one kidney.

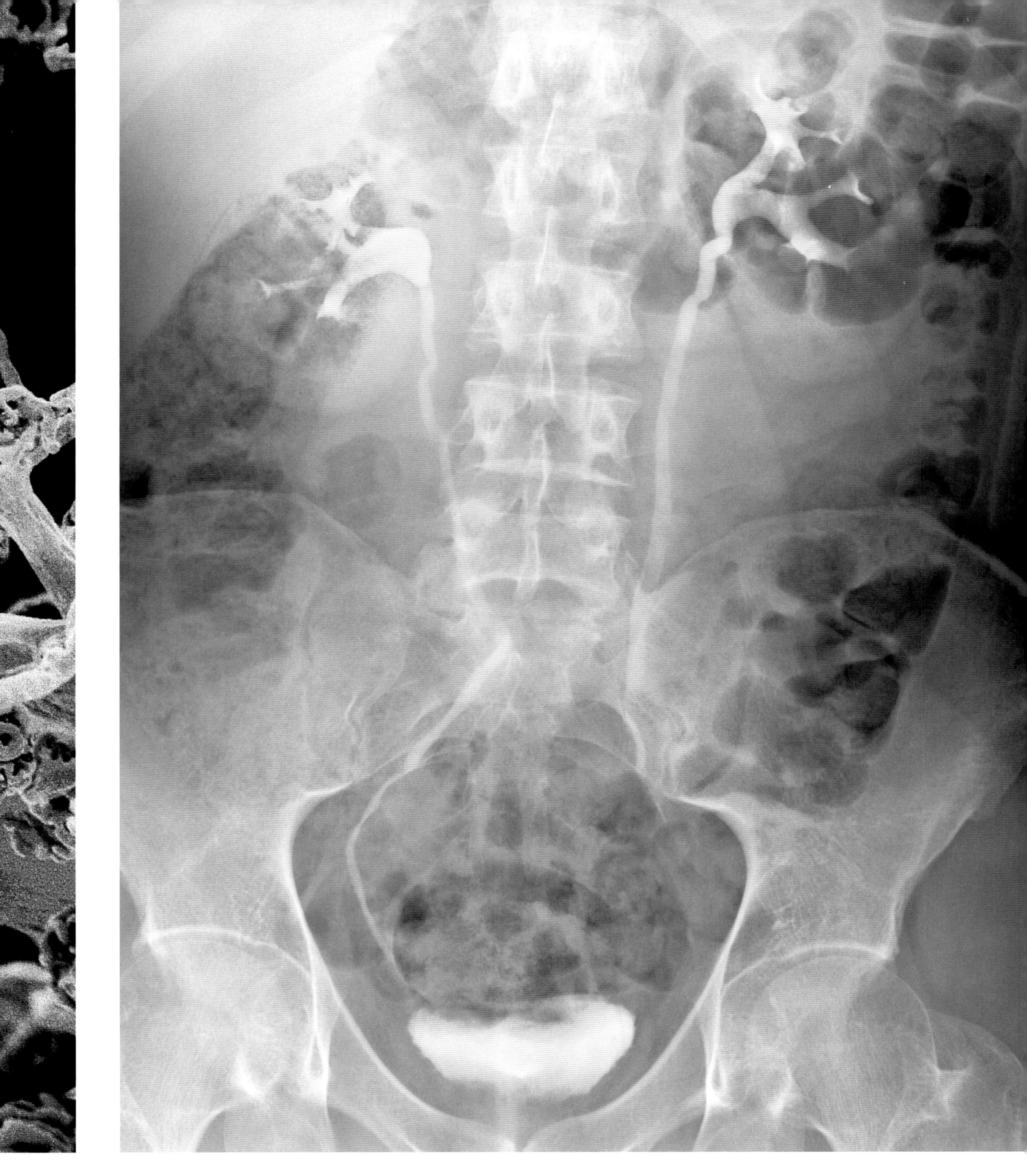

Urinary system [X-ray]

URINE PRODUCED IN THE KIDNEYS (orange, top left and right) is channelled down the ureters (pink) to the bladder (yellow and pink). When the bladder begins to get full, stretch receptors in its walls inform the brain. The sphincter muscles around the urethra are then voluntarily relaxed to let urine drain from the bladder, through the urethra, and out of the body.

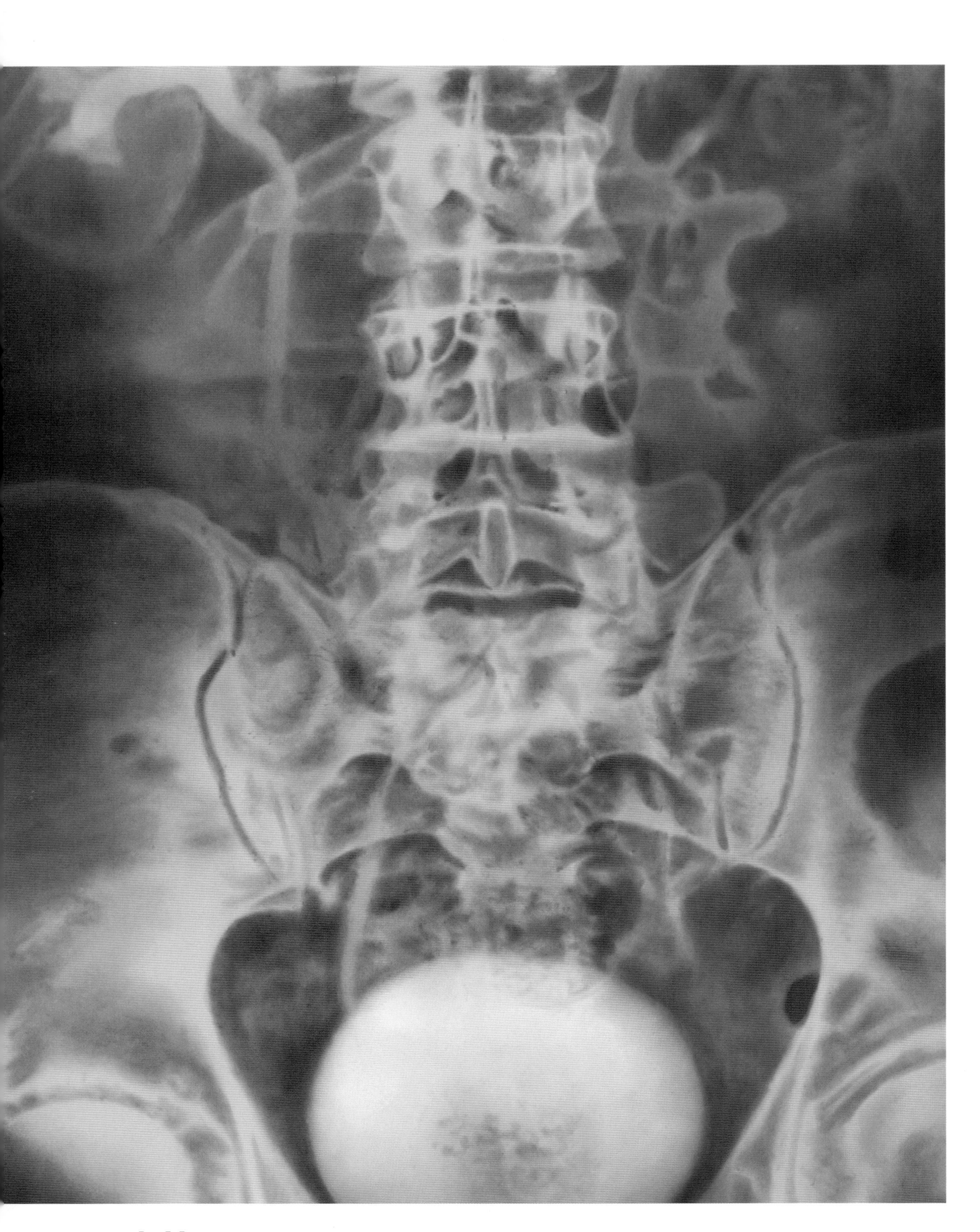

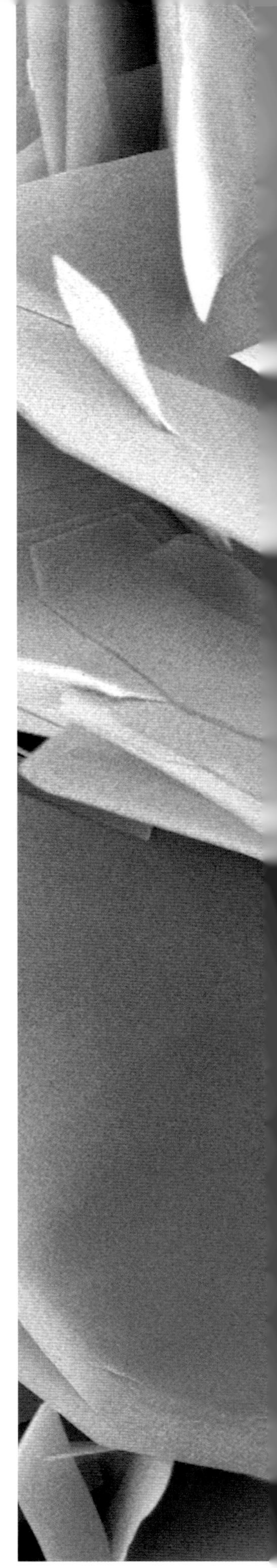

Bladder [X-ray]

THE BLUE SPHERE at bottom, between the pelvic hip bones, is a full bladder. The branched collecting ducts of a kidney can be seen at upper left. From here, urine passes down the tube-like ureter to the bladder. A human bladder can hold more than 0.5 litre/1 pint of urine. Its muscular walls relax as it fills, keeping the tension constant.

Kidney stone crystals [SEM]

KIDNEY STONES are usually formed by the precipitation of the mineral salt calcium oxalate from the urine. The hard stones may cause severe pain, especially as they pass down the urinary tract. Large stones may need to be surgically removed or broken down using ultrasound treatment.

The circulatory system has many functions. It delivers oxygen, nutrients and hormones to cells. It removes the waste products of metabolism and carries them to the lungs and kidneys for excretion. And it distributes heat generated mainly by the muscles, keeping the body at a constant temperature.

The centre of this system is the heart, the best-designed pump in the world. In a healthy adult human at rest the heart beats 60 to 100 times a minute, which adds up to 2.5 billion times in the course of an average lifetime. During this period it pumps about 150 million litres/33 million gallons of fluid through the arteries. In a single day the heart generates enough power to lift an average-sized car about 15m/50ft. Yet this incredible machine weighs only about 200g/7oz, is completely self-regulating and is so energy-efficient that its fuel consumption for seventy years could be covered by 250kg/550lb of sugar.

The heart is really two pumps, arranged side by side in a single organ. Each half is equipped with cunningly designed valves that ensure the blood flows in the correct direction. The left side of the heart receives blood rich in oxygen from the lungs and pumps it under high pressure through the arteries and capillaries to all parts of the body. Oxygen-poor blood under low pressure returns through

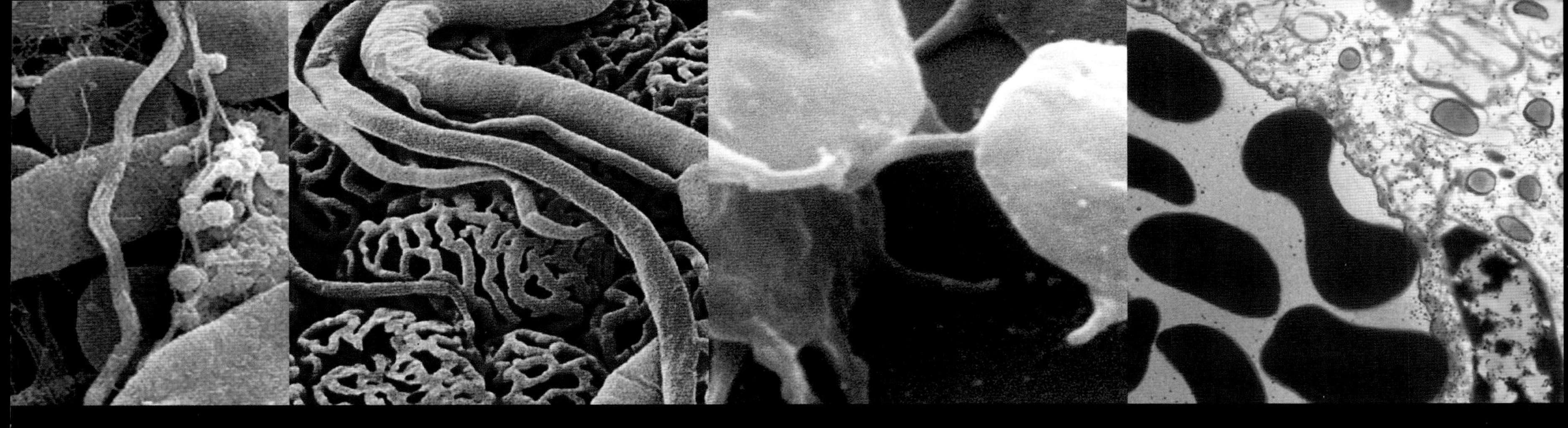

veins to the right side of the heart, which pumps it back to the lungs for new oxygen. A given portion of blood circulates in less than a minute.

Blood itself is a complex fluid. More than half its volume is made up of plasma, a straw-coloured solution of proteins, sugars, salts and other chemicals. The actual cell content of blood is about 38 per cent in women, 46 per cent in men. By far the most numerous cells in blood are the red blood cells – 5 million in each millilitre/0.04fl oz of blood. Their role is to carry life-giving oxygen to cells and transport waste carbon dioxide back to the lungs. The other cells found in blood are the various white blood cells, which fight infection, and cell fragments called platelets, which help repair damaged blood vessels and aid in blood clotting.

Major blood vessels [Artwork, see right]

THE AORTA (red, down centre of torso) is the body's main artery, receiving about 0.08 litre/3fl oz of blood each time the heart beats. Branches of the aorta lead to the internal organs (not seen). At lower centre the aorta divides, forming the femoral arteries that deliver blood at high pressure to the lower body and limbs.

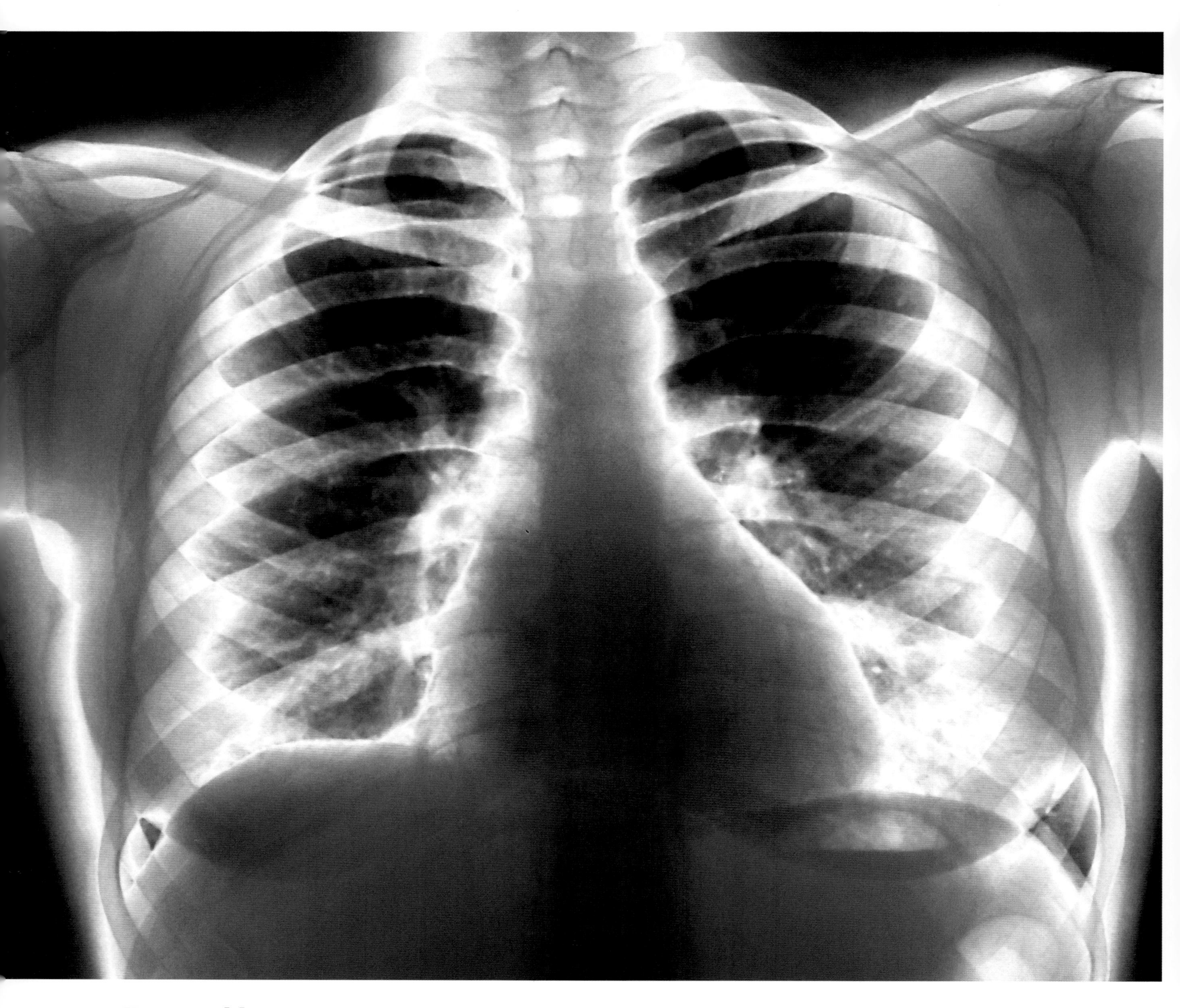

Heart and lungs [X-ray]

AS THE CENTRE OF THE CIRCULATORY SYSTEM, the heart (red) must be well protected. It lies between the lungs (black), enclosed by the ribs (blue bands) and shielded at front by the sternum, a hard bony plate.

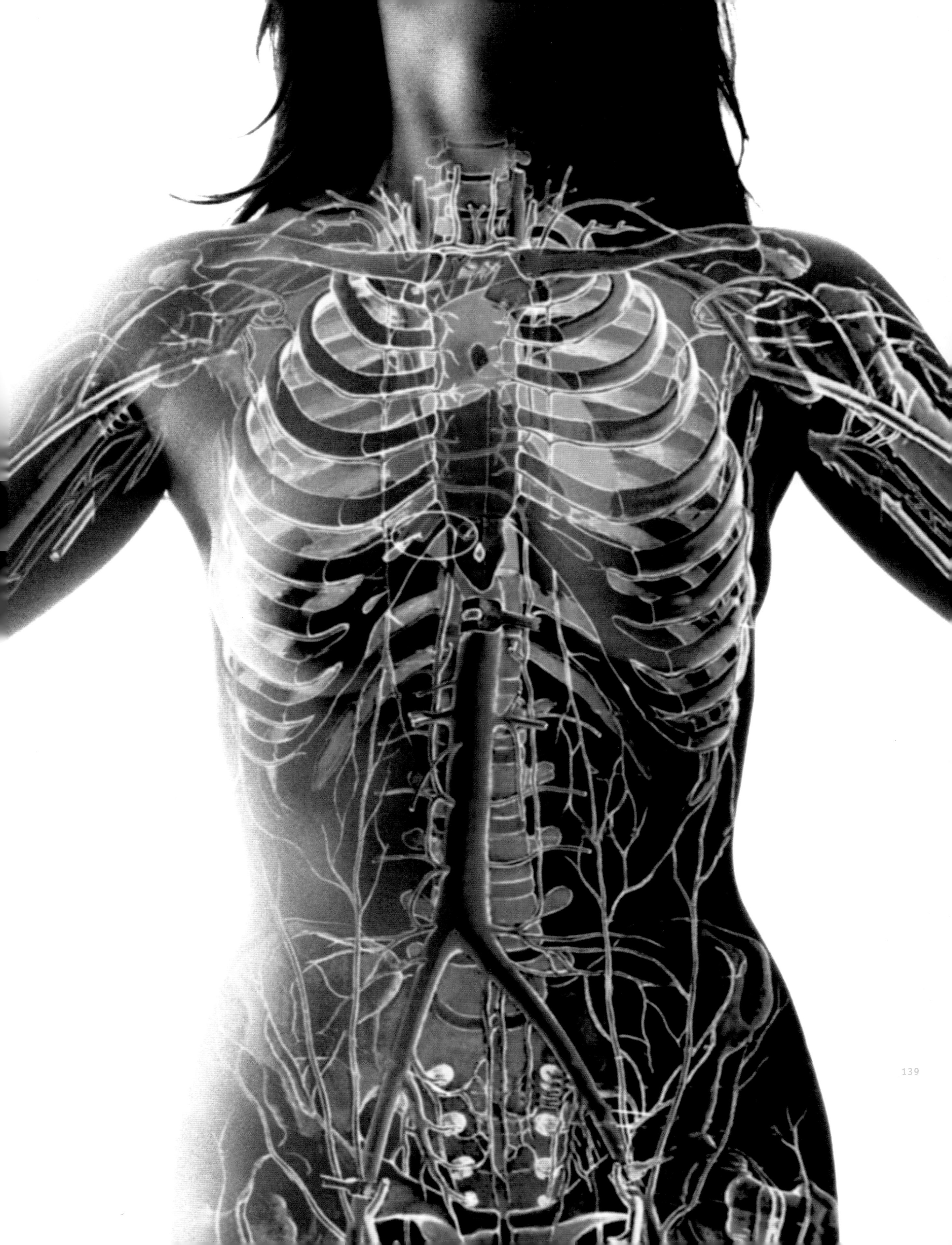

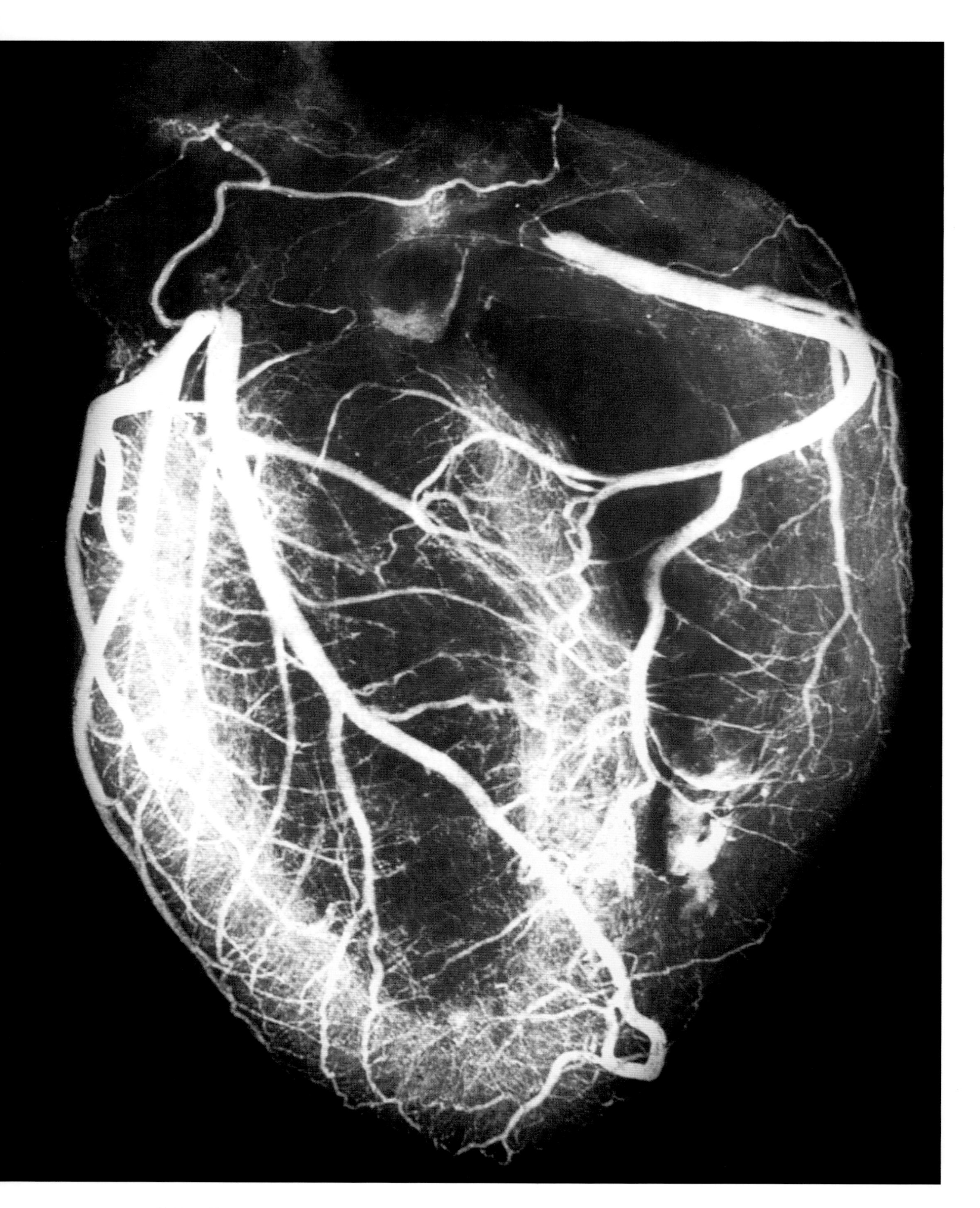

Healthy heart [Angiogram]

SHOWN IN WHITE, the left and right coronary arteries provide the heart with its own blood supply. They are sensitive to changes in the blood's oxygen levels, and increase the blood supply to the heart to meet demand. Angiograms are used to detect heart disease or blockages to the coronary arteries that may lead to a heart attack.

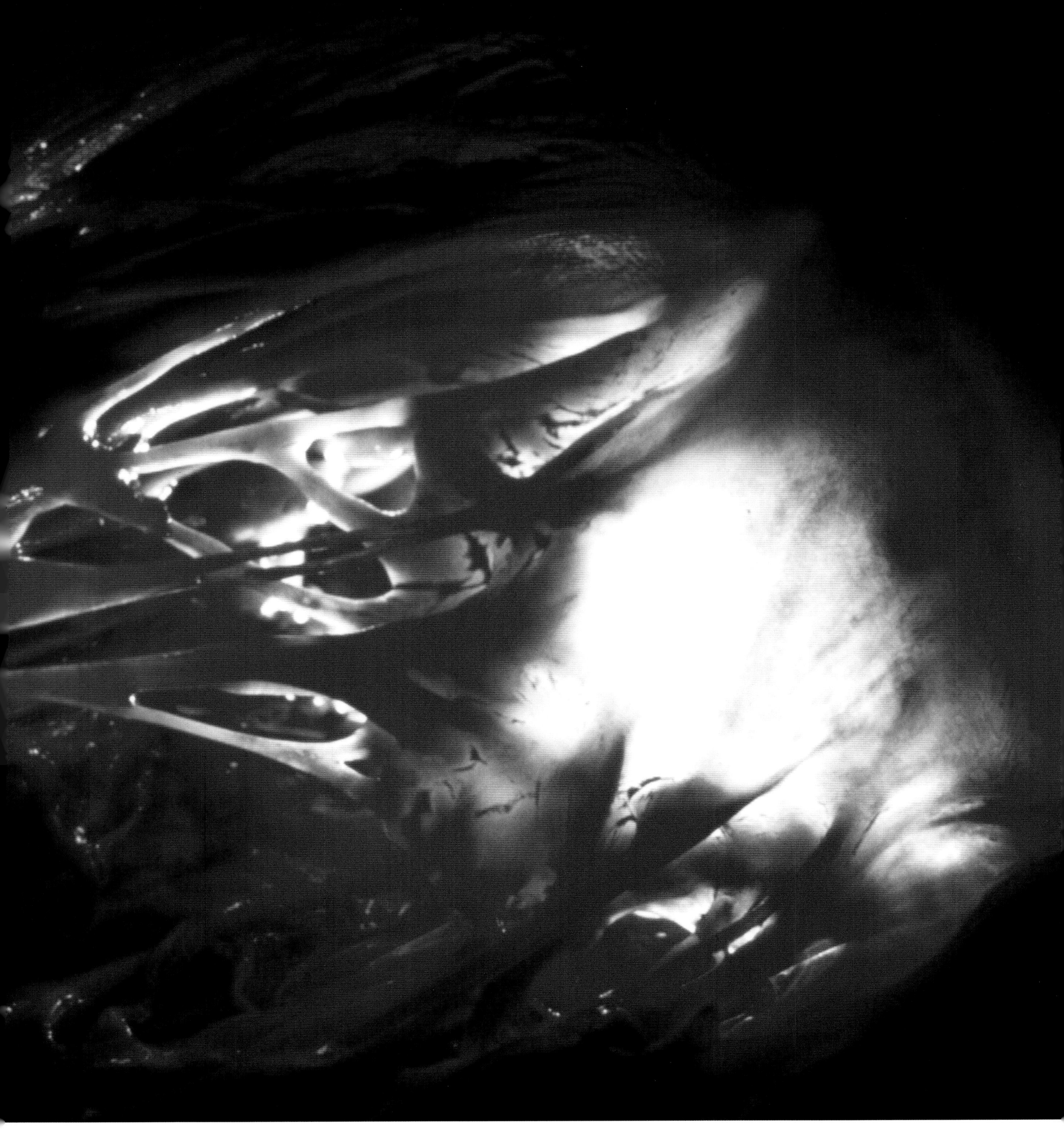

Mitral heart valve [Macrophoto]

THE OPENINGS BETWEEN THE CHAMBERS of the heart are guarded by a system of valves that ensure blood flows in one direction only. This is the mitral valve, which controls the flow of blood from the left atrium into the left ventricle. The tendons seen at bottom act as guy ropes, preventing the valve from turning inside out under the great pressure they are subjected to.

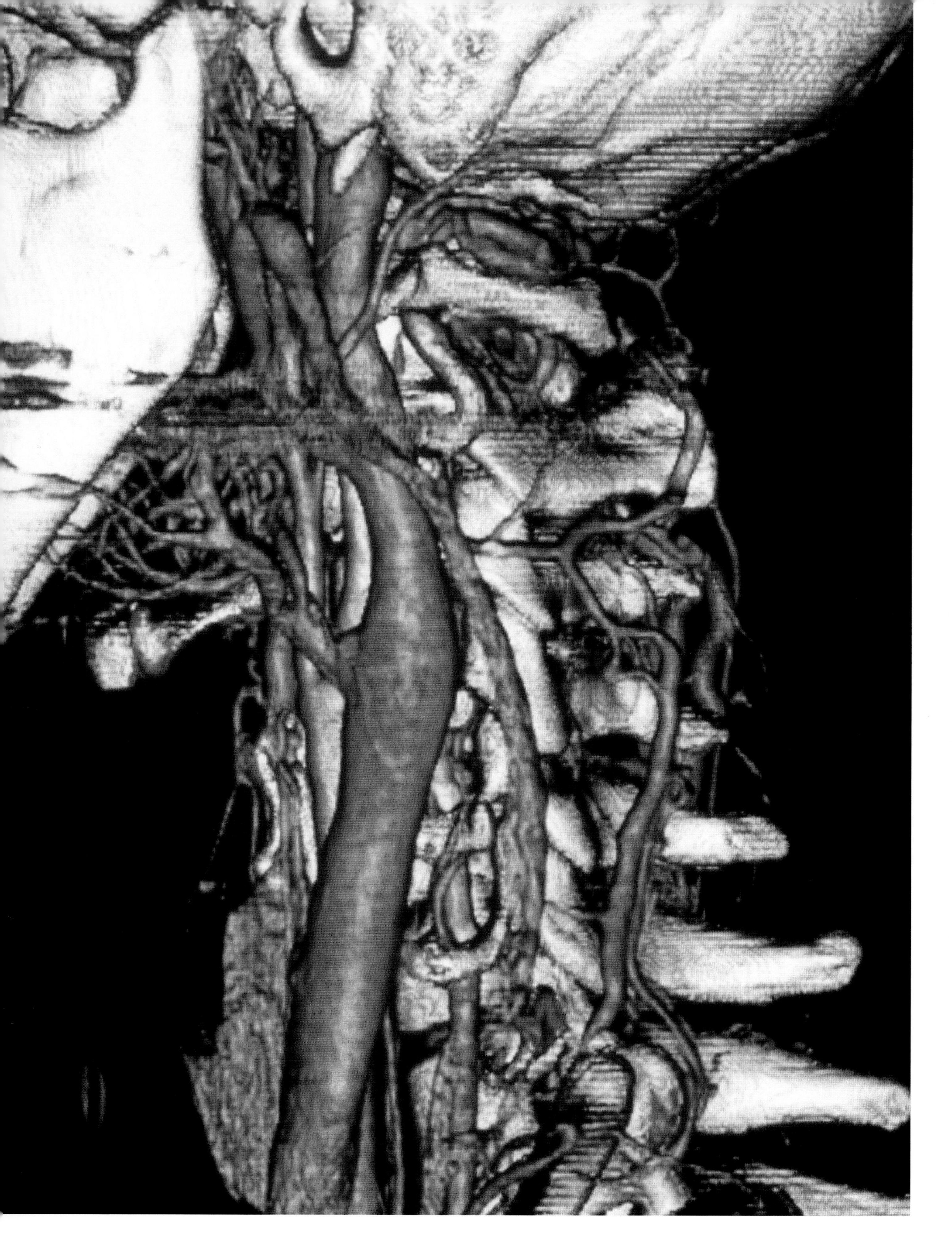

Blood vessels in the human neck [EBT]

ARTERIES (RED) AND VEINS (BLUE) are revealed in this scan of the neck. The large artery is one of the two carotids that distribute blood to the head and neck. Stretch receptors in the walls of the carotid arteries help control blood pressure. If blood pressure rises, the stretch receptors send information to the brain, which responds by sending nerve signals that slow heart rate and dilate the arteries in peripheral tissue, thereby resetting blood pressure.

Blood vessel in spinal cord [TEM]

A BLOOD VESSEL IN THE SPINAL CORD is packed with red blood cells that carry oxygen, nutrients and hormones around the body. Red blood cells are bi-concave discs that have a dumbbell shape when viewed end-on – an arrangement that increases their oxygen-carrying capacity. The green ovals are nuclei of cells that form the blood vessel's inner lining. The vessel is surrounded by nervous tissue called white matter.

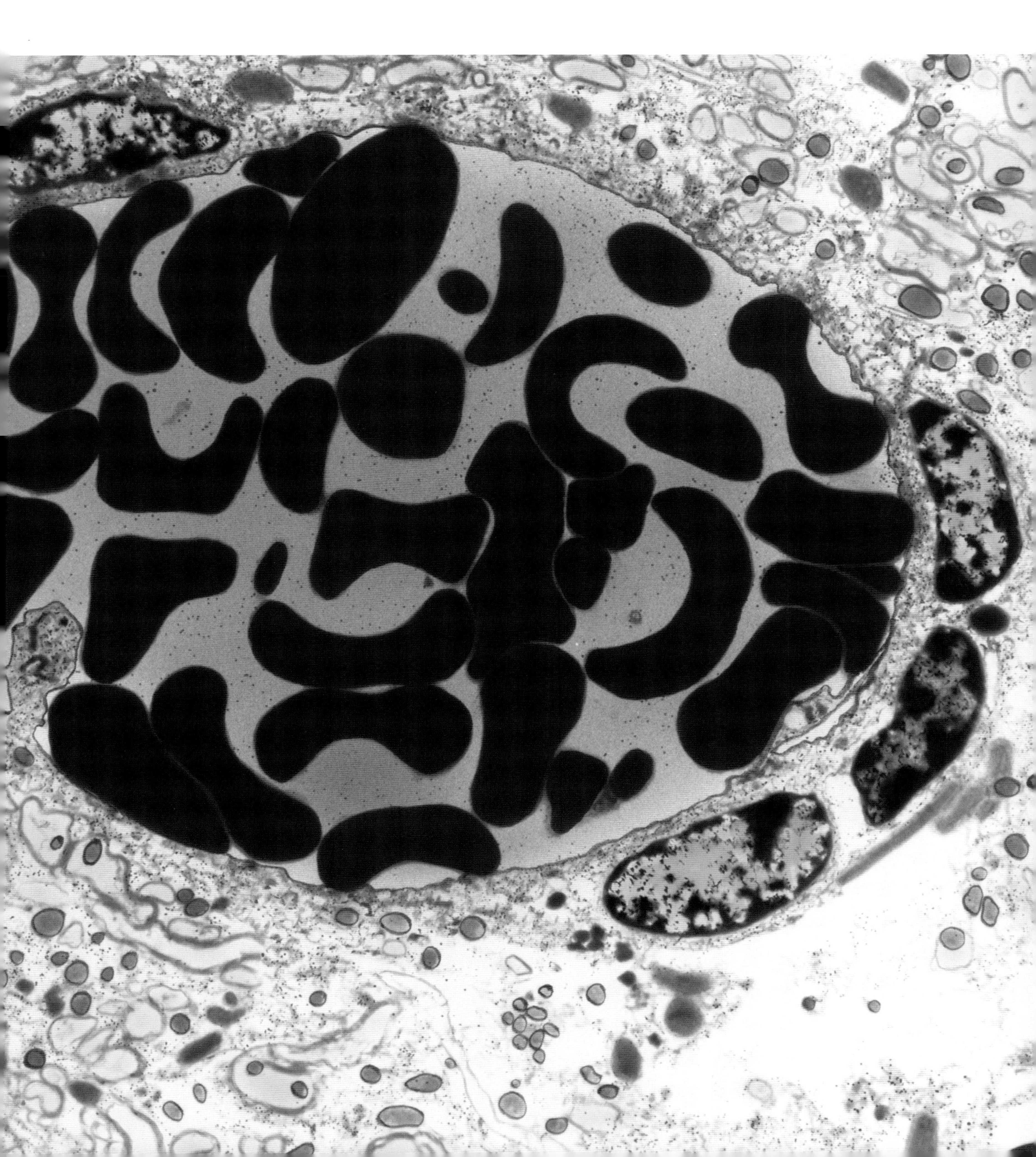

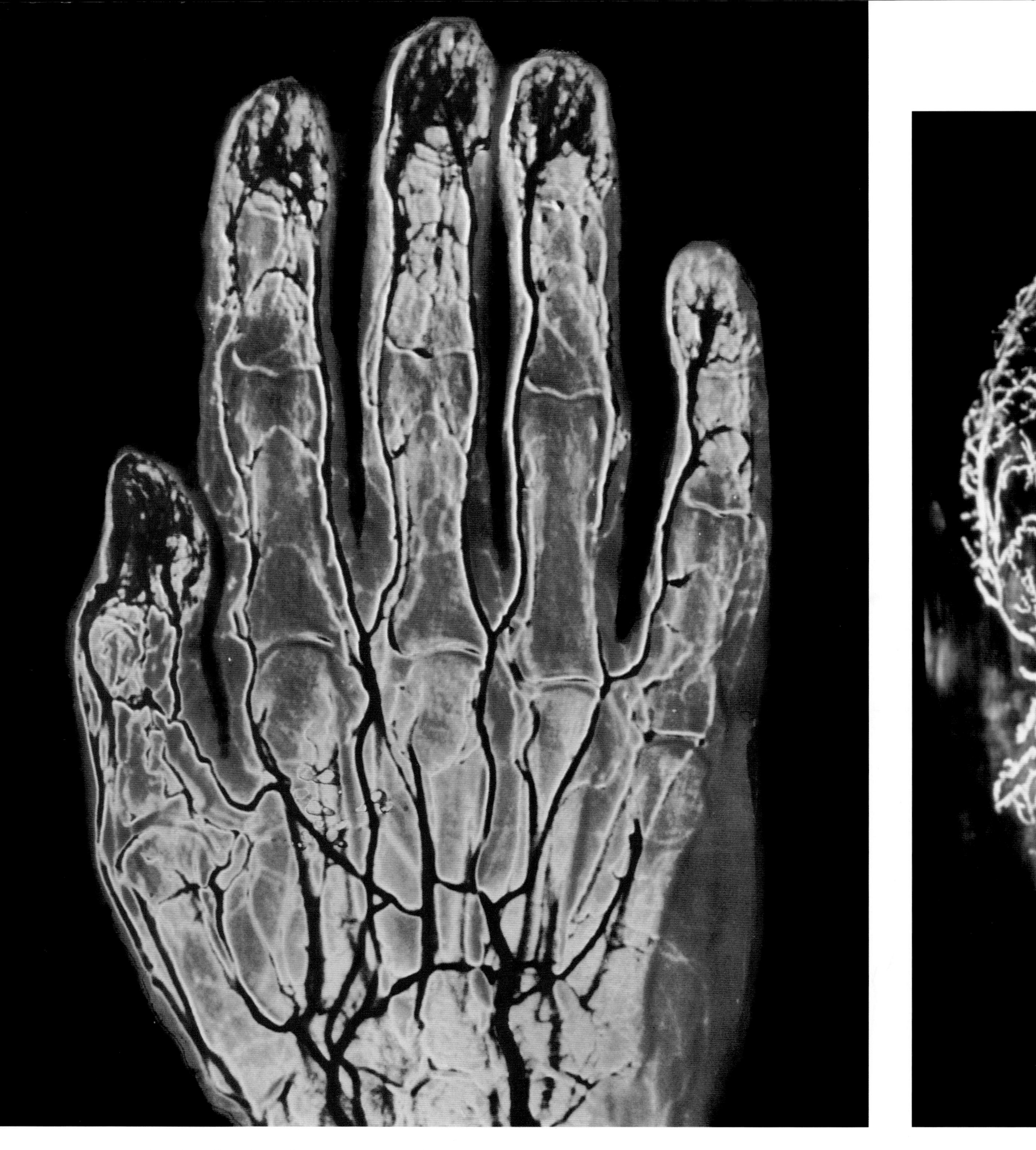

Arteries in the hand [Angiogram]

THE CAPILLARIES OF THE NAIL BEDS are prominent in this image showing the network of arteries in the human hand. As highly active tissues, the nail beds need a good blood supply. Arteries have thick muscular walls, and their contractile power maintains blood pressure at a distance from the heart. Angiogram imaging is an X-ray technique used to diagnose numerous circulatory disorders, including obstructions in the blood vessels and swellings of the arterial walls.

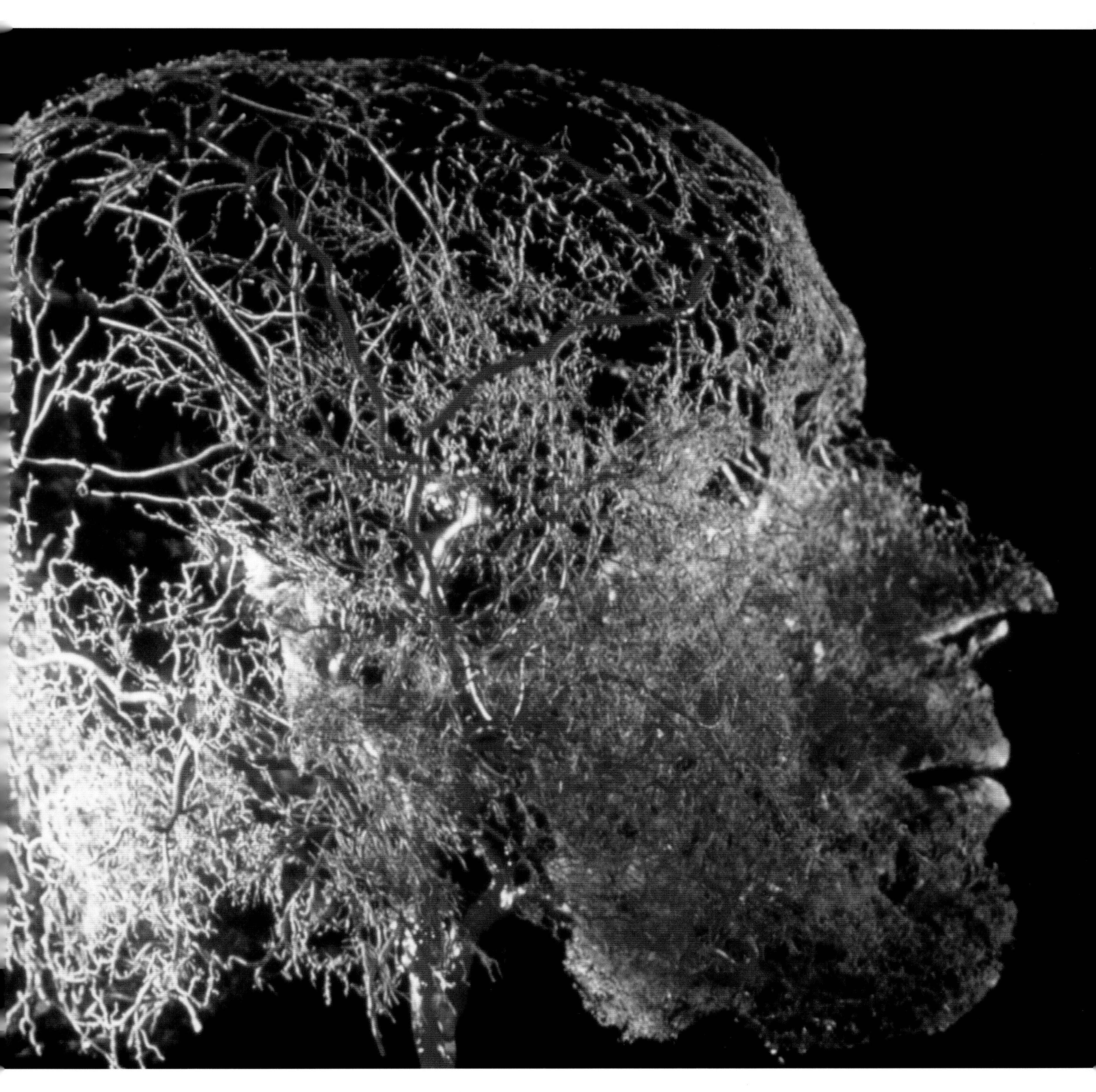

Blood vessels in the head [Resin cast]

THE BRAIN HAS A HIGH OXYGEN demand and needs a large blood supply. If the supply is cut off for even a few seconds, irreversible brain damage may occur. The main blood vessels are the two internal carotid arteries, which enter the skull base. The dense network of capillaries in the cheeks helps to cool the body by transferring heat to the external environment. When they dilate, causing a blush, they also help to communicate emotions – anger, embarrassment or sexual arousal.

Blood vessels in thyroid [SEM]

BLOOD VESSELS SNAKE through the thyroid gland, a hormone-producing gland that lies at the base of the neck. The gland has no ducts, but simply releases its hormone directly into the blood, which carries it to target organs and cells elsewhere in the body.

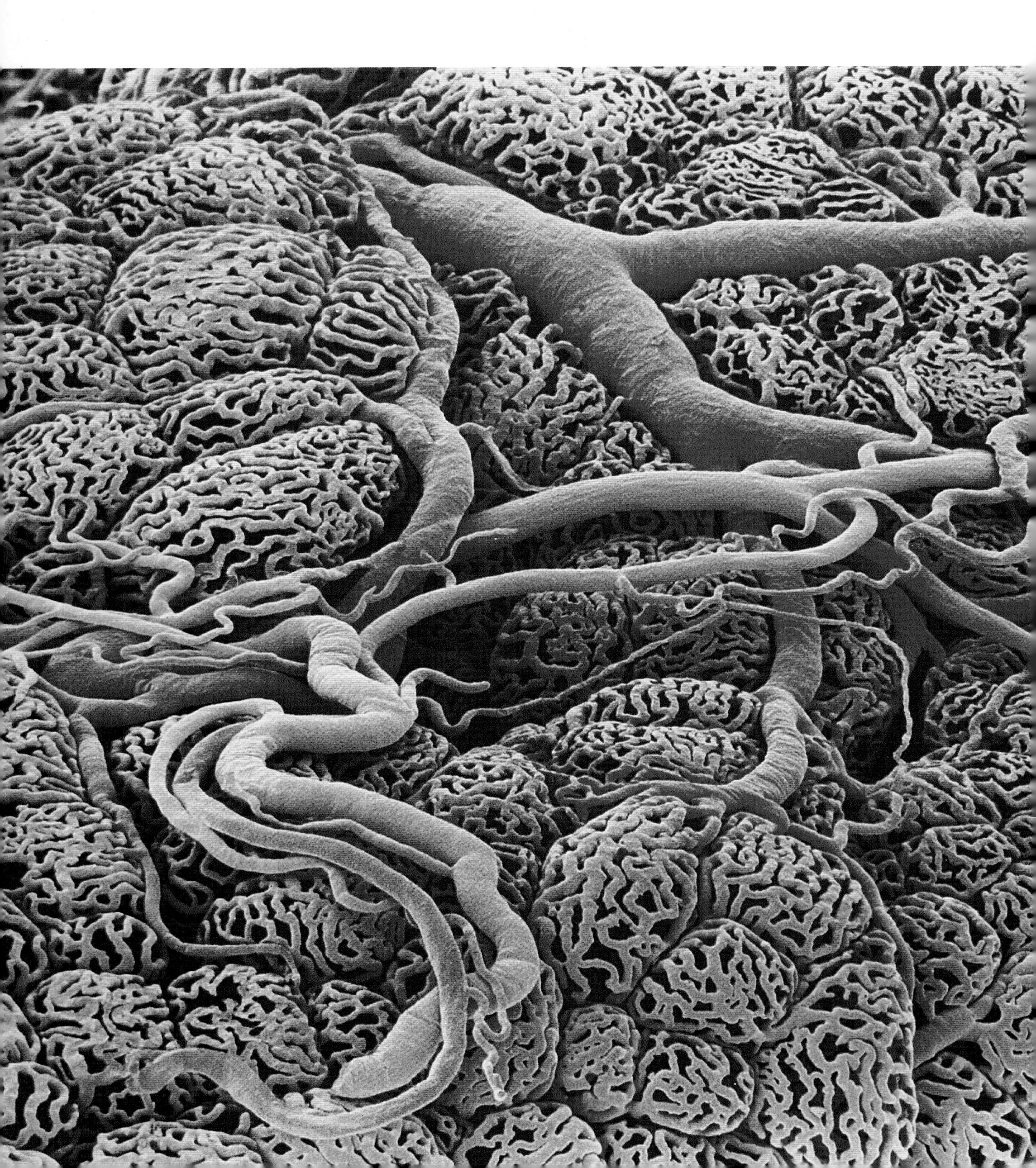

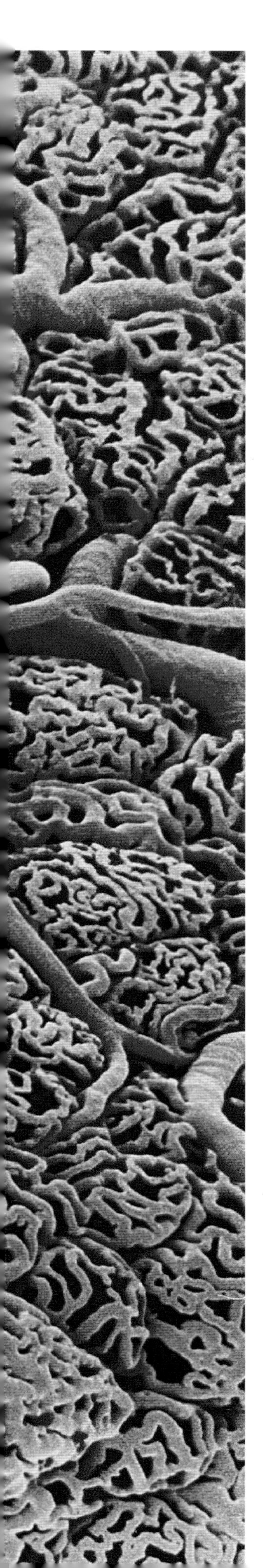

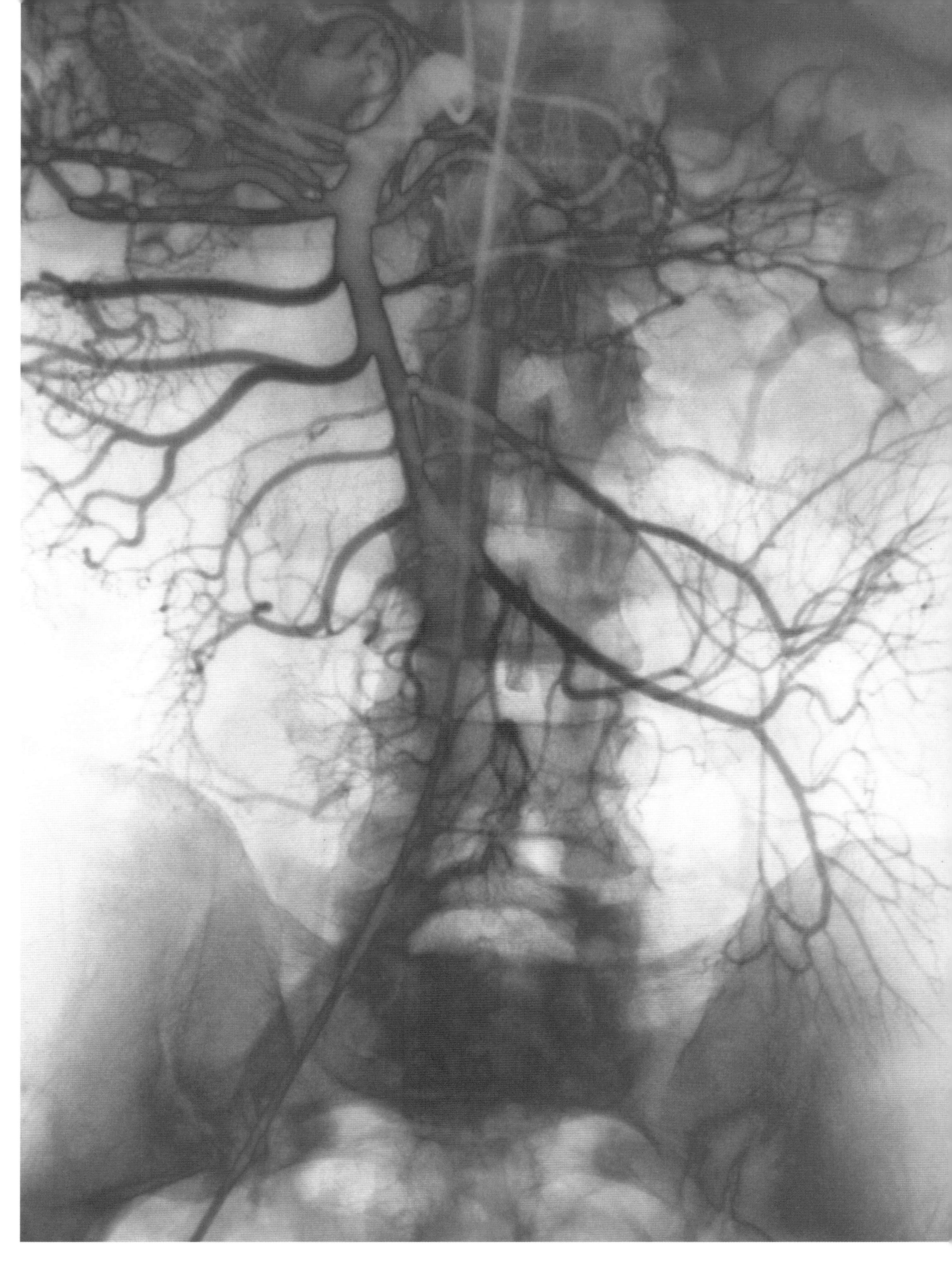

Mesenteric arteries [Angiogram]

THESE BRANCHING MESENTERIC ARTERIES supply the intestines with blood. The dark line running along an artery from bottom left to top centre is a catheter, a fine tube via which contrast medium is injected into arteries to enable them to show up on the X-ray. Whereas blood leaving most parts of the body returns directly to the heart, blood from the stomach and intestines is filtered by the liver before it returns to the heart.

Capillary on muscle [SEM]

RED BLOOD CELLS are so tiny that up to six million are found in a millilitre/0.04fl oz of blood. Yet some capillaries are so thin that a single red blood cell can only just pass through. There are so many capillaries in the body that their total sectional area may be one thousand times that of the aorta, the body's main artery.

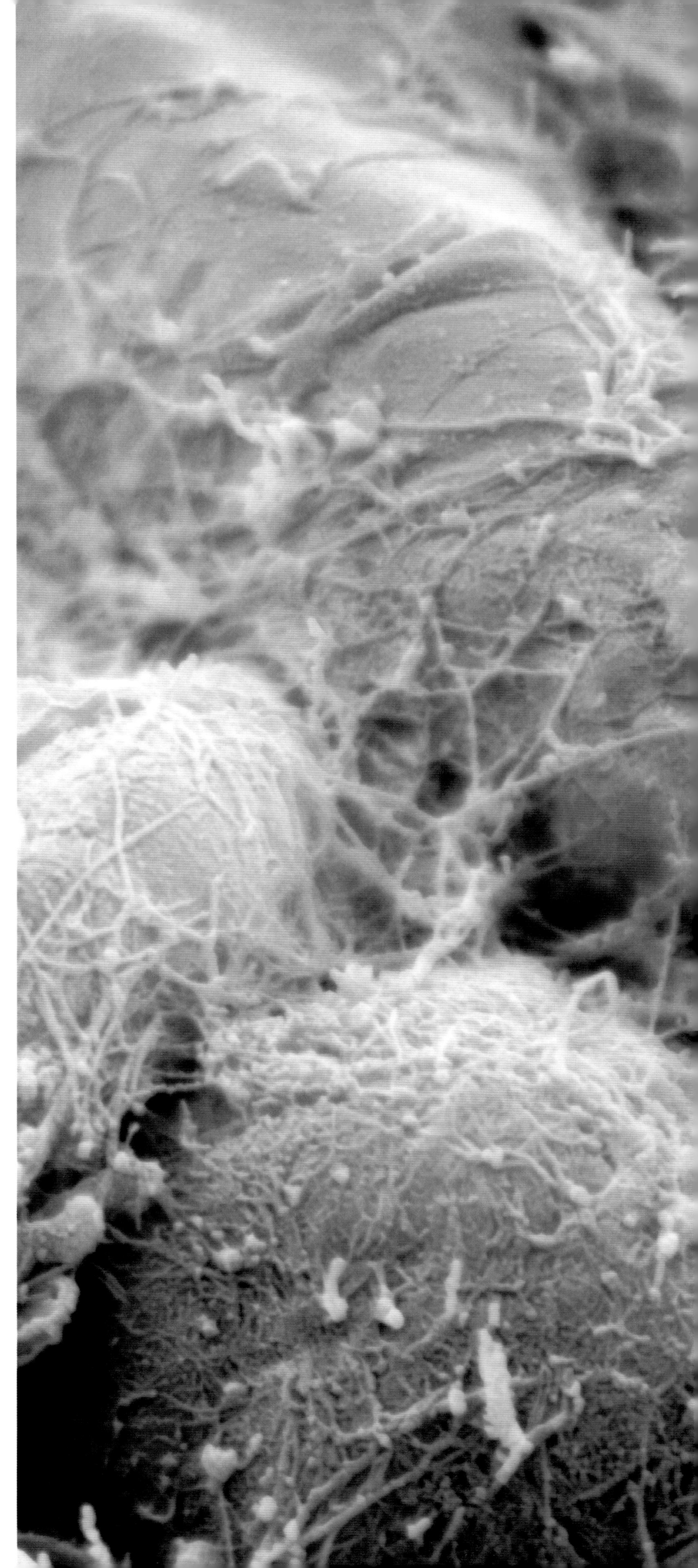

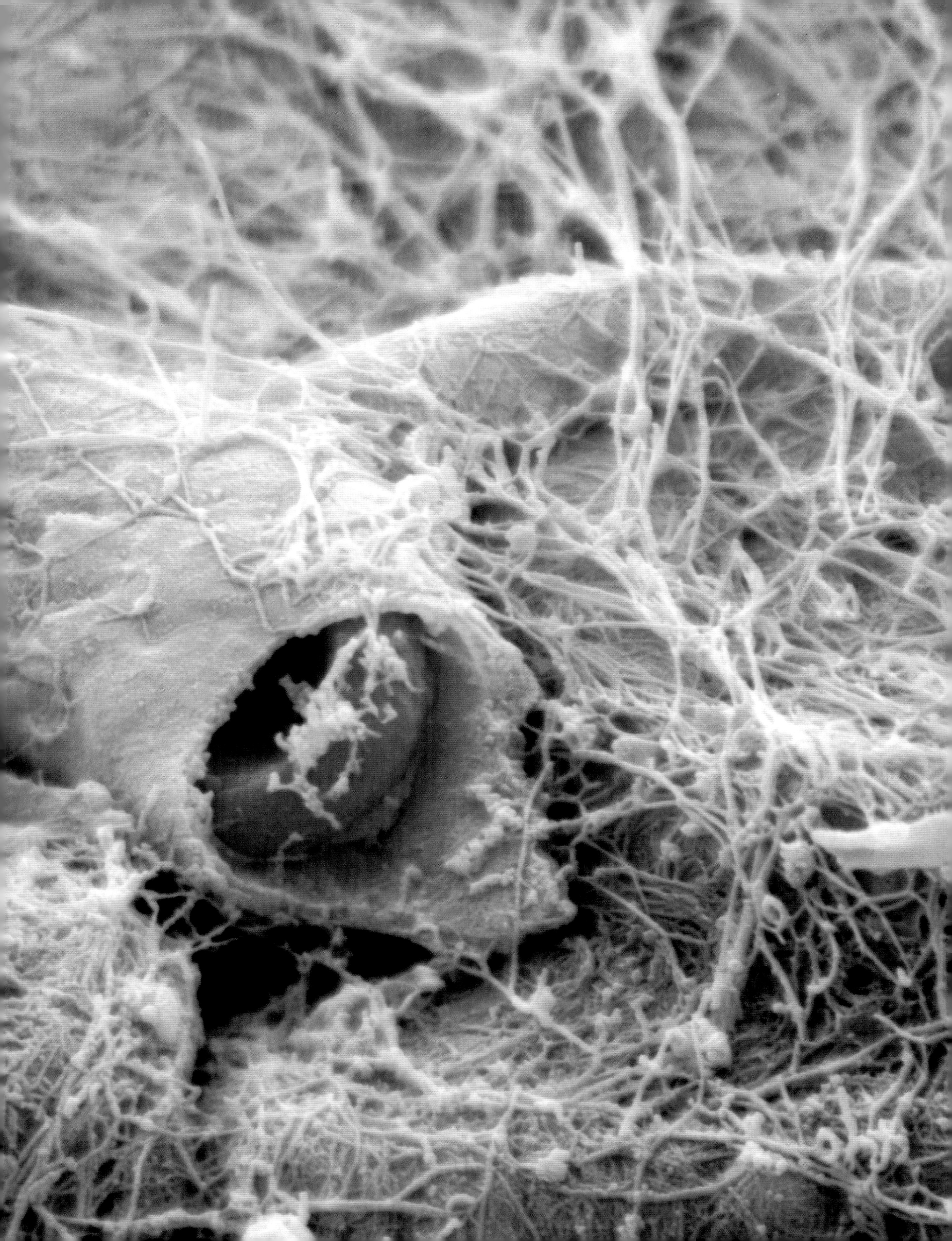

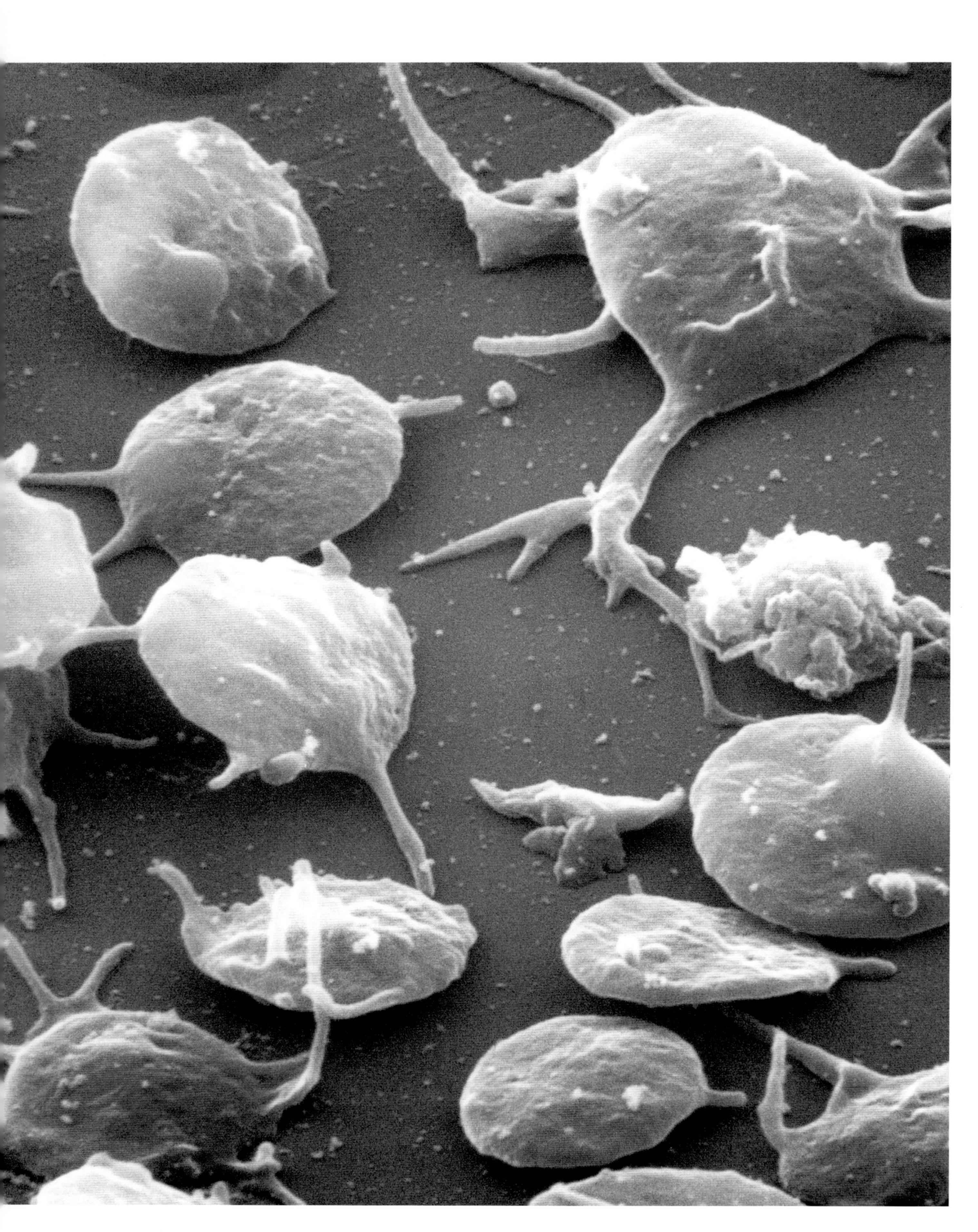

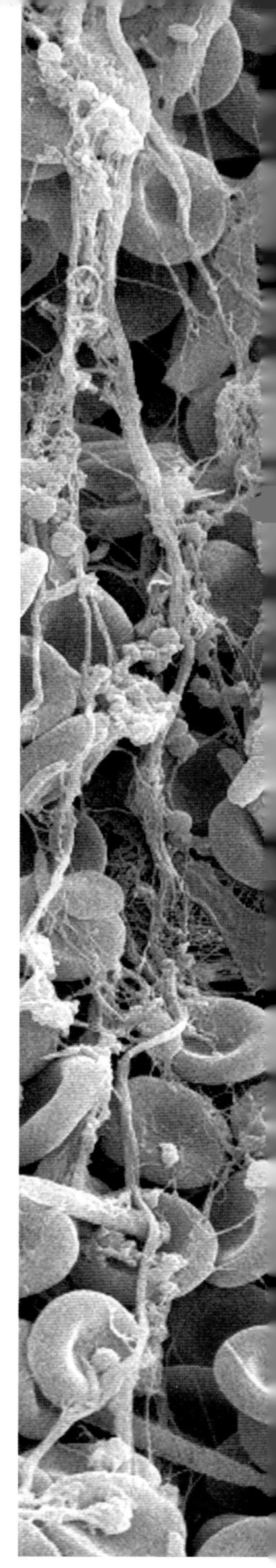

Active blood platelets [SEM]

PRESENT IN THE BLOODSTREAM in large numbers, platelets are not true cells but fragments of bone marrow cells shed into the circulation. When inactive, they are oval or round, but when activated, as here, they develop extensions that help plug defects in the walls of damaged blood vessels. They also release a substance called serotonin, which constricts blood vessels, thereby reducing blood loss.

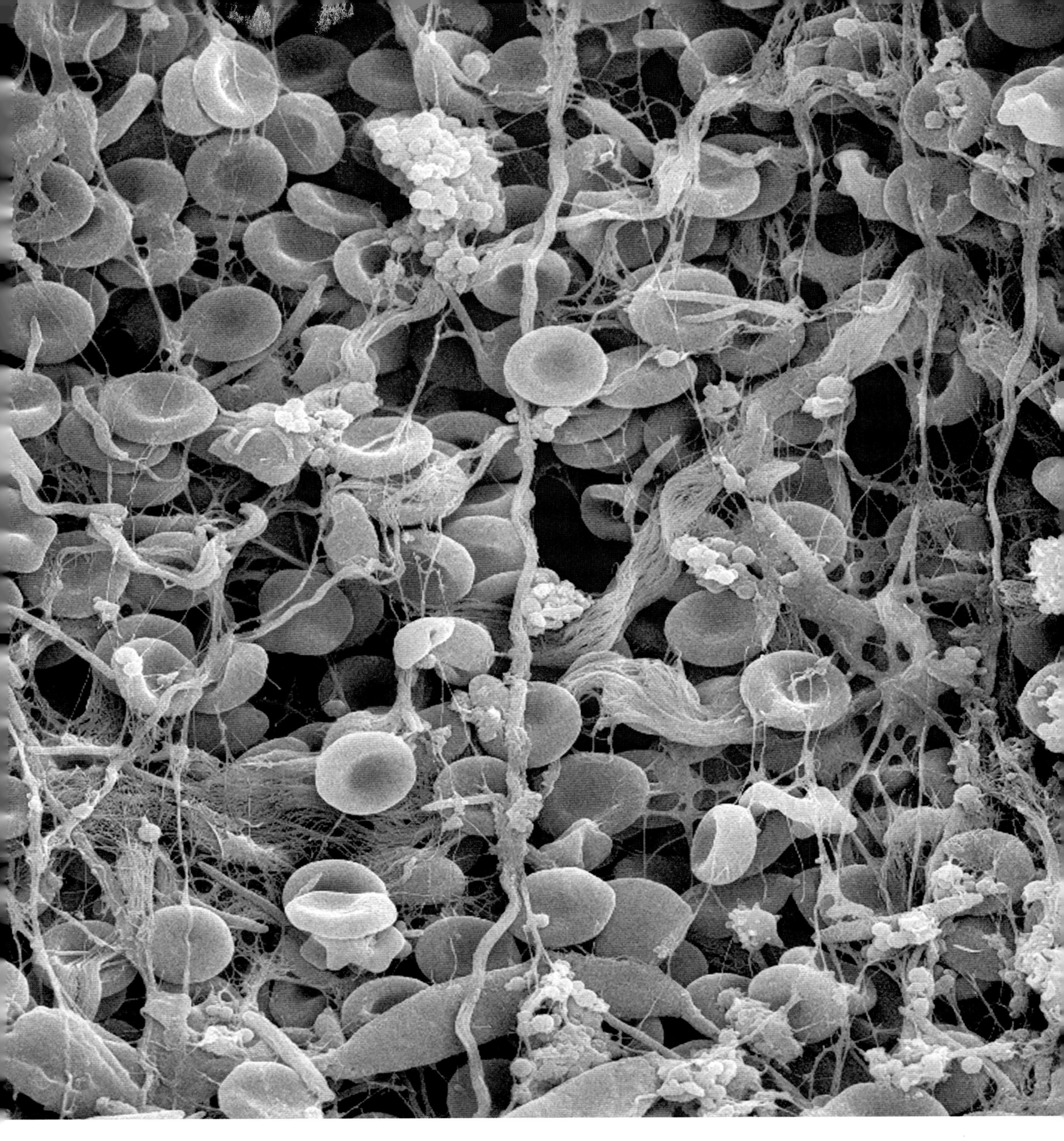

Blood clot [SEM]

A MASS OF RED BLOOD CELLS has been trapped in filaments of protein (brown) in damaged connective tissue, forming a clot. Blood clotting reduces blood loss and helps prevent infection. It is a complicated process involving blood platelets, which stimulate the conversion of soluble proteins to insoluble fibres that form a mesh.

Oxygen is the fuel that releases energy from glucose in our cells. Every minute on average we use about 400cc/25cu in of oxygen to burn up food material. This oxygen comes from the air by way of the lungs, which comprise a pair of spongy, lobed organs located behind the ribs in the thorax.

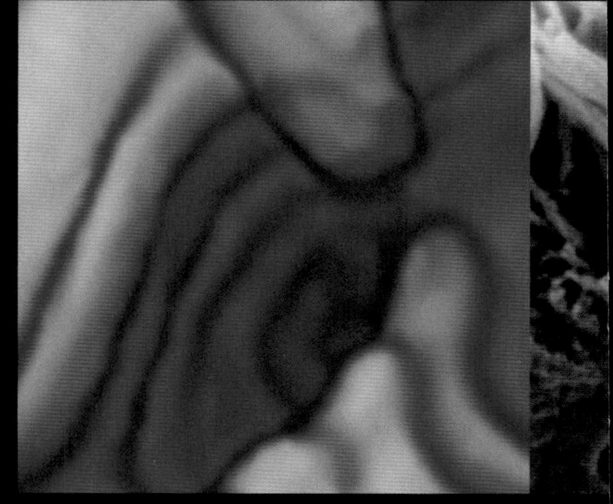

Because lungs contain no muscle, they cannot move themselves. Instead, they are inflated and deflated by movements of the ribcage and diaphragm, to which they are attached by membranes. Each day we breathe about 23,000 times and move about 12,000 litres/425cu ft of air.

When we breathe in, air is drawn through the nose and passes down the trachea (windpipe) into two bronchi, one for each lung. Inside the lungs the bronchi branch again and again until they are thinner than hairs. At the end of these bronchioles the fresh air enters bubble-like air sacs called alveoli. The total area covered by the alveoli is about 100sq m/1,000sq ft.

Oxygen from the inhaled air diffuses rapidly through the thin walls of the alveoli and into the blood, where it binds to the protein haemoglobin in red blood cells. People who live or work at high altitudes, where the air is thin, have up to 30 per cent more red blood cells than people who live near sea level. The oxygenated blood is carried to the heart and then distributed by the circulatory system

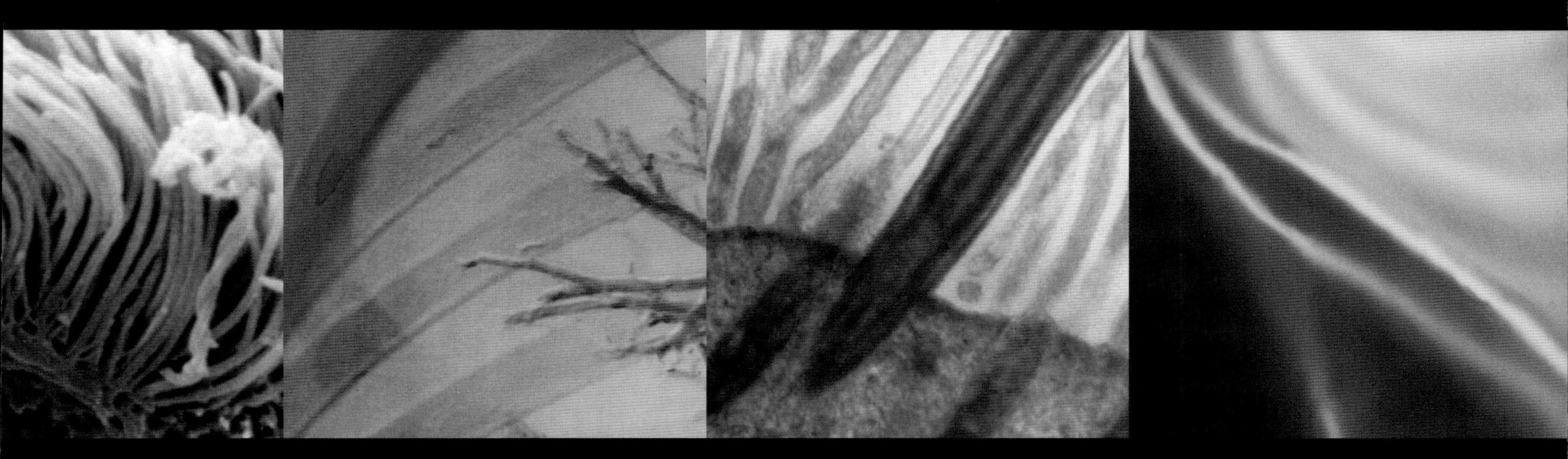

to every cell in the body. On its return journey, blood carries waste carbon dioxide back to the lungs, where it is expelled. Although we can alter our breathing rate at will, breathing usually requires no conscious effort, the rate adapting automatically to meet respiratory demand at any given time. Breathing rate is regulated by a part of the brain that reacts to the amount of carbon dioxide in the blood, rather than to the amount of oxygen. Chemical receptors in the brain and the heart region detect higher than normal levels of carbon dioxide and send nerve impulses to the respiratory centre in the brain. The brain responds by sending messages to the muscles that control the ribcage and diaphragm, telling them to work harder. As a result we breathe faster and deeper, automatically increasing oxygen delivery at the same time as removing the carbon dioxide.

Respiratory system

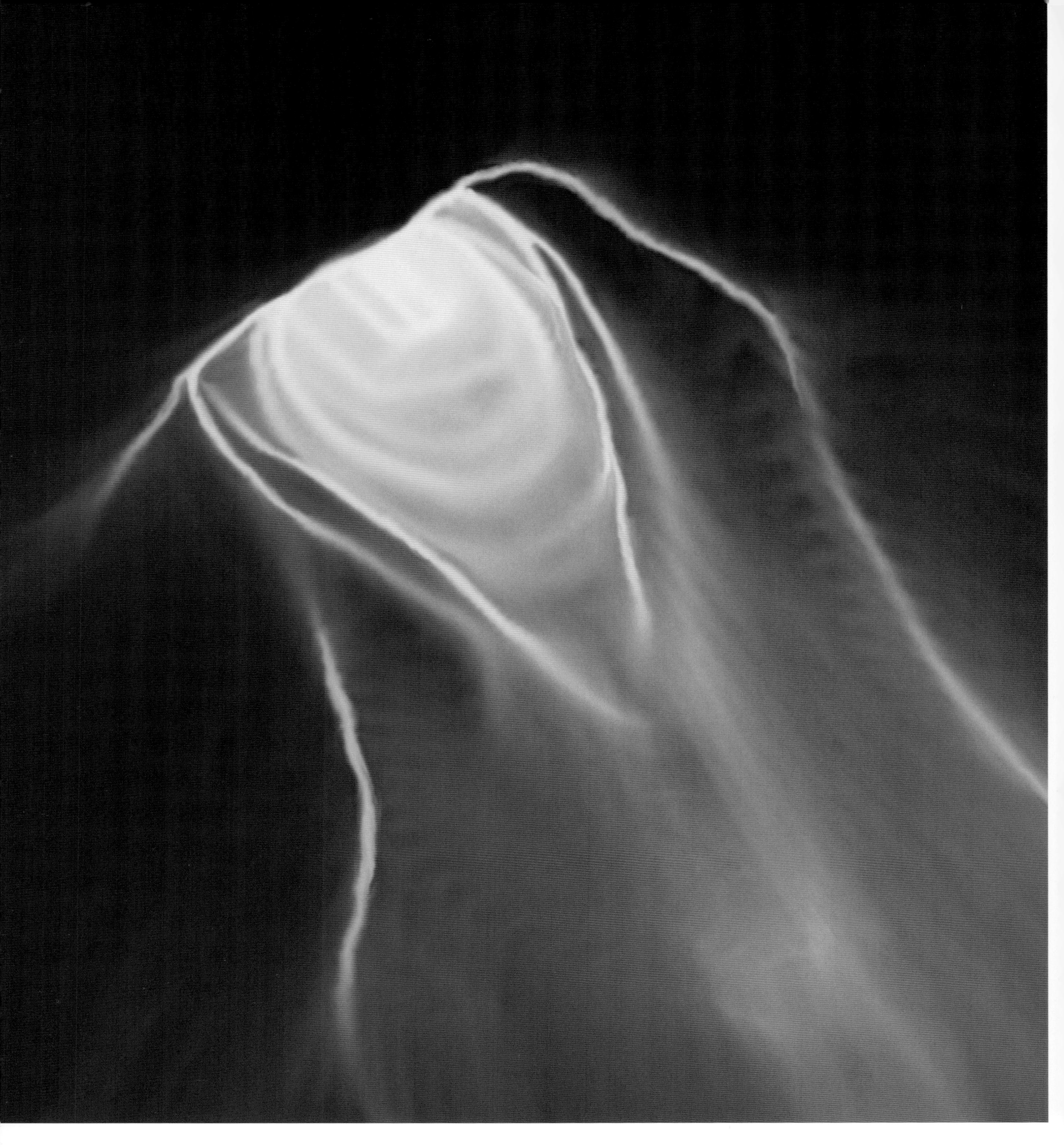

Vocal cords in the larynx [CT]

THE LARYNX AT THE BACK of the throat is part of the respiratory system and also the organ of voice, containing the vocal cords (V-shaped fold, centre). Here, the cords are shown relaxed at the sides of the larynx, and air can pass soundlessly between them. But when muscles pull them taut, air is forced between them and they vibrate, making sounds. Their name is slightly misleading, however, as the majority of sound production is carried out by the tongue and lips.

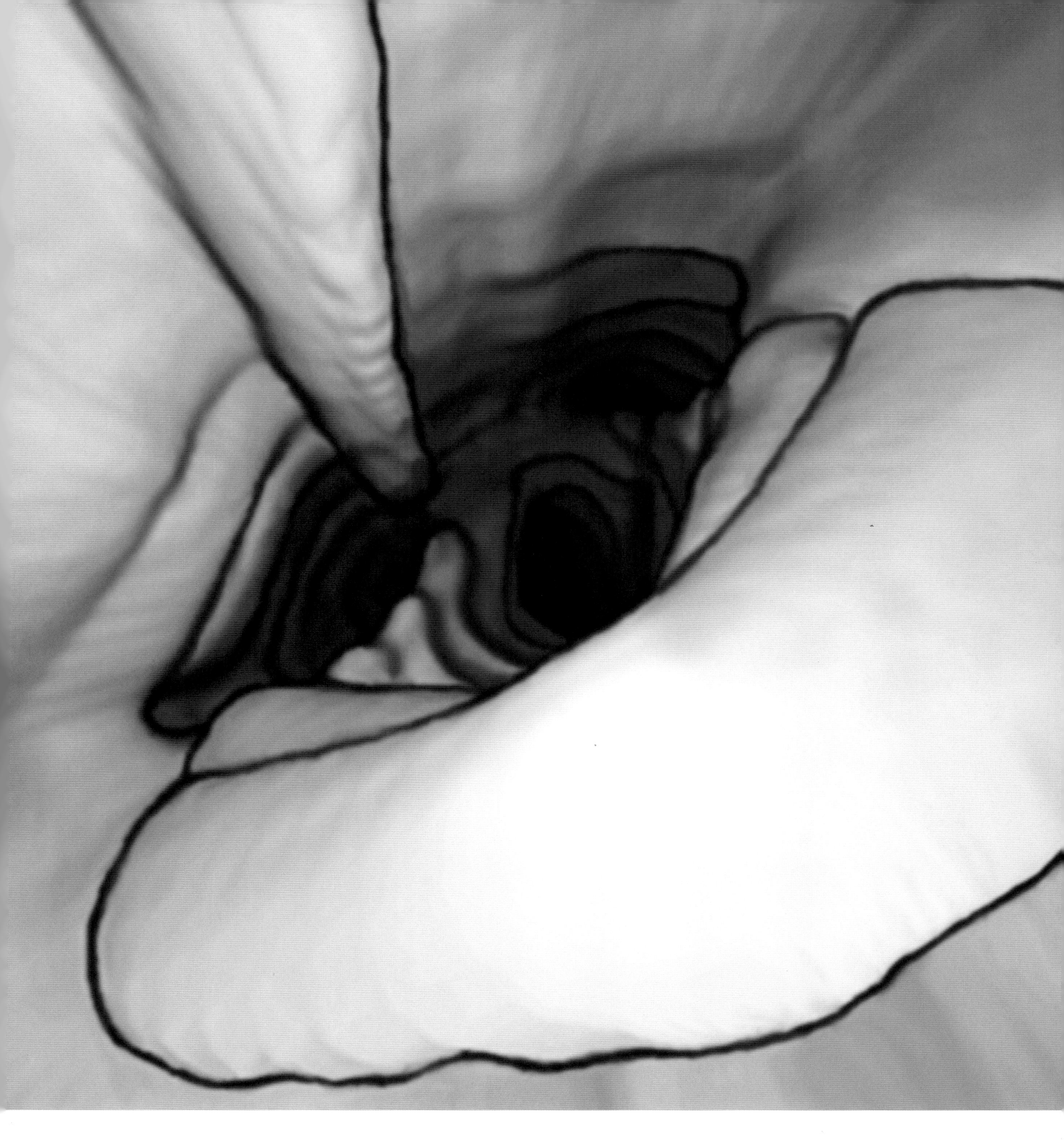

Epiglottis in larynx [CT]

AT THE TOP OF THE DESCENDING LARYNX (centre) is the epiglottis (lower right), a flap of elastic cartilage that separates the trachea (windpipe) from the oesophagus (gullet). It flips between two positions to prevent food from entering the lungs, and air from entering the stomach.

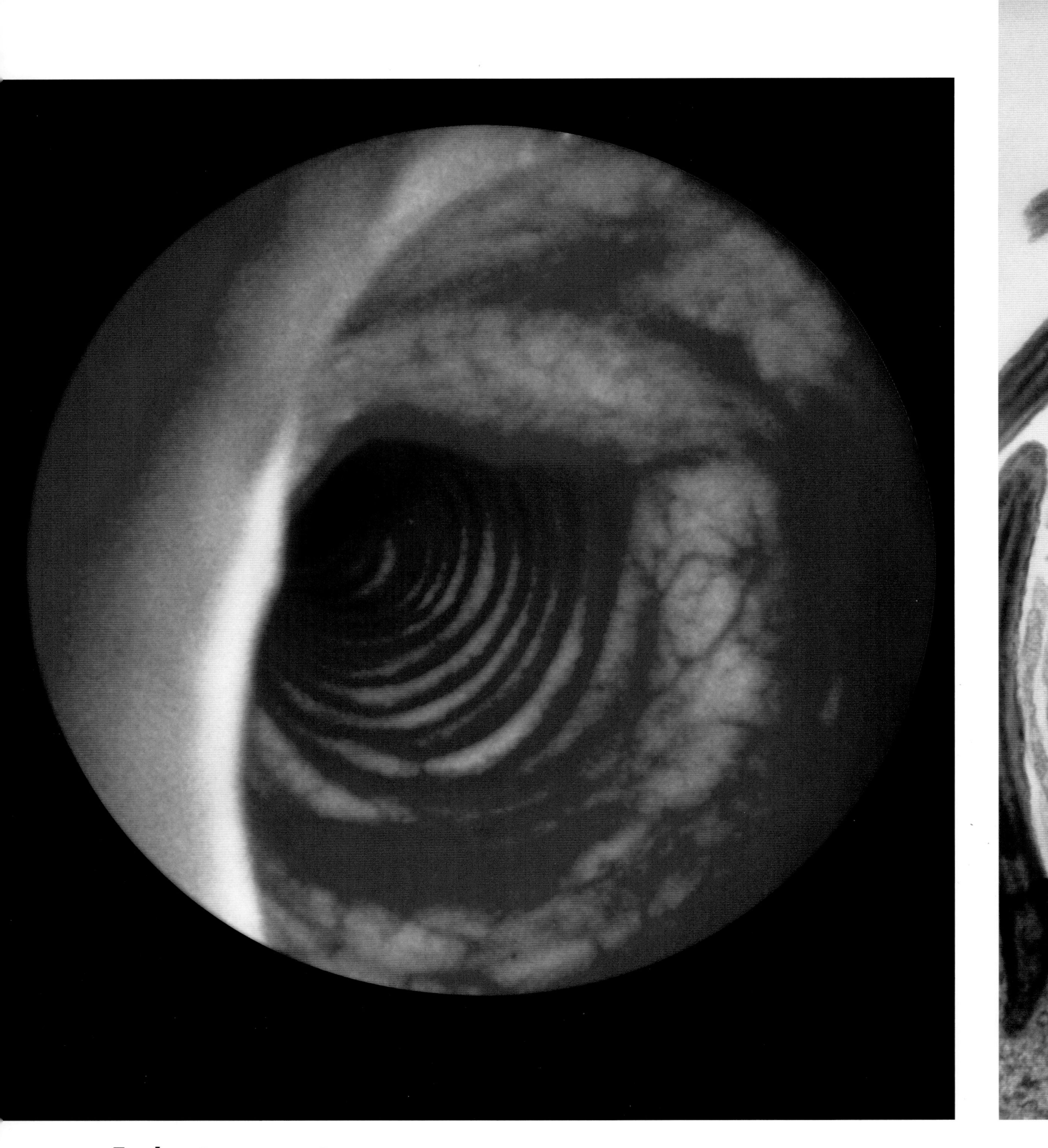

Trachea [Bronchoscope]

BELOW THE LARYNX IS THE TRACHEA, the airway between throat and lungs. The trachea is a hollow tube about 10cm/4in long supported by cartilaginous rings, clearly visible in this image.

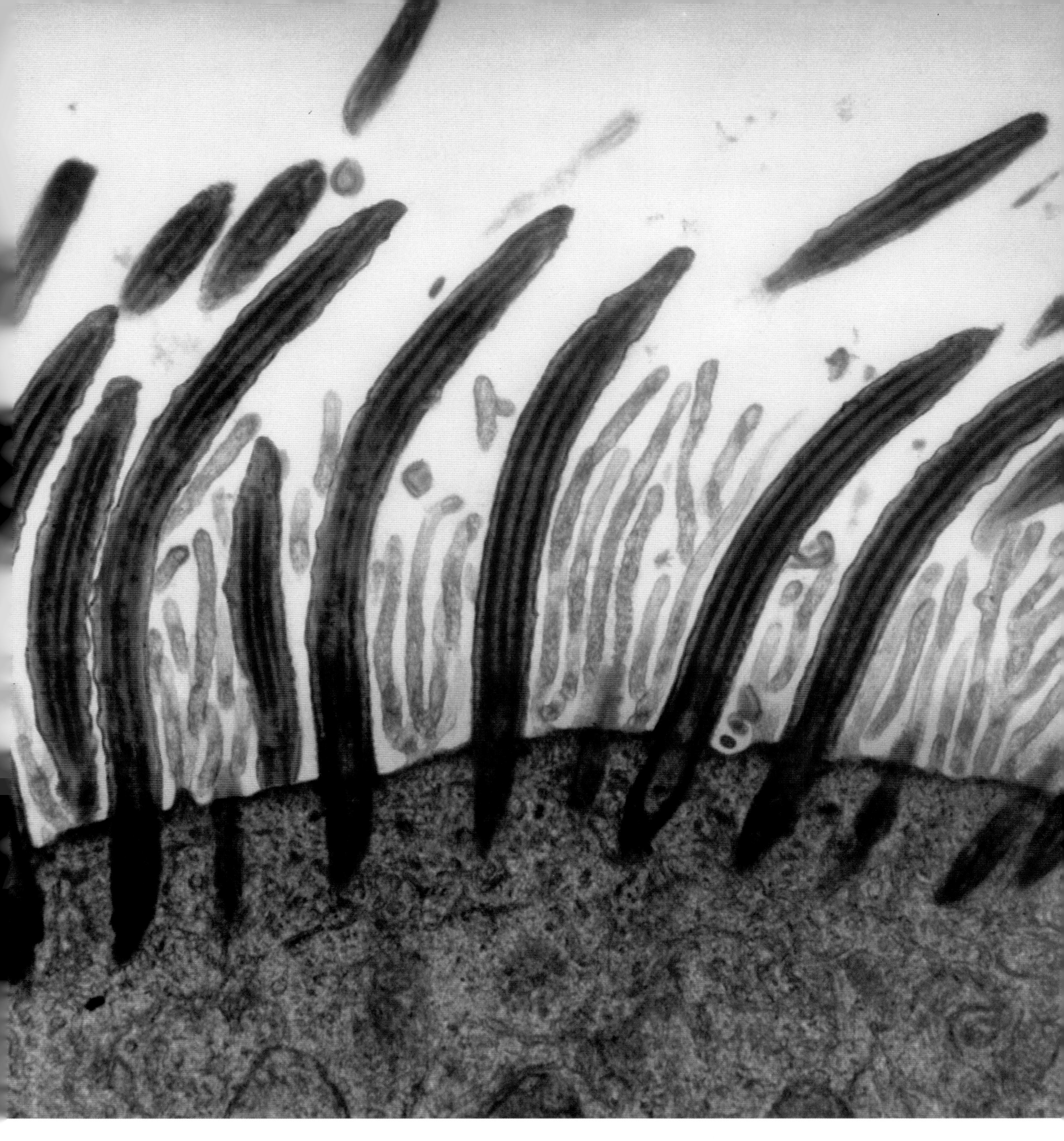

Ciliated cell on trachea [TEM]

AN EPITHELIAL CELL (green) from the trachea has many cilia (red) – finger-like extensions that beat rhythmically, carrying dust and other irritants back up towards the throat, where they can be swallowed or coughed up. The smaller blue extensions between the cilia are microvilli, which increase the cell's surface area and aid in the exchange of substances.

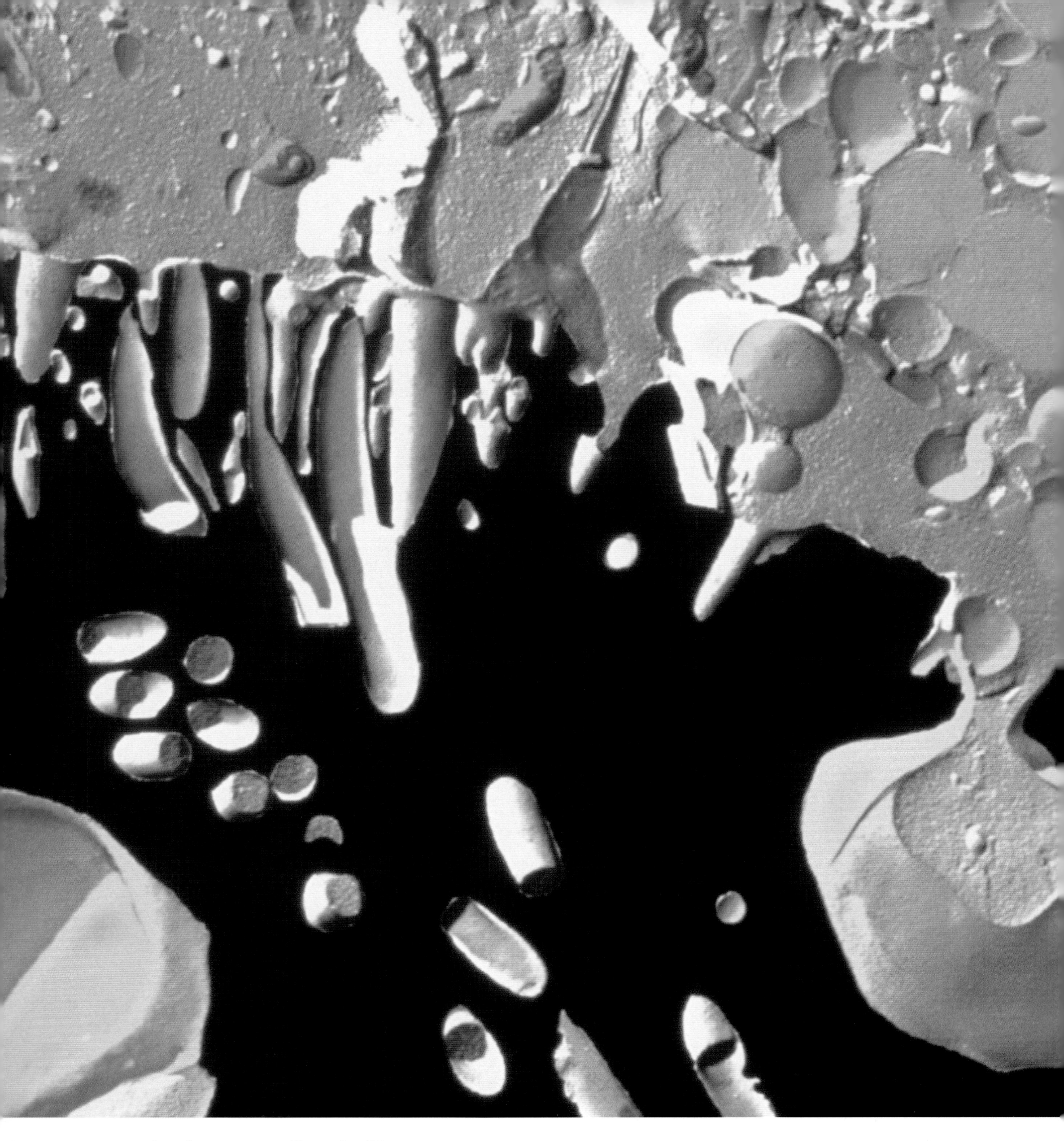

Tracheal mucus cell and cilia [TEM]

AT UPPER CENTRE IS A GOBLET CELL (purple) which produces mucus (green) that lines the walls of the trachea. Micro-organisms and other airborne irritants breathed into the windpipe are trapped by the mucus, which is then transported back up the throat by the rhythmic beating of the hair-like cilia on each side of the goblet cell.

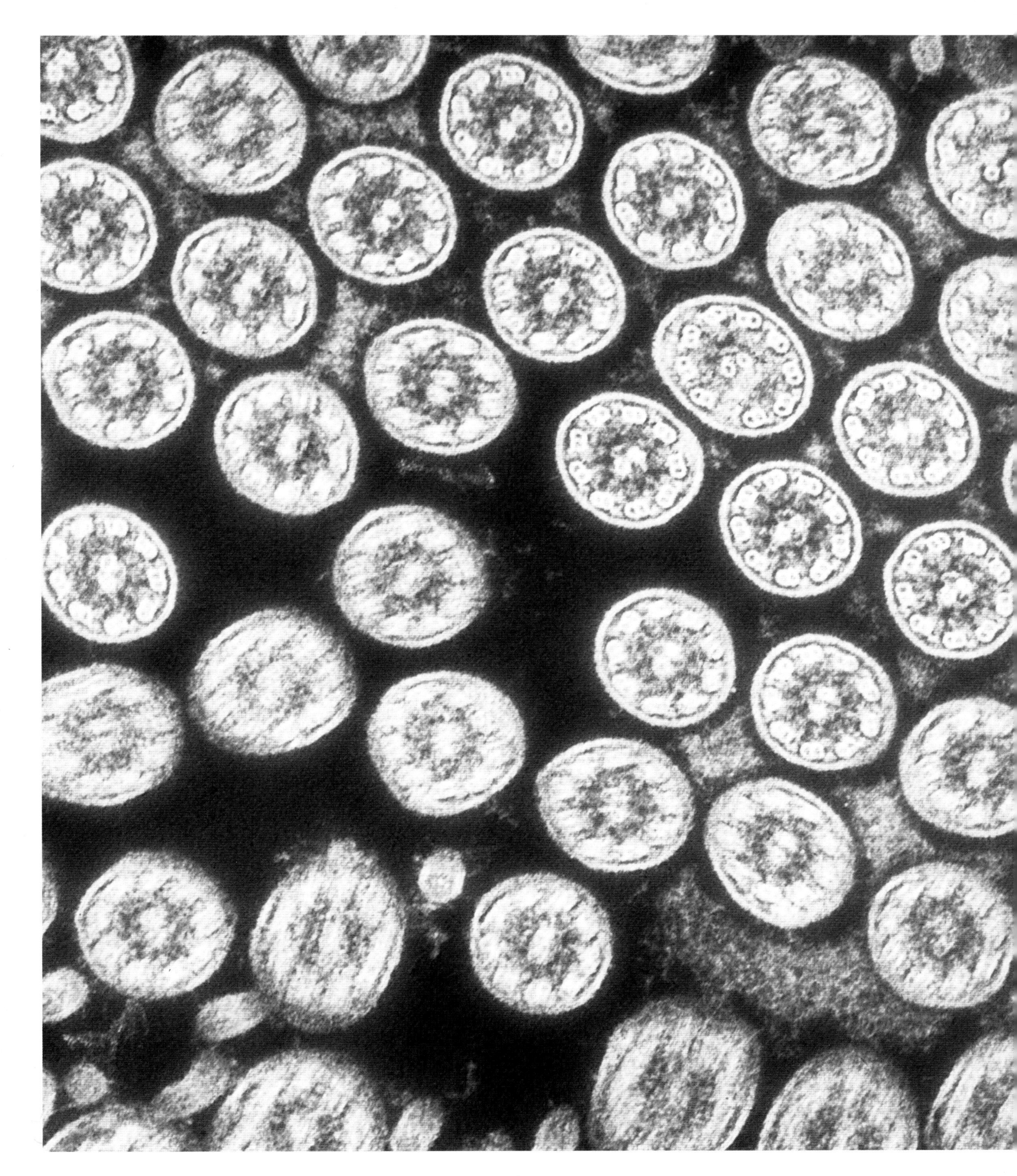

Cilia in cross section [TEM]

THE CELLS THAT LINE THE TRACHEA (windpipe) are bombarded with inhaled foreign particles, every time we breathe. Each cell has about 300 cilia. These are hair-like projections which sweep the airways, moving particles back up and out of our lungs. The cilia in this picture are shown in cross section. Their microscopic anatomy shows the same arrangement as the motile apparatus of all cells, from protozoa to human sperm.

Dividing bronchioles in the lung [CT]

AT ITS BASE THE TRACHEA divides into two tubes called the bronchi, one for each lung, which in turn divide like the branches of a tree into very narrow airways called bronchioles. This image shows a section through dividing bronchioles in the right lung. The smallest bronchioles have muscles in their walls, which can make them wider or narrower, thereby altering the rate of air flow in and out of the lungs.

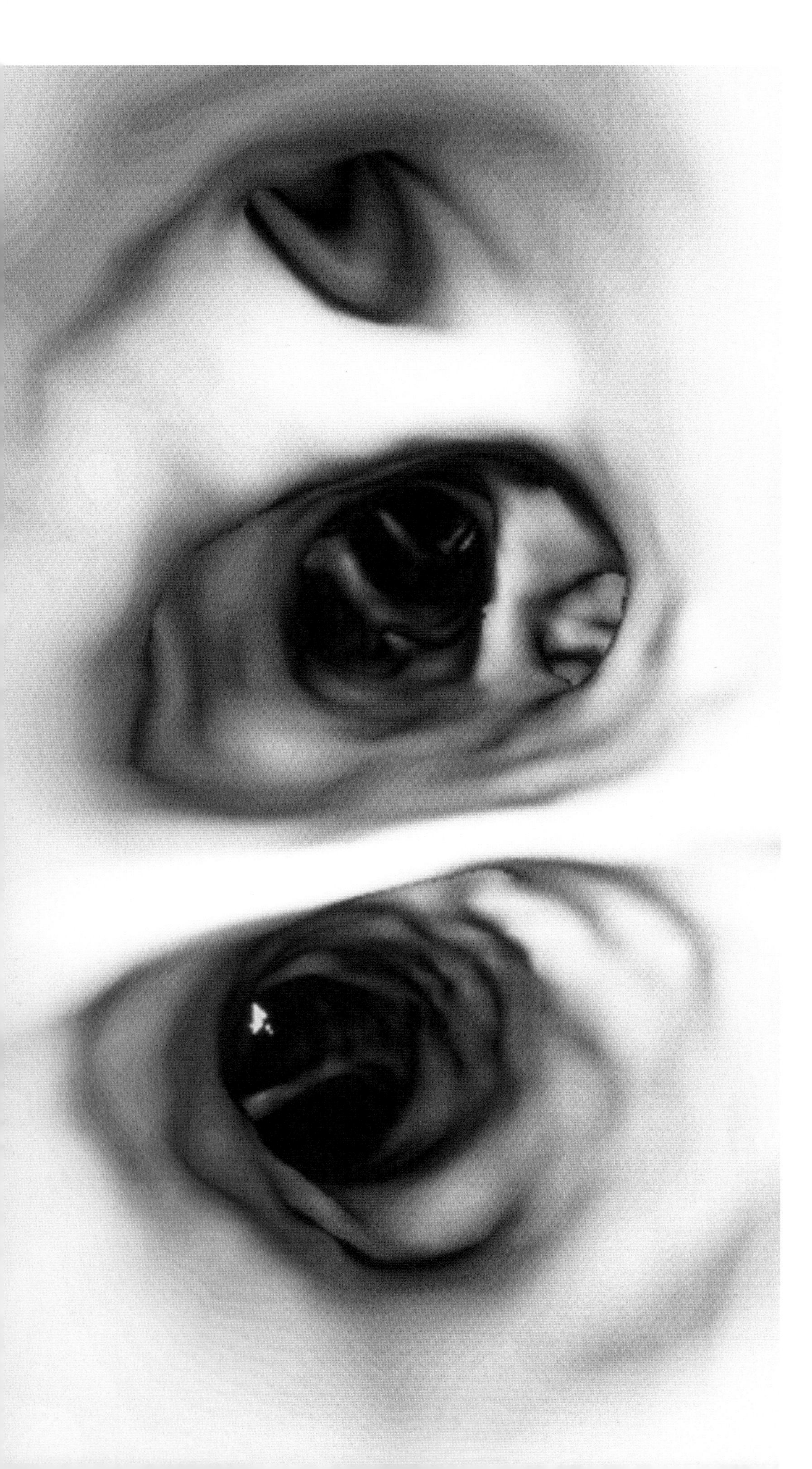

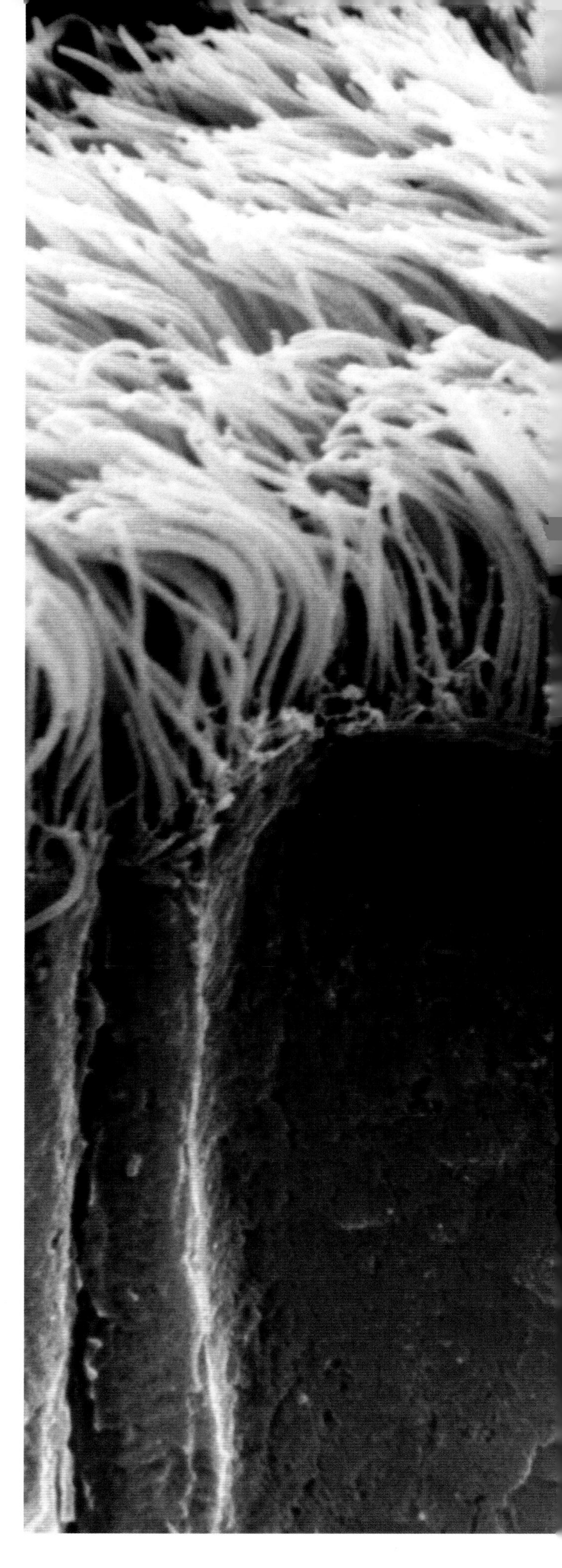

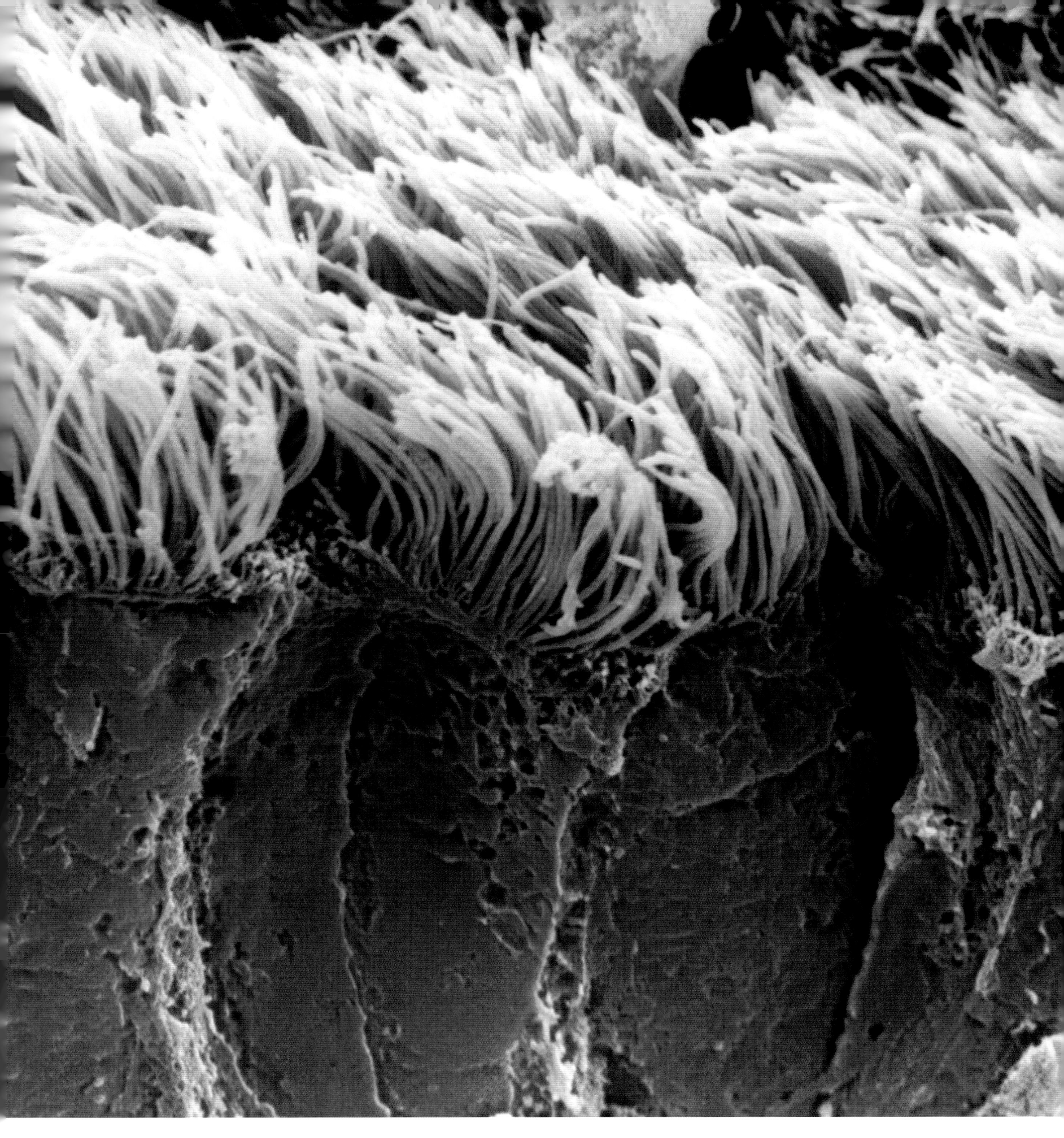

Bronchial epithelial cilia [SEM]

LIKE THE TRACHEA, THE BRONCHIOLES of the lung are lined with mucus membrane made up of epithelial cells (brown) with hair-like cilia (green, pink) protruding from their exposed surfaces. Rhythmic sweeping movements of the cilia collect bacteria and other particles trapped by the mucus and move it towards the throat, where it can be expelled.

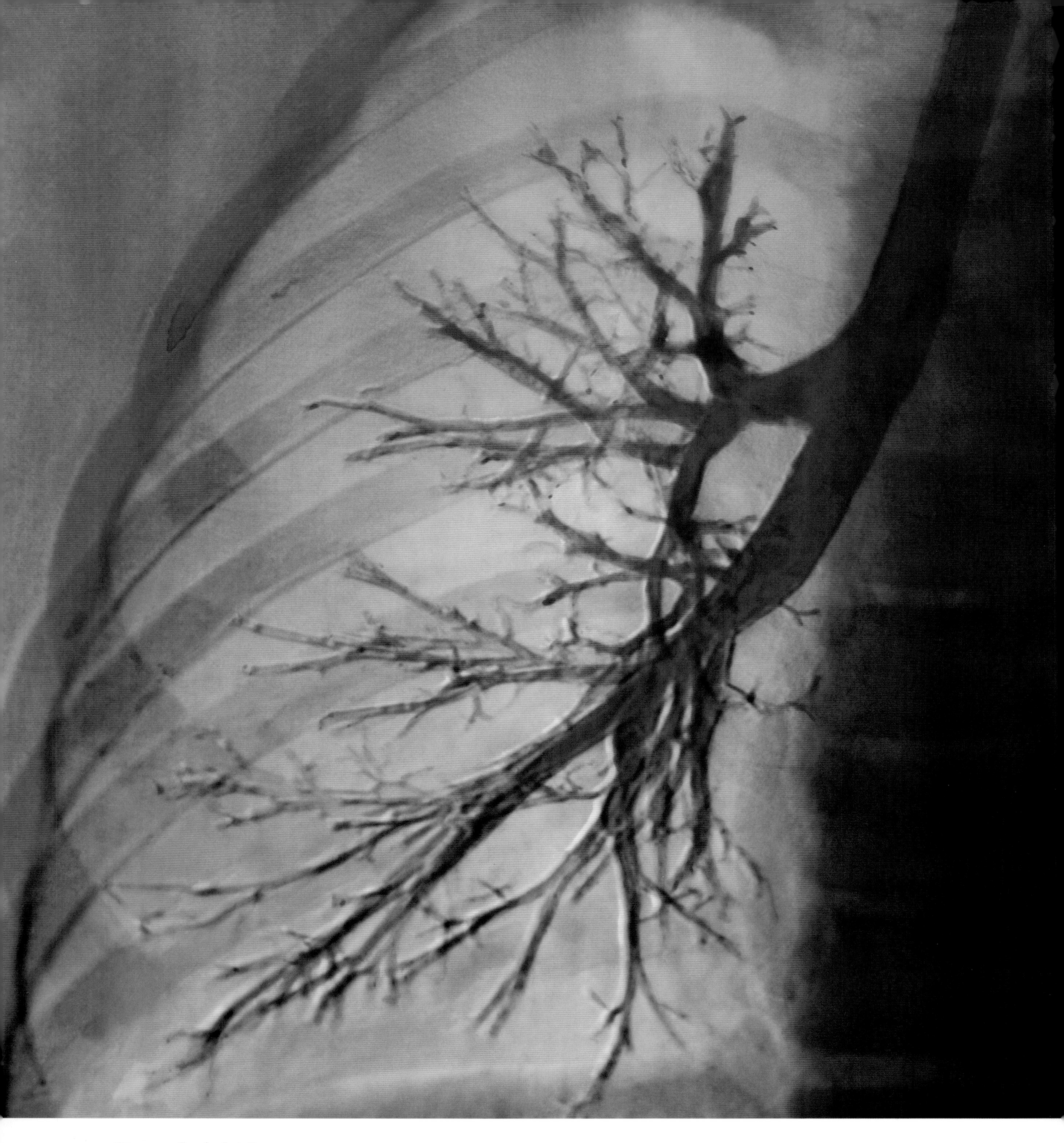

Normal right lung [Bronchogram]

THIS IMAGE SHOWS THE BRANCHING 'tree' of bronchioles arising from a division in the bronchus that delivers air to each lung. The bronchioles have muscular walls that automatically constrict or relax to alter the volume of air entering or leaving the lung. During an asthma attack, the muscles go into spasm, making it difficult for the sufferer to breathe out.

Bronchiole and alveoli in lung [SEM]

A BRONCHIOLE (lower left), one of the tiny airways in the lung, ends in a cluster of air sacs called alveoli. It is in the air sacs that interchange takes places between the gases of the inhaled air and gases dissolved in the blood.

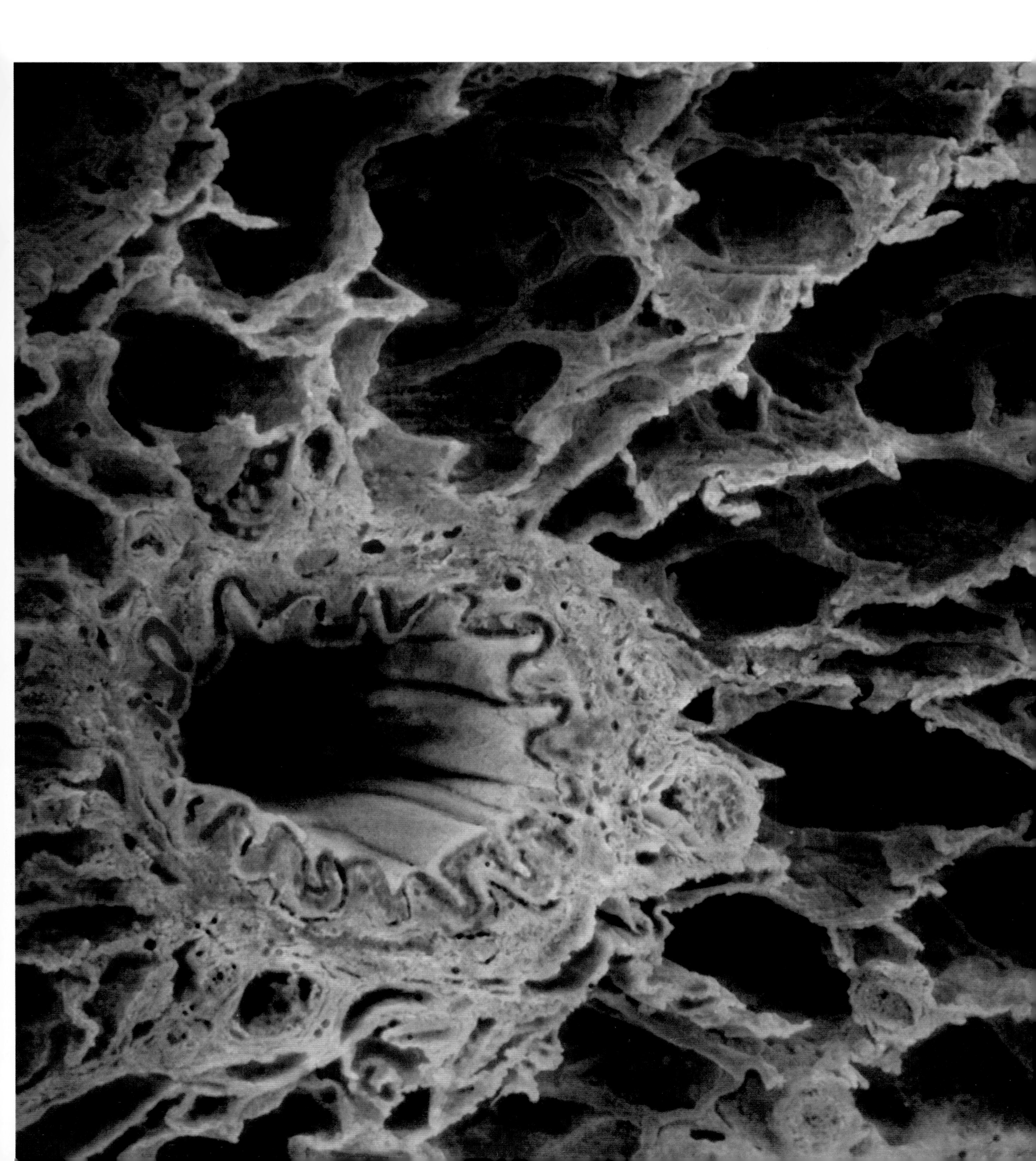

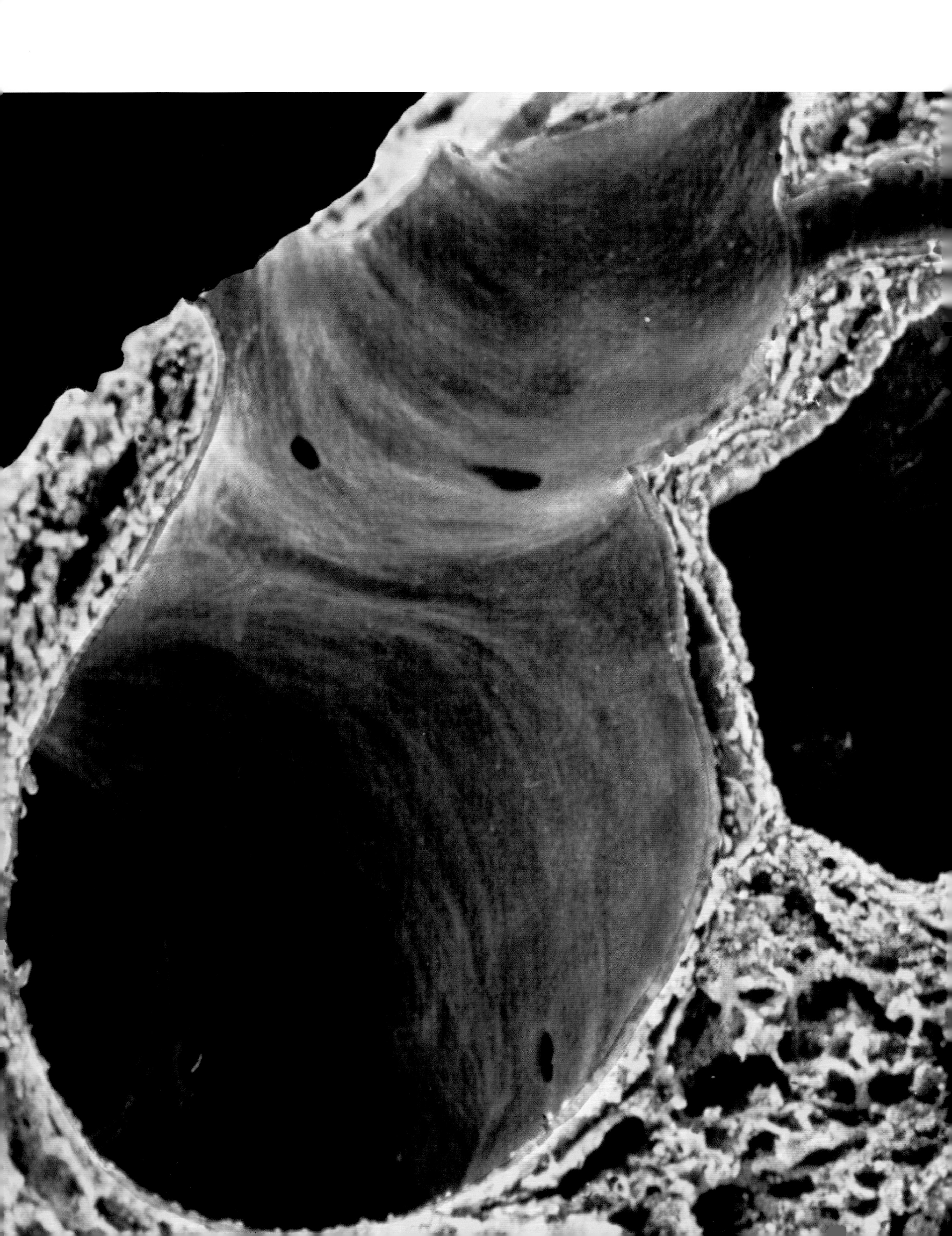

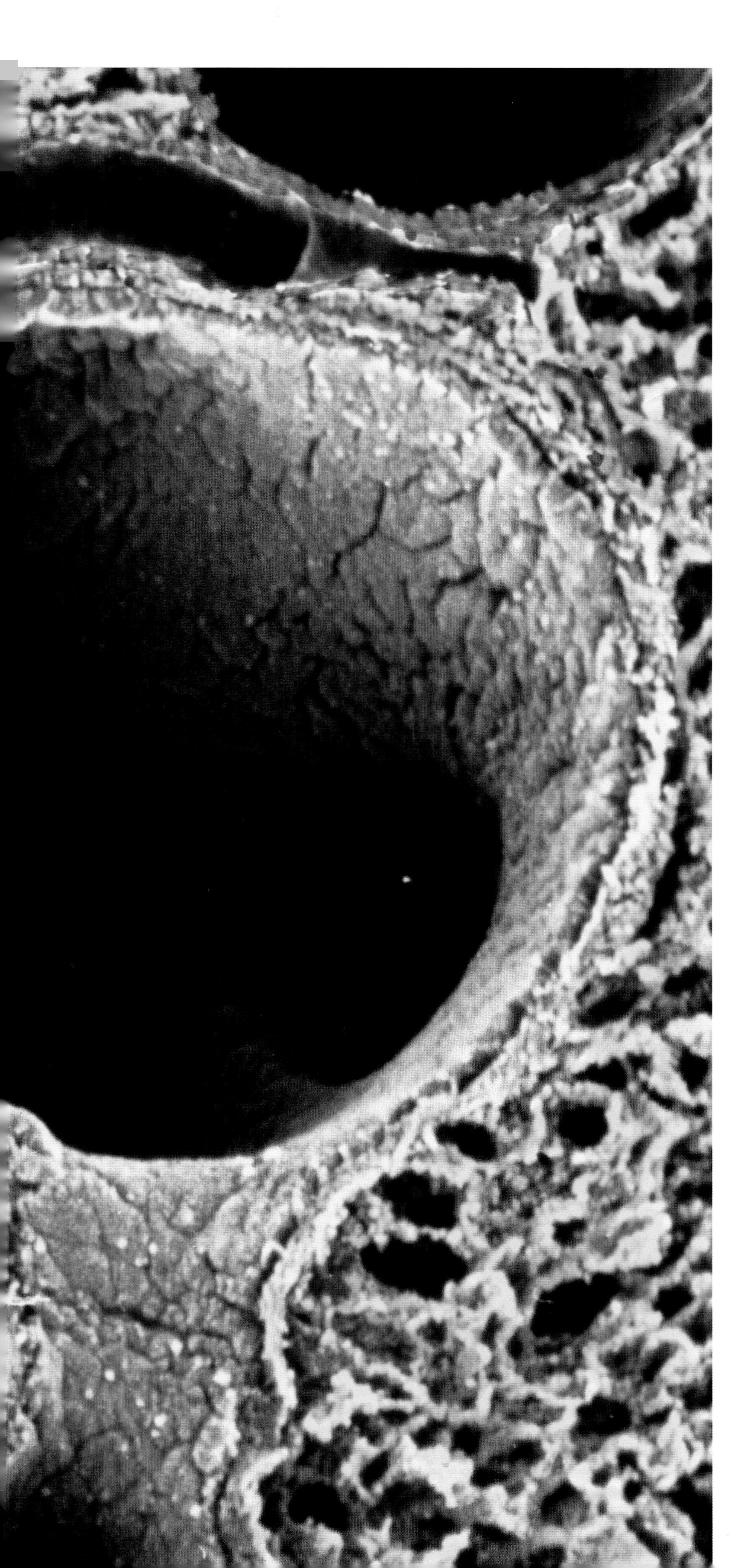

Lung alveoli and bronchus [SEM]

A BRONCHUS (blue), one of the main airways in the lungs, is seen next to a pulmonary blood vessel (pink) that pumps blood to the lungs. The spongy tissue (yellow) around them is made up of air sacs whose walls are lined with blood capillaries where interchange occurs between the contained air and gases dissolved in the blood. The millions of air sacs in a lung provide an interchange area the size of a tennis court.

Sex is a subject of endless fascination for humans, yet the basic principles of reproduction are simple. A sperm (the male sex cell) unites with an egg (the female sex cell). If fertilization takes place, about nine months later a baby is born. What complicates the process is the fact that two individuals and two reproductive systems are involved. To make the outcome more hit-and-miss, fertilization can only take place for a few days during the female's monthly sexual cycle.

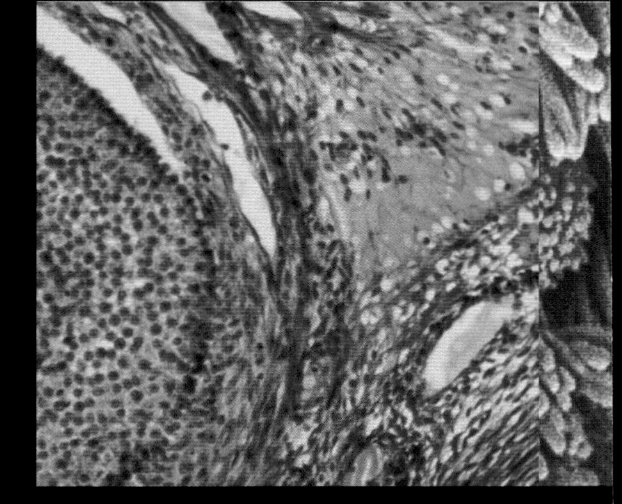

Humans become sexually mature at puberty, which usually occurs a year or two earlier in girls than in boys. The two reproductive systems are anatomically analagous, the female clitoris being a much smaller counterpart of the male's erectile organ, the penis, and the egg-bearing ovaries corresponding to the sperm-producing testes.

Each month, about halfway through the female's menstrual cycle, an ovarian follicle matures in the ovary, discharging an egg into the Fallopian tube, where it is propelled by muscular contractions towards the uterus (womb). Fertilization takes place within the Fallopian tube and can only occur if intercourse has taken place within the previous three to four days. During intercourse, the man inserts his erect penis into the woman's vagina and ejaculates sperm. Millions of sperm begin swimming towards the egg, but only a tiny proportion survives the journey up from the vagina, through the uterus, and into the tube. The feat has been compared to that of a man swimming the Atlantic – in treacle. But it has to be said that, in purely biological terms, the provision of sperm is the male's only

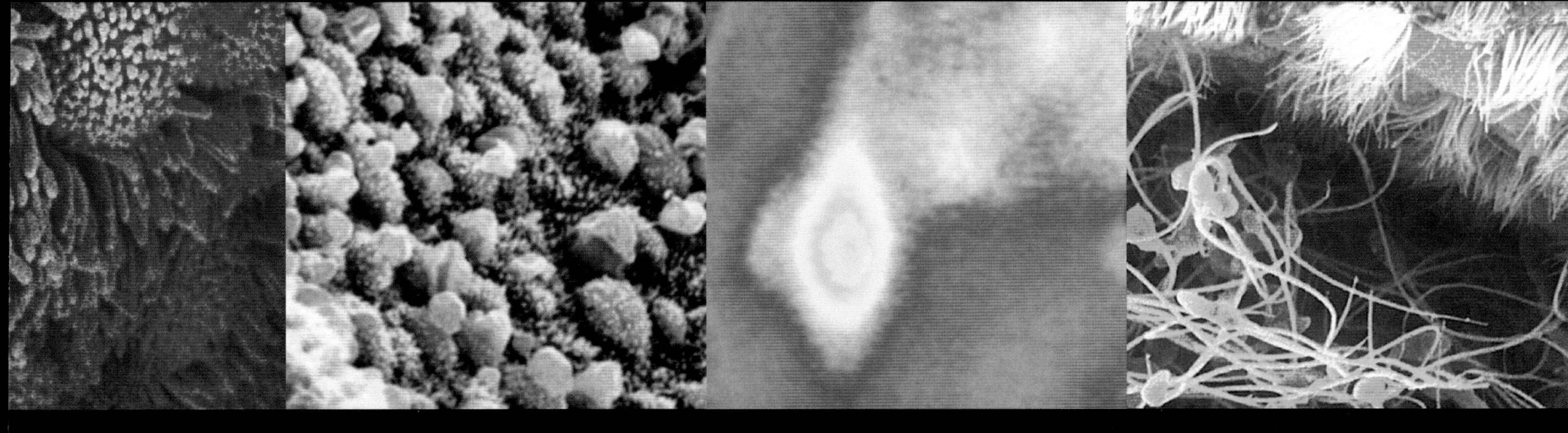

contribution to the reproductive process. If a sperm penetrates an egg and their nuclei fuse, the egg begins to divide. Occasionally, one of the early cleavages may divide the embryo into equal parts, which develop into identical twins. Pregnancy begins when the embryo, by now a ball of cells, reaches the uterus and embeds itself in the wall of the uterus. At first the embryo is nourished by its own food reserves and glandular secretions from the uterus. As it develops, the future baby obtains food and oxygen from its mother's blood through an organ called the placenta. Nutrients and waste products pass to and from the baby through a tube called the umbilical cord.

After two months the embryo is known as a foetus, which becomes recognizably human at about twelve weeks. After a total of about forty weeks, all being well, the fully formed baby will emerge into the world. He or she will have inherited characters from both parents, but the genetic shuffling that took place during the formation of the sex cells will have ensured that he or she is unique.

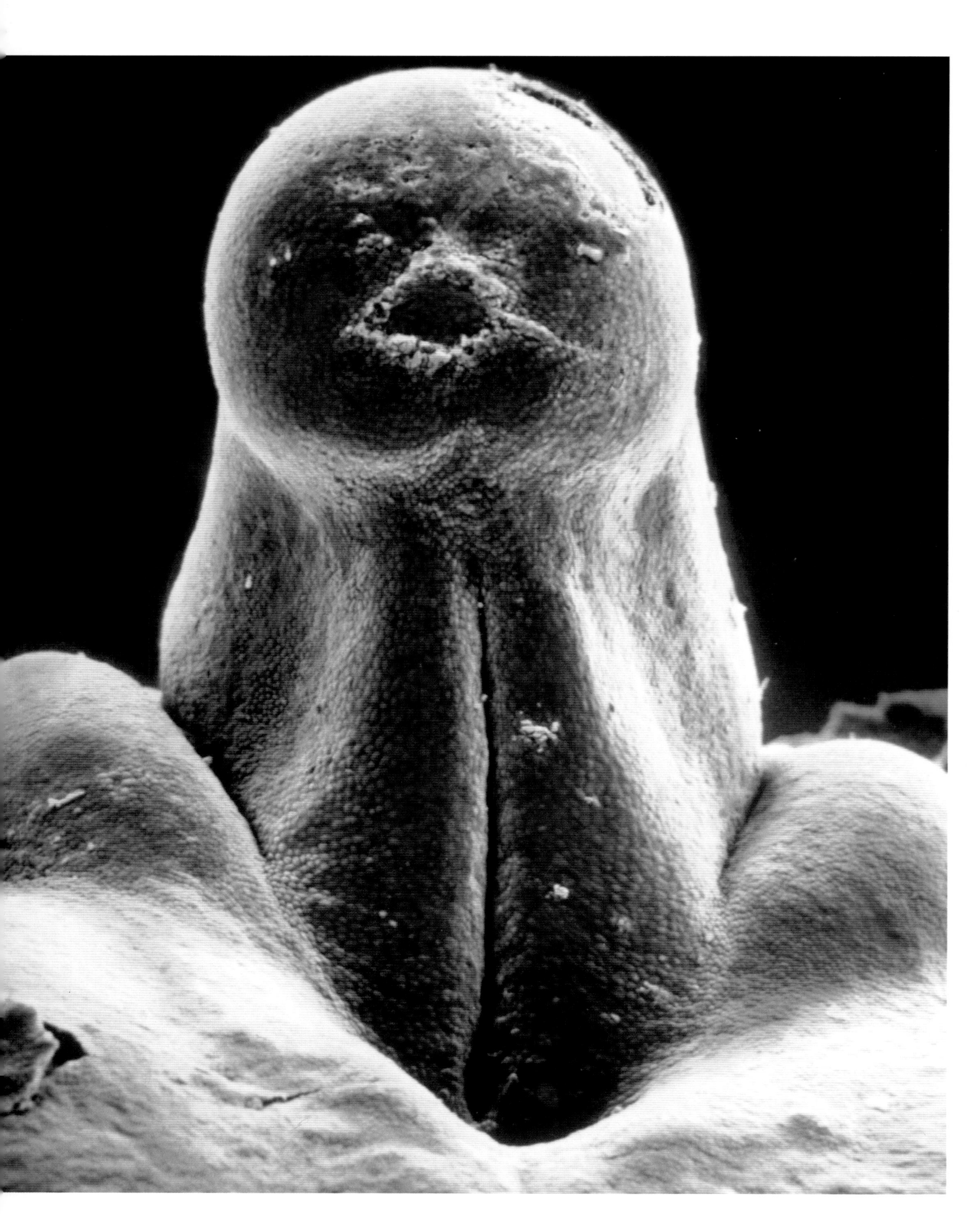

External genitalia in embryo [SEM]

IN THE EARLY EMBRYONIC STAGE, the male and female genitalia are not differentiated. The system here, in an embryo about ten weeks old, is destined to become the female genitalia, or vulva. The round tip at top centre will elongate slightly to form the clitoris, the female equivalent of the penis. The long cavity at centre will form the vaginal opening, while the folds on each side and at bottom will develop into the labia minora and labia majora respectively.

Blood vessels of penis [SEM]

THE NETWORK OF FINE VESSELS branching off from the blood vessel at centre infiltrates the spongy tissue of the penis. When males become sexually aroused by mental or physical stimulation, the blood vessels to the penis become dilated and the penis receives more blood than can drain away, making it erect.

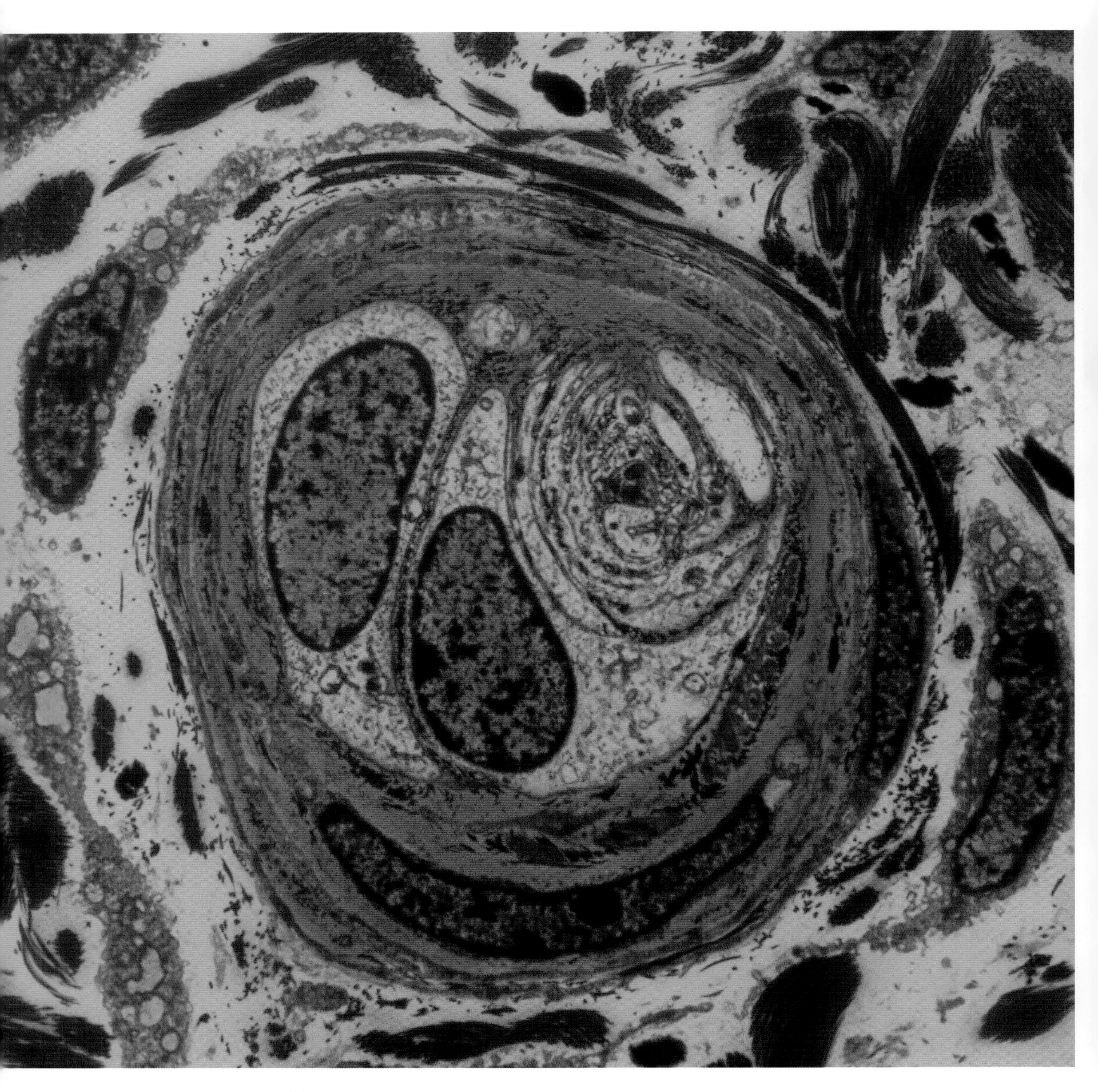

Sensory nerve endings in penis [TEM]

THE END OF THE PENIS, the glans, contains many sensory nerves, like the one shown here (centre). Stimulation of these nerves by rhythmic movement produces a series of involuntary reflexes – orgasm – and ejaculation of sperm.

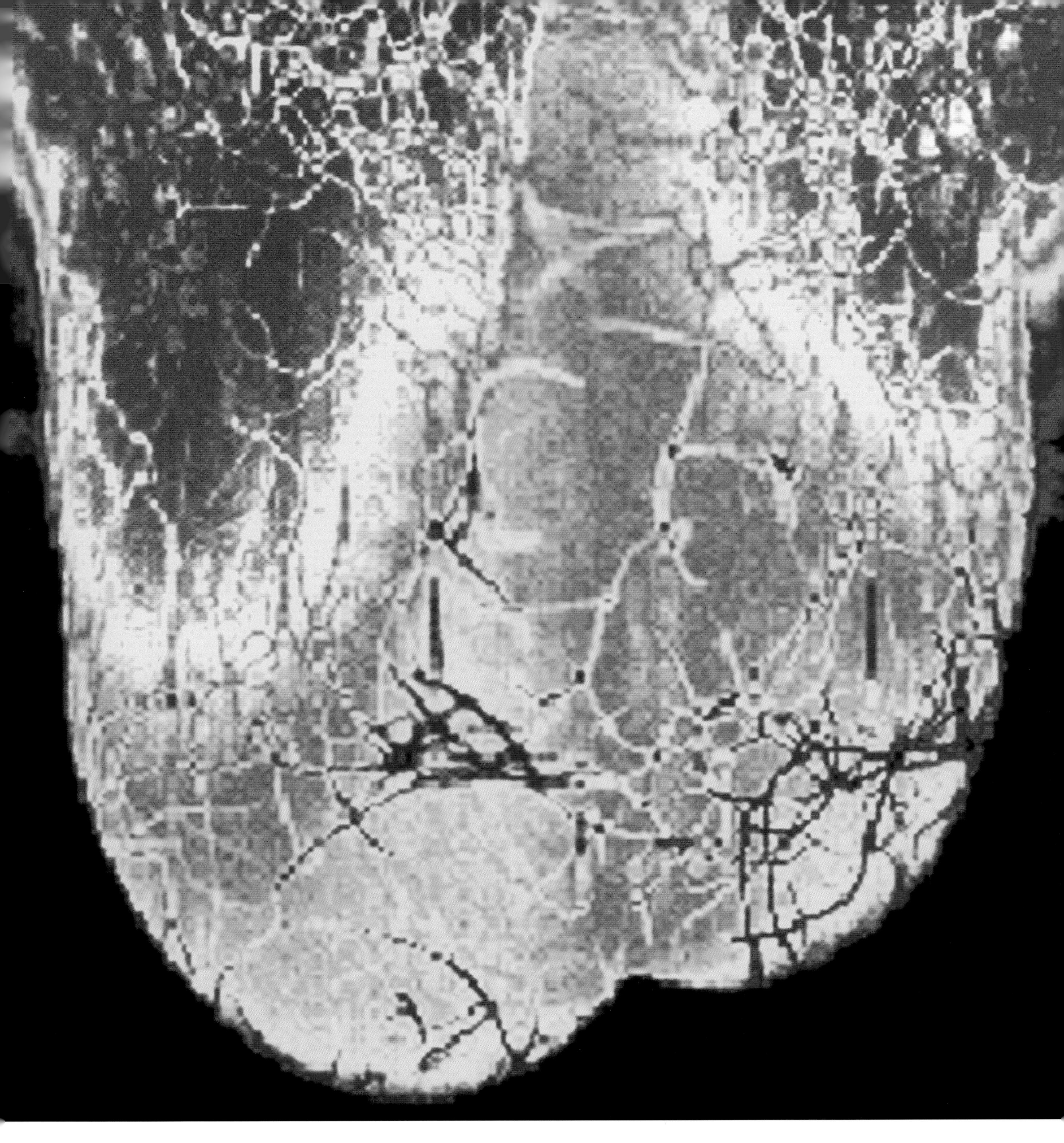

Scrotum [Thermogram]

THE DIFFERENT COLOURS REFLECT the different temperatures on the scrotum, the pouch that houses the testes. The colours range from red (warmest) through yellow and green to blue (coldest). As can be seen, the testes occupy the coolest part of the scrotum, which is about 2°C/4°F cooler than the body core and the most efficient temperature for production of sperm.

Sperm production in a testis [SEM]

A SECTION THROUGH A TESTIS shows the tails of developing sperm (blue/pink) in a seminiferous tubule, the site where sperm is produced. Sperm production occurs at a remarkable rate, over a thousand being made every second. They develop from cells in the lining of the tubule, which also contains cells that nourish the developing sperm.

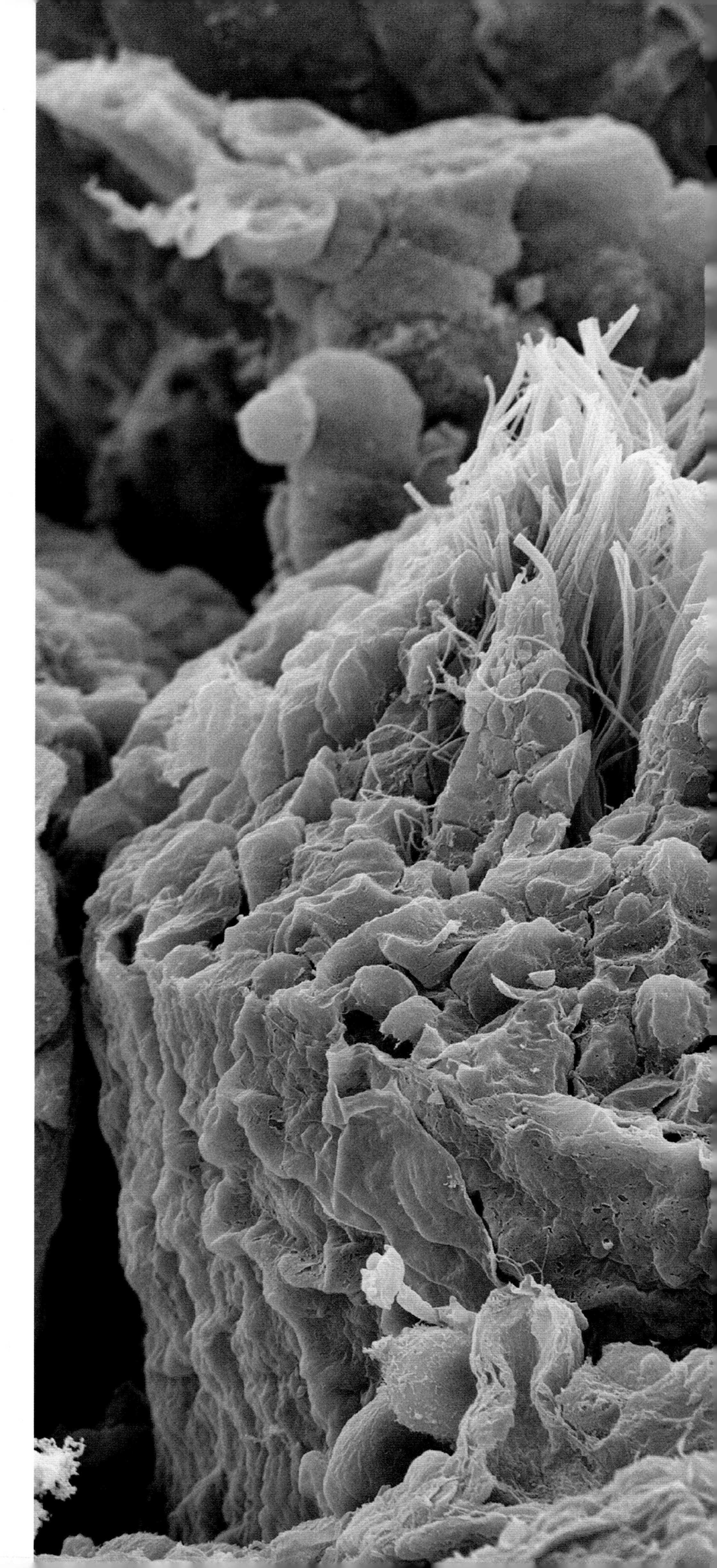

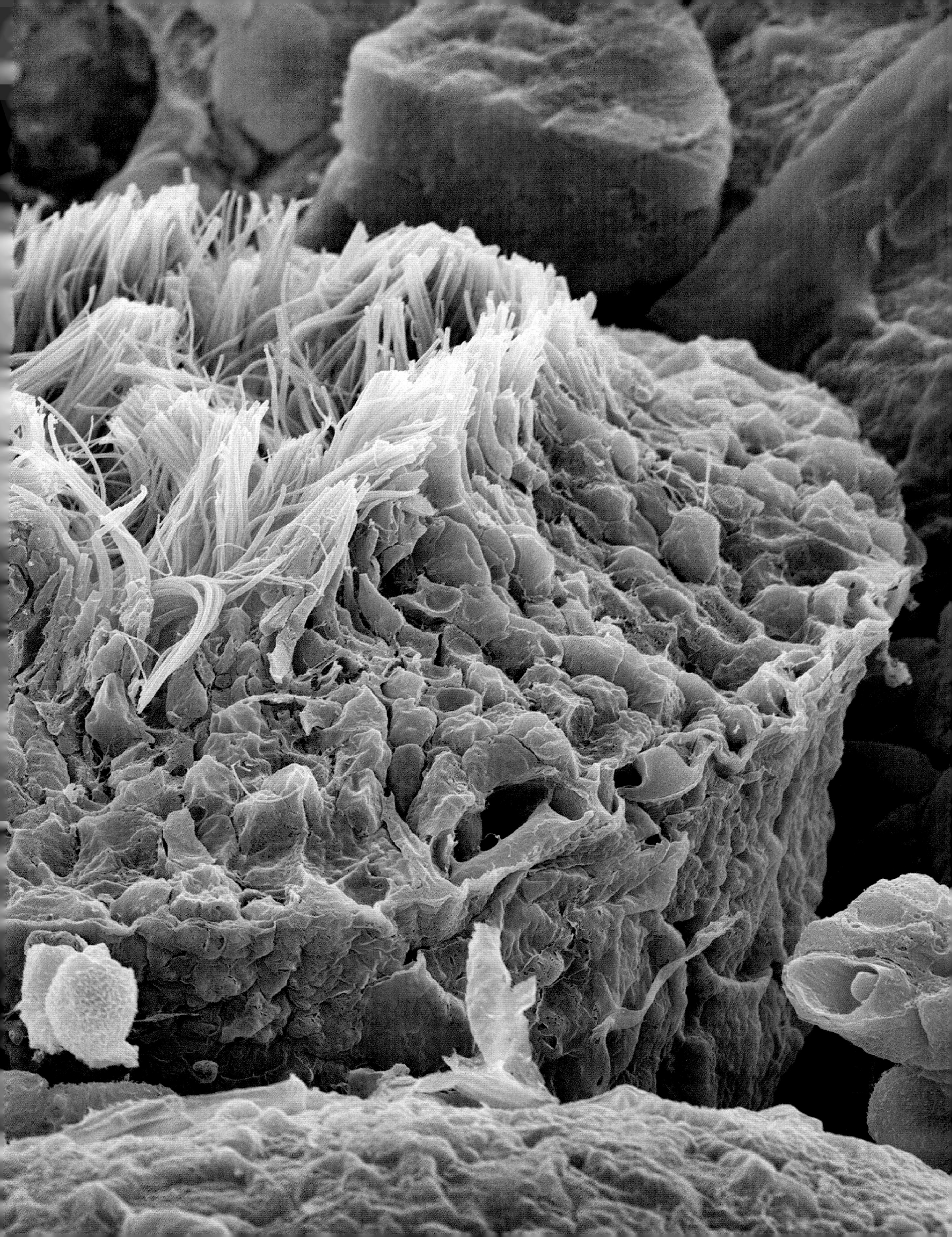

Sperm in a testis [SEM]

MATURE SPERM (blue) occupy a storage site in a testis. Unlike mature human eggs, which are produced at the rate of one a month from puberty to the menopause, sperm are produced constantly from puberty into old age.

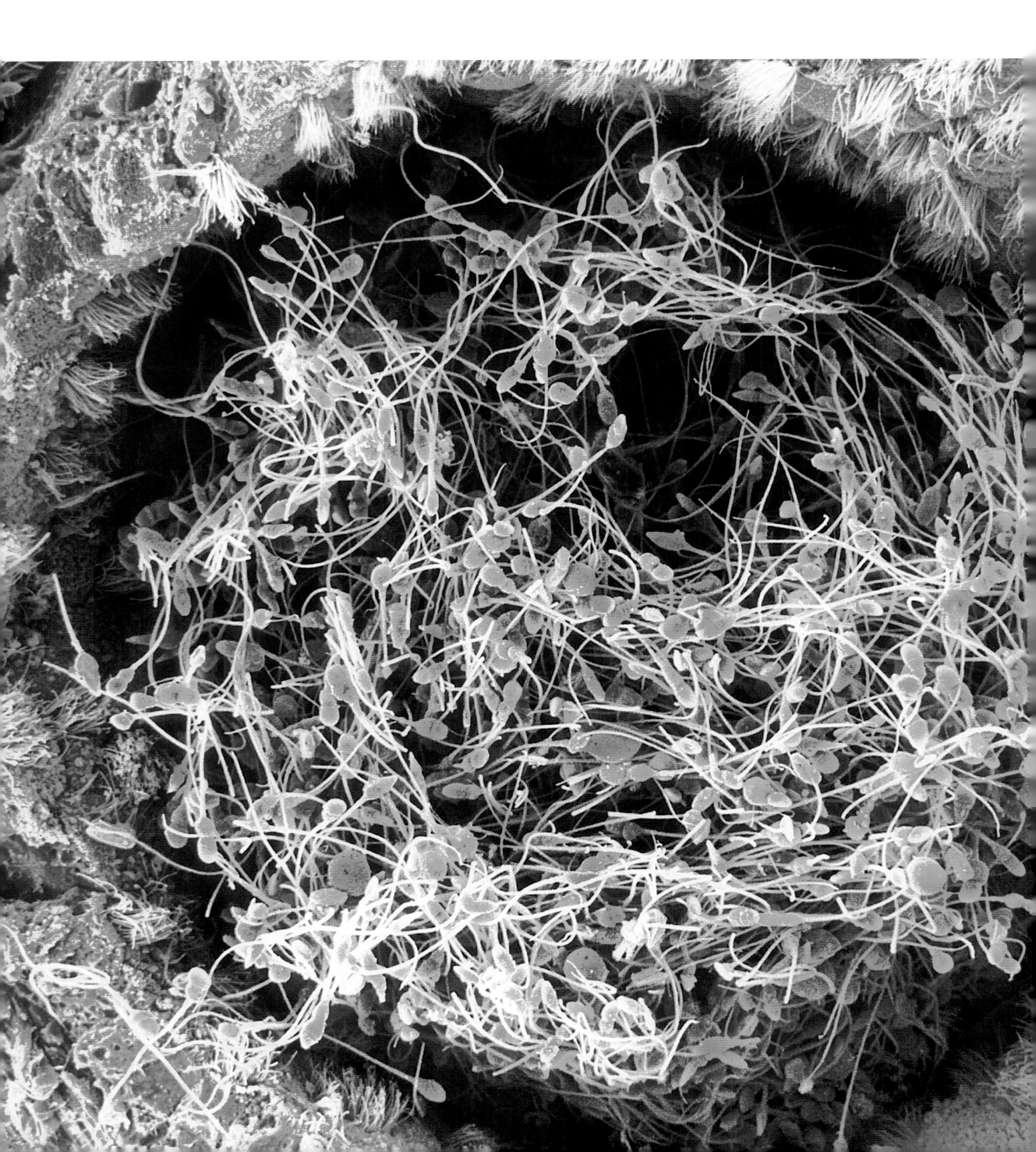

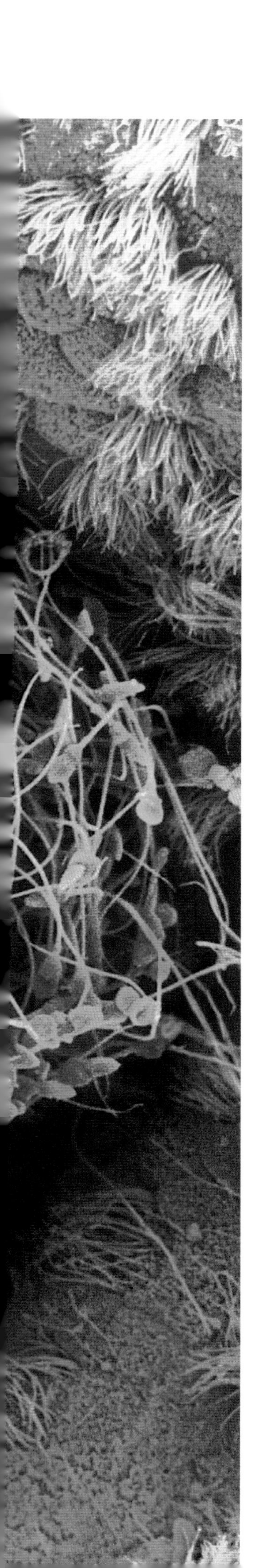

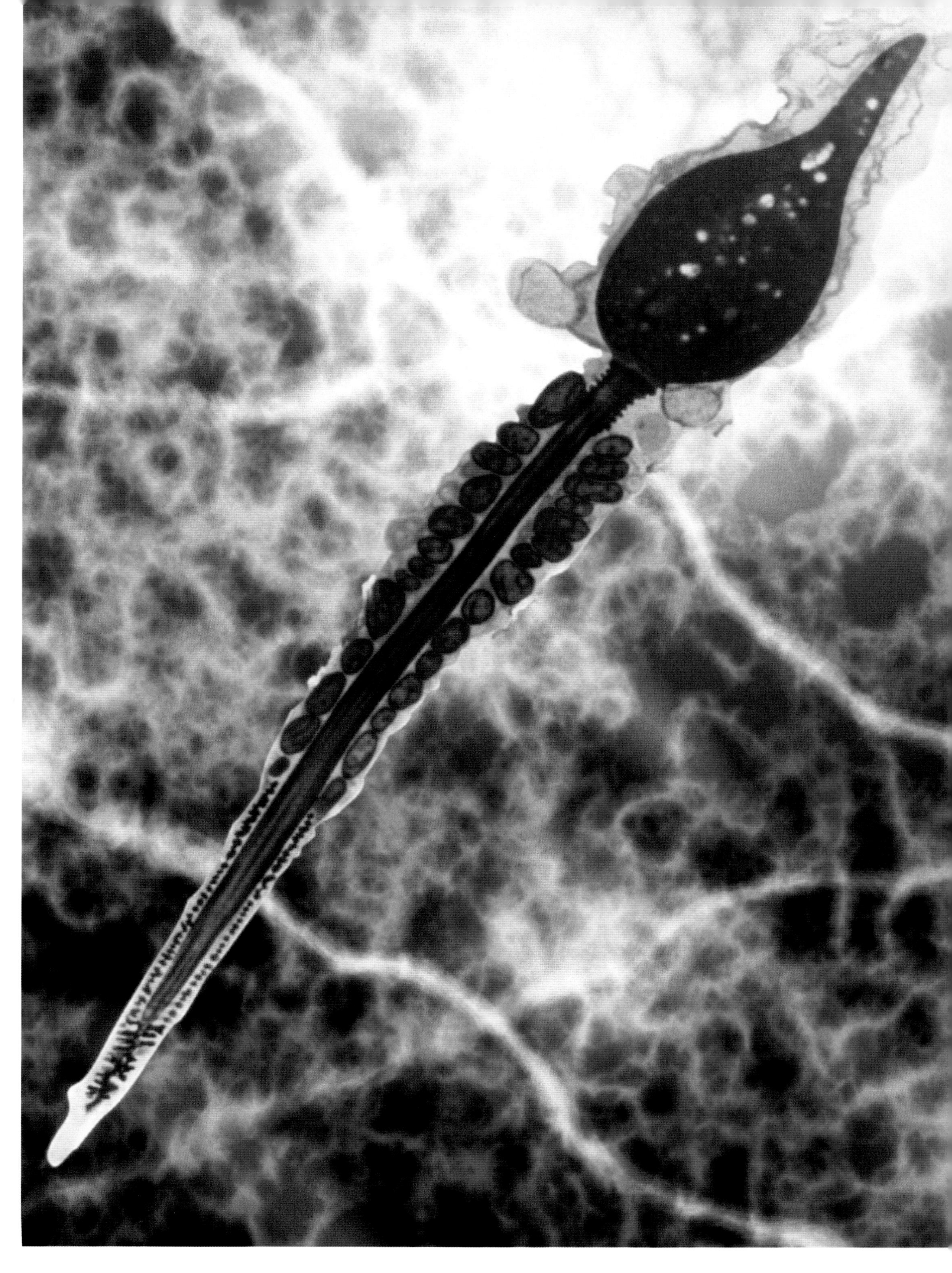

Human sperm [TEM]

A HUMAN SPERM has a head containing the twenty-three paternal chromosomes. It is capped by the acrosome (pink, surrounding head), which releases enzymes that help the head of the sperm to penetrate the egg cell. The middle section below the head contains many mitochondria (purple circles), which provide the energy required for the sperm to swim along the female's reproductive tract. It does this by rapidly moving its tail.

Vaginal membrane [SEM]

HERE, THE SURFACE OF THE VAGINA is seen to be a much folded mucus membrane. The folds allow the walls of the vagina to expand during intercourse and childbirth. During sexual arousal, the membrane secretes mucus that lubricates the vagina, making it easier for the penis to enter.

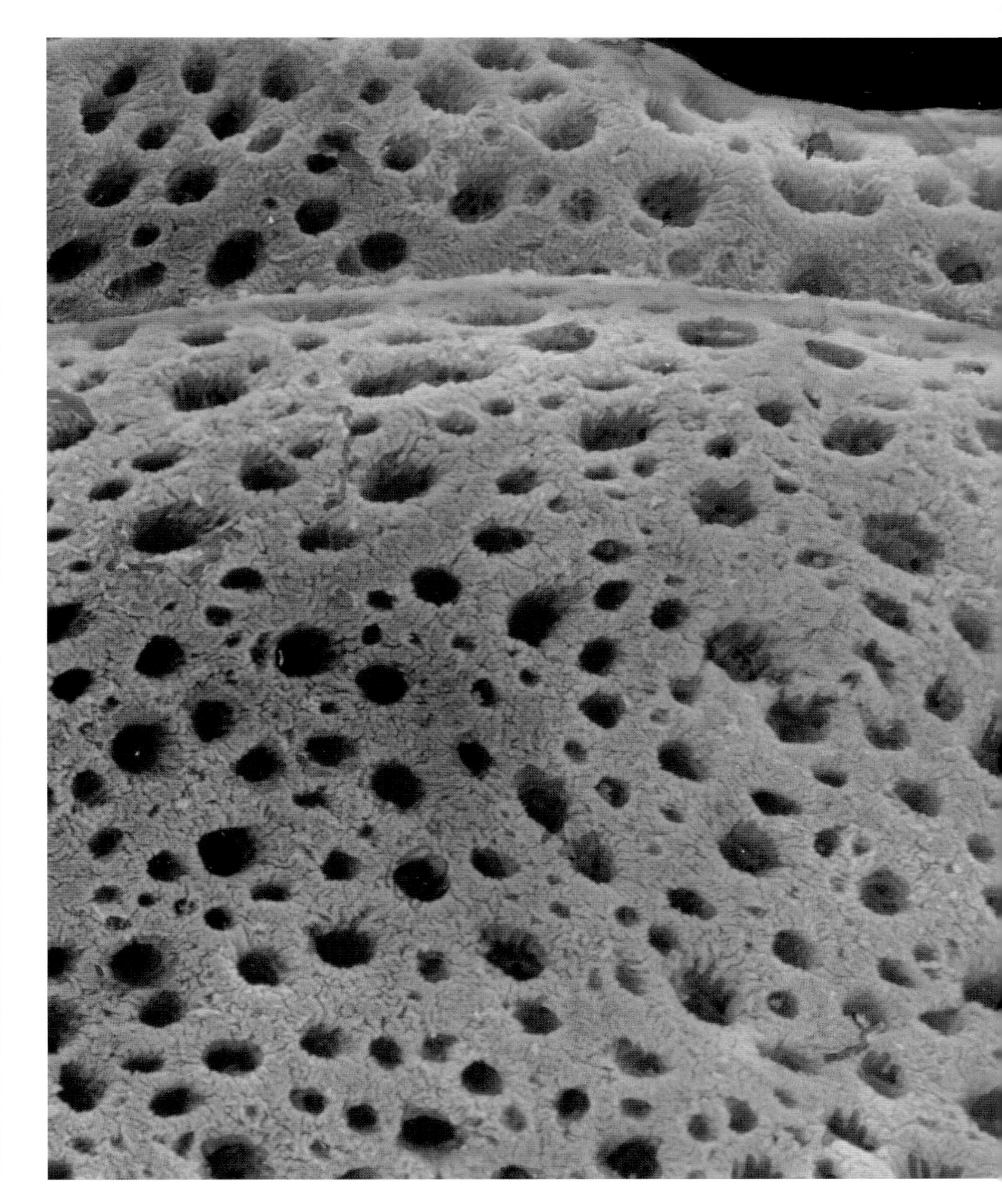

Uterus at ovulation [SEM]

AROUND THE TIME OF OVULATION (the release of an egg from an ovary), the inner layer of the uterus (womb) develops this honeycomb appearance, with numerous dark glandular openings. After ovulation, the glands secrete substances such as glycogen and fats and oils, which provide nutrients for the fertilized egg.

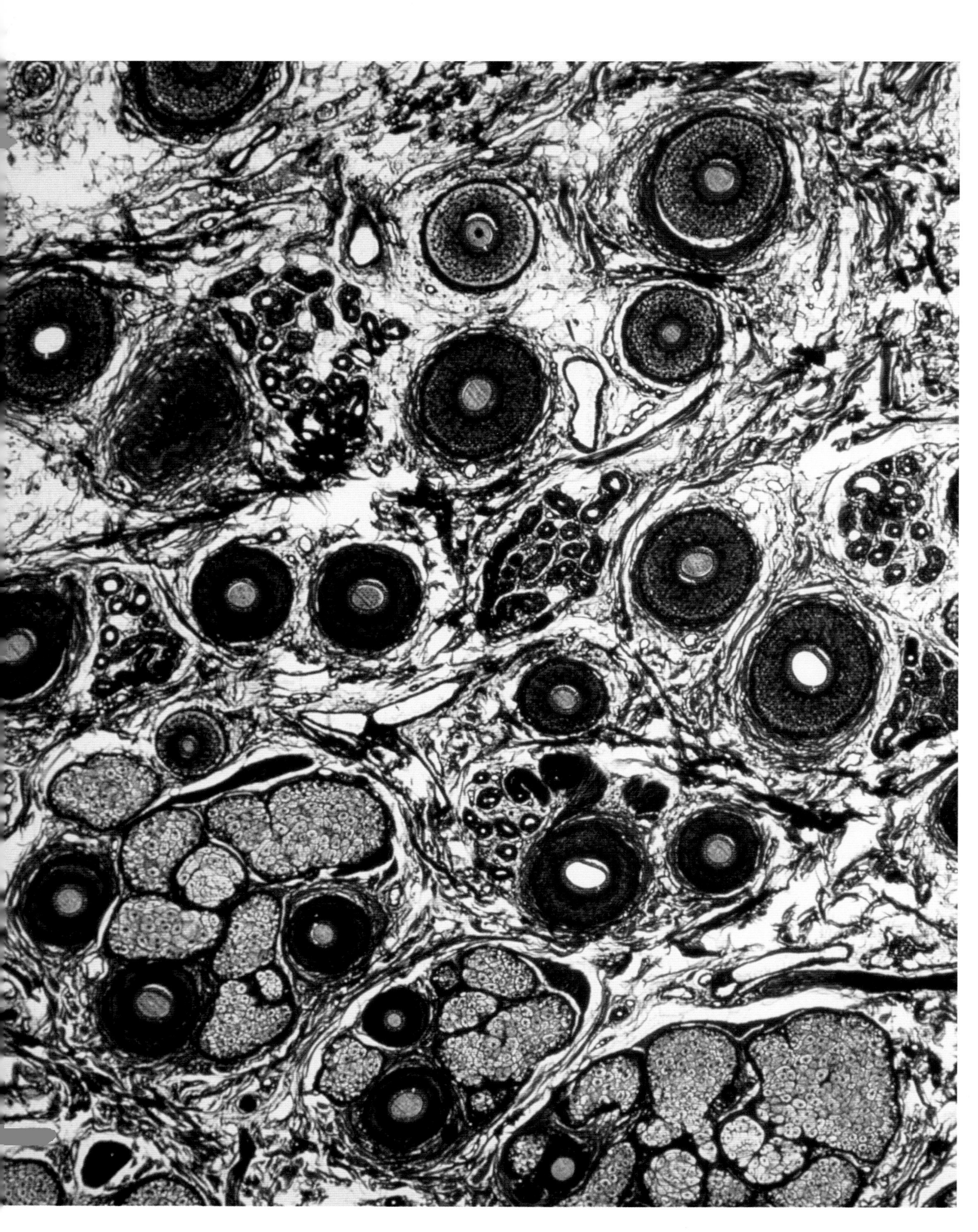

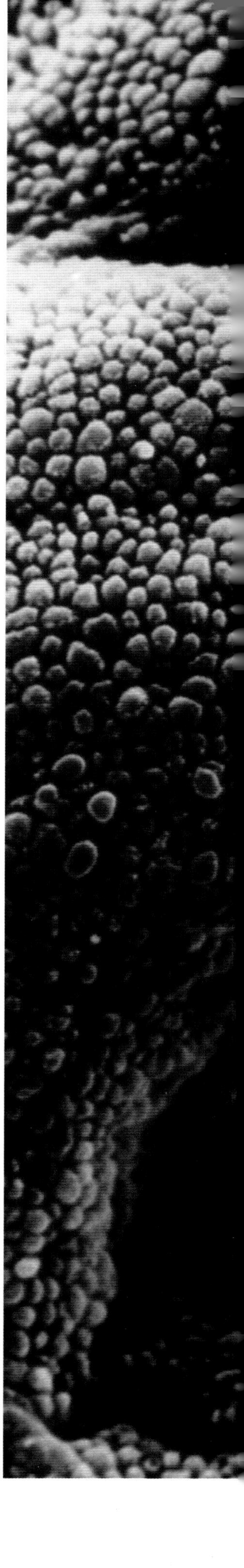

Uterus during the proliferative stage [LM]

DURING THE FIRST FOURTEEN DAYS of the female menstrual cycle, the lining of the uterus (womb) undergoes growth or proliferation to replace the layer lost during the previous menstruation. This image, taken during the late proliferative stage, shows the glands (purple with pink centres) that will nourish the egg after its release. If pregnancy does not occur, the uterus lining breaks down and is shed at menstruation after about twenty-eight days. It varies for every woman.

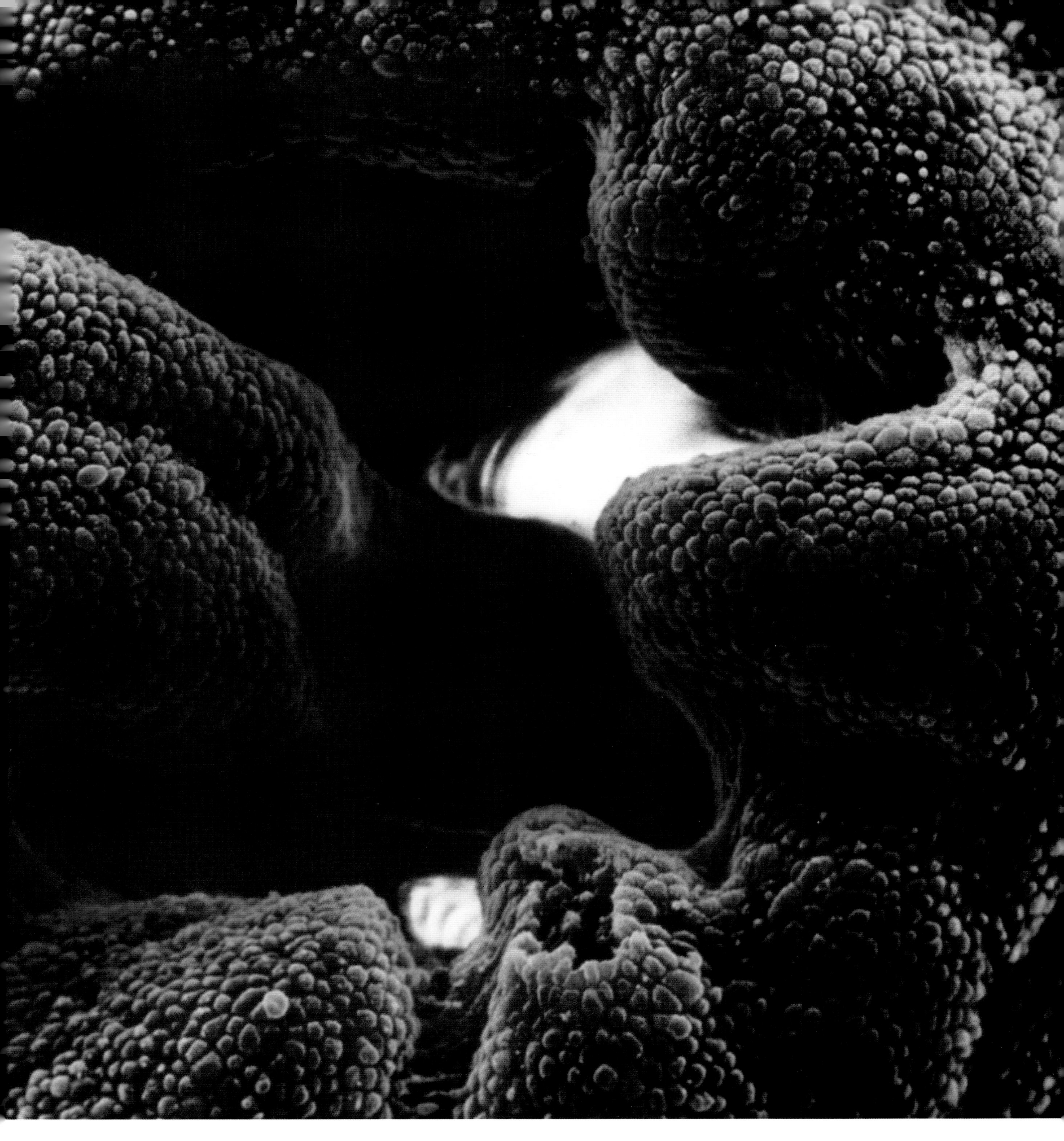

Uterus and entrance to Fallopian tube [SEM]

PART OF THE UTERUS (womb) is seen around the frilled entrance to one of the Fallopian tubes, the ducts by which fertilized eggs reach the uterus. The journey takes about seven days, and during this time the lining of the uterus develops in readiness to receive an embryo. If conditions are right when the embryo leaves the Fallopian tube, it implants in the wall of the uterus.

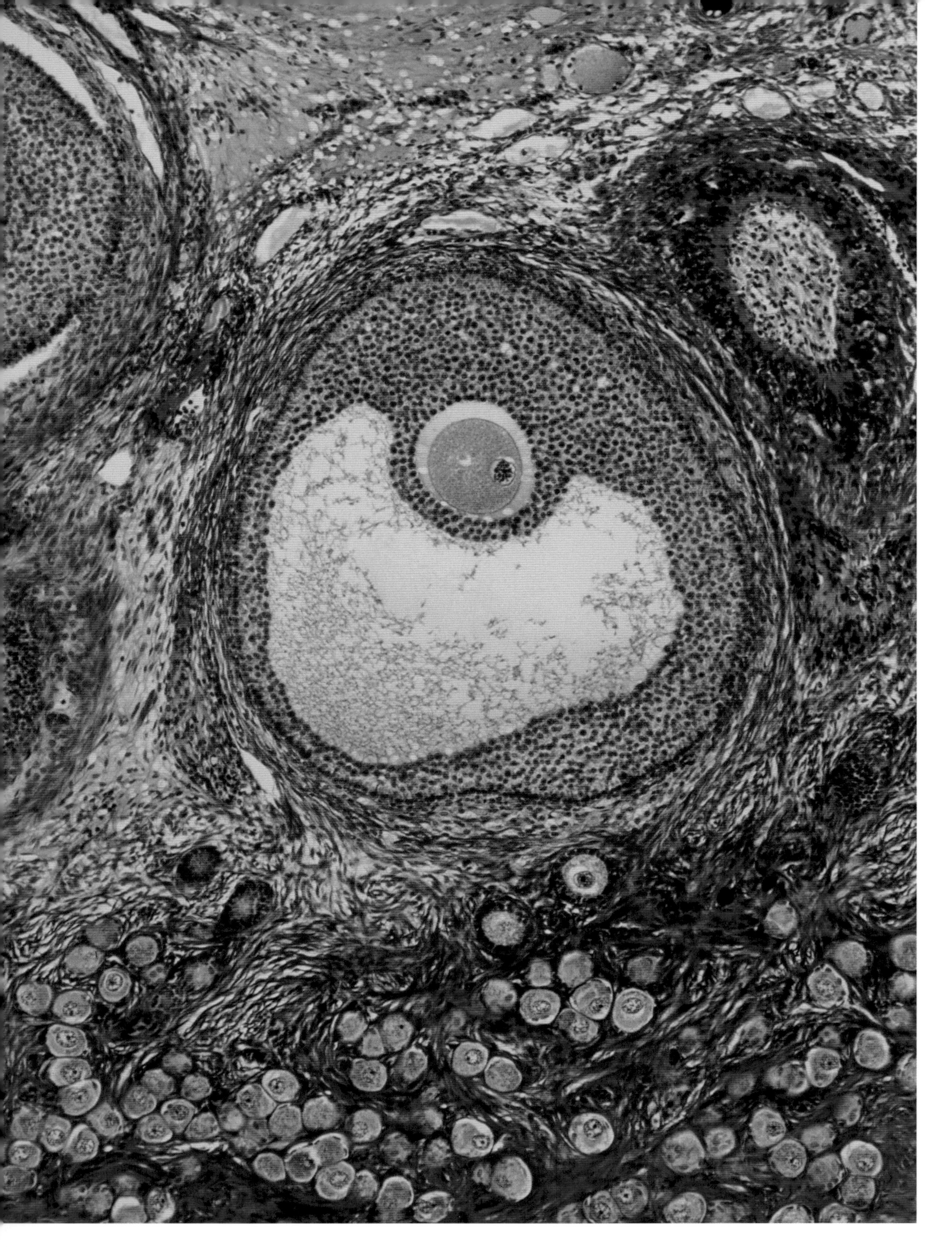

Maturing secondary follicle in an ovary [LM]

THE OVARIES CONTAIN AN ENORMOUS NUMBER of egg cells in sacs called primary follicles (small purple circles, lower frame). About ten of these begin to mature during each menstrual cycle, but only one (upper centre) completes its development and becomes a secondary follicle. Inside this follicle, the egg cell (small purple sphere) begins to enlarge and is surrounded by a space filled with follicular fluid (white marbled with blue). The mature follicle, called a Graafian follicle, releases its egg at ovulation.

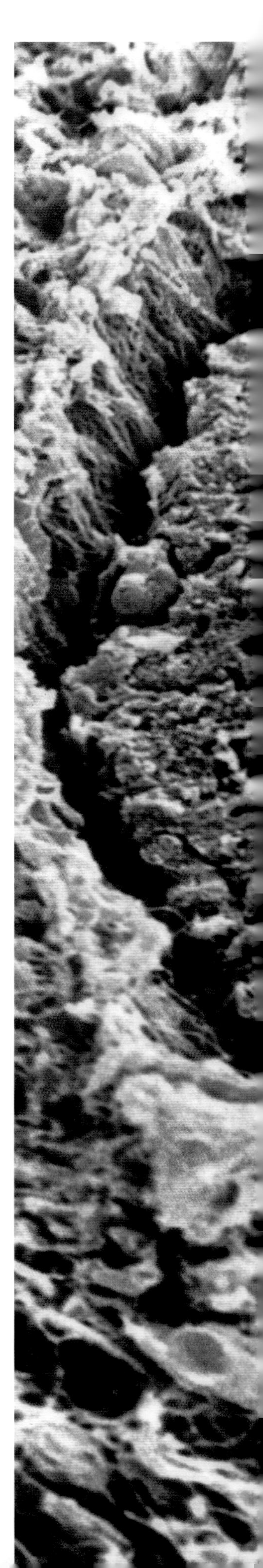

Uterine gland [SEM]

WHEN AN EGG REACHES THE UTERUS (womb) and implants in the uterus wall, it is nourished by secretions from glands like the one shown here in section. The glands are stimulated by progesterone, a hormone produced in the ovaries. If the egg is not fertilized, progesterone secretions stop after about ten days and the uterus wall breaks down. If a fertile egg is successfully implanted, the uterus wall forms the placenta, the organ through which the foetus obtains food from its mother's blood.

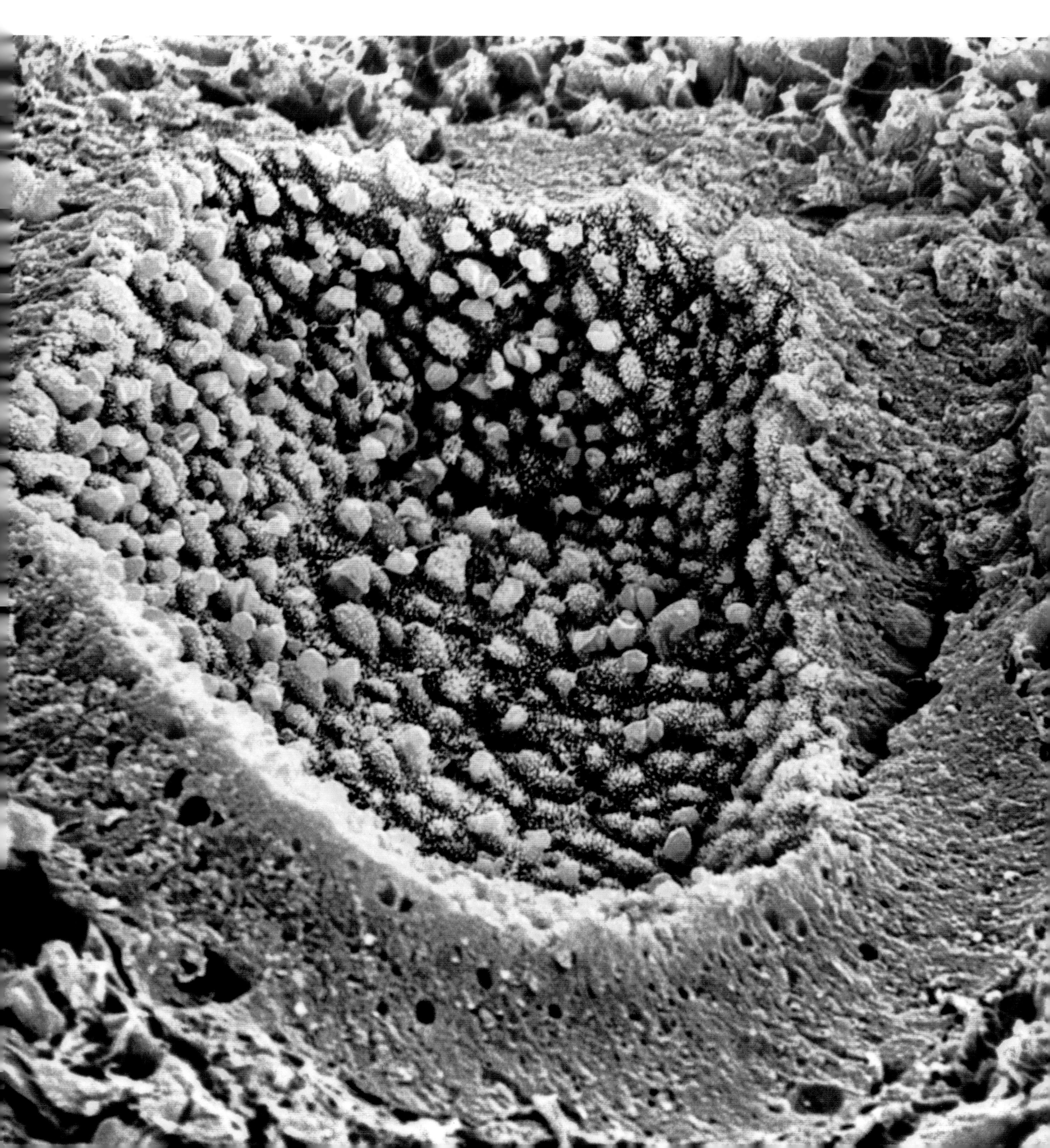

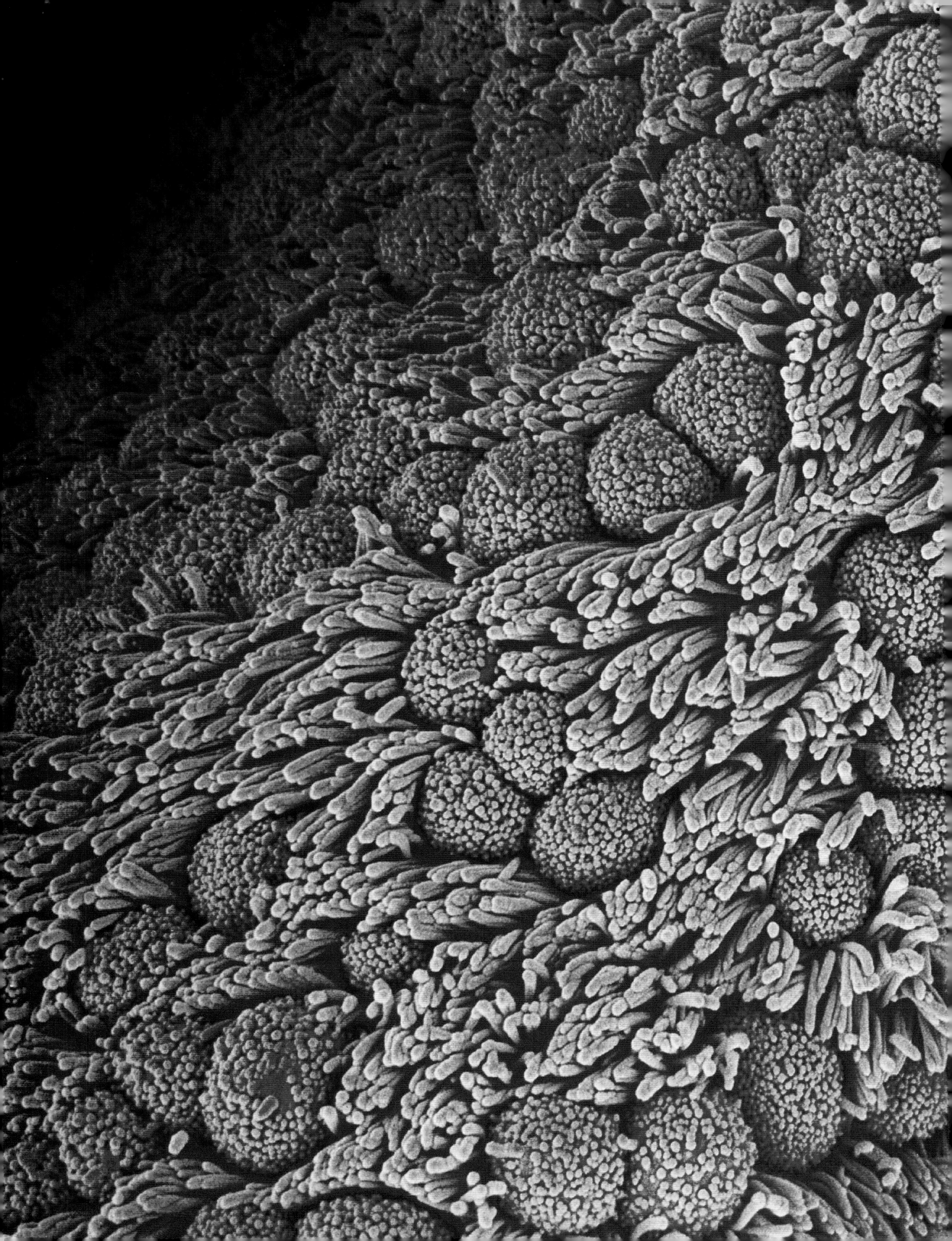

Lining of Fallopian tube [SEM]

FALLOPIAN TUBES ARE DESIGNED to carry eggs from the ovaries to the uterus (womb). They are lined with mucus-secreting cells (red) and covered with hair-like cilia (brown) that propel the egg along the tube. The journey takes seven days, and during this time a fertilized egg divides several times to become an embryo.

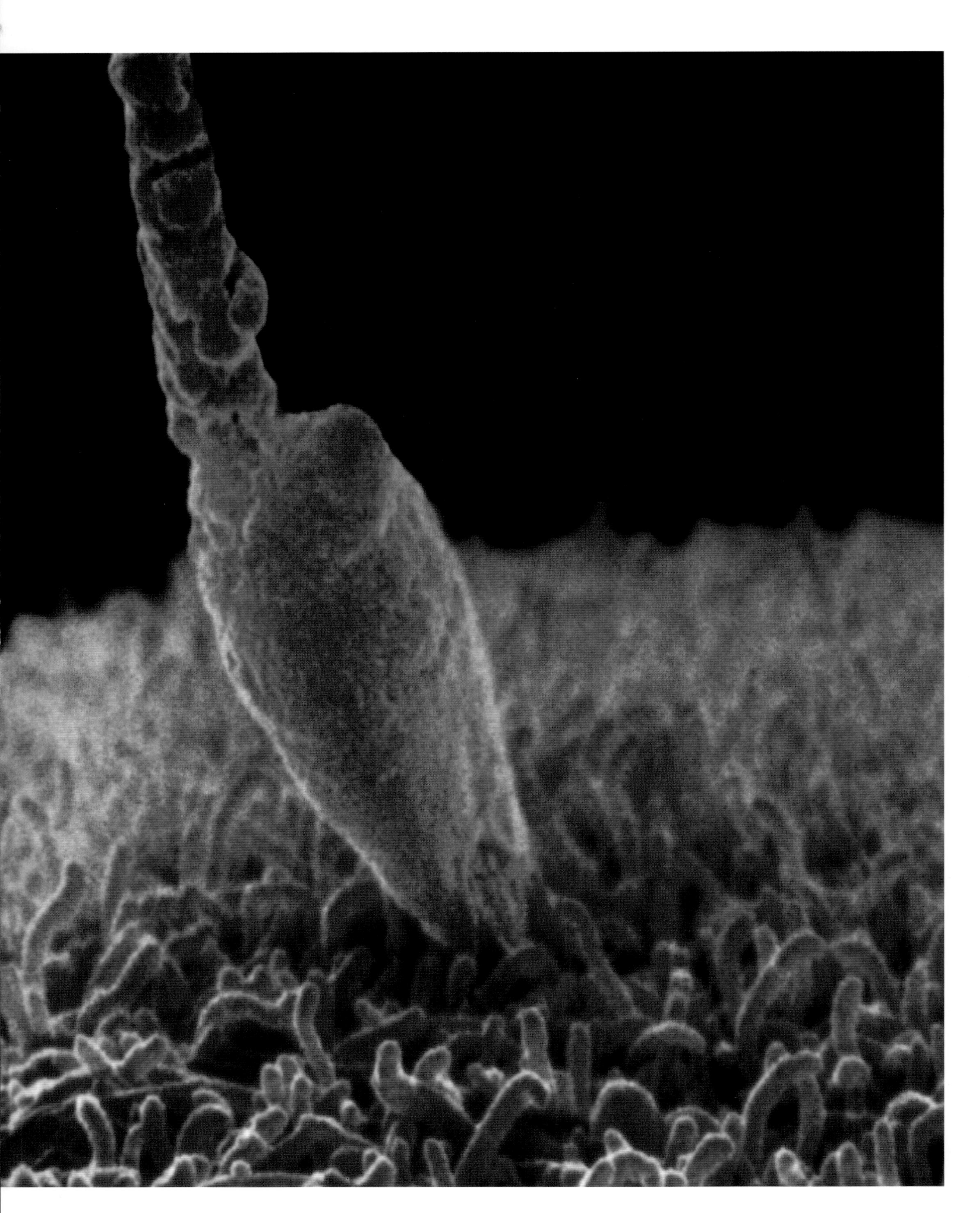

Sperm fertilizing an egg [SEM]

OF THE 200 MILLION or so sperm released during intercourse, only a few hundred survive the journey and reach the egg in the Fallopian tube. Of these, only one sperm will actually fertilize the egg. If a sperm succeeds in penetrating the egg's wall to fuse with the egg nucleus, a membrane forms around the egg, creating a barrier to other sperm.

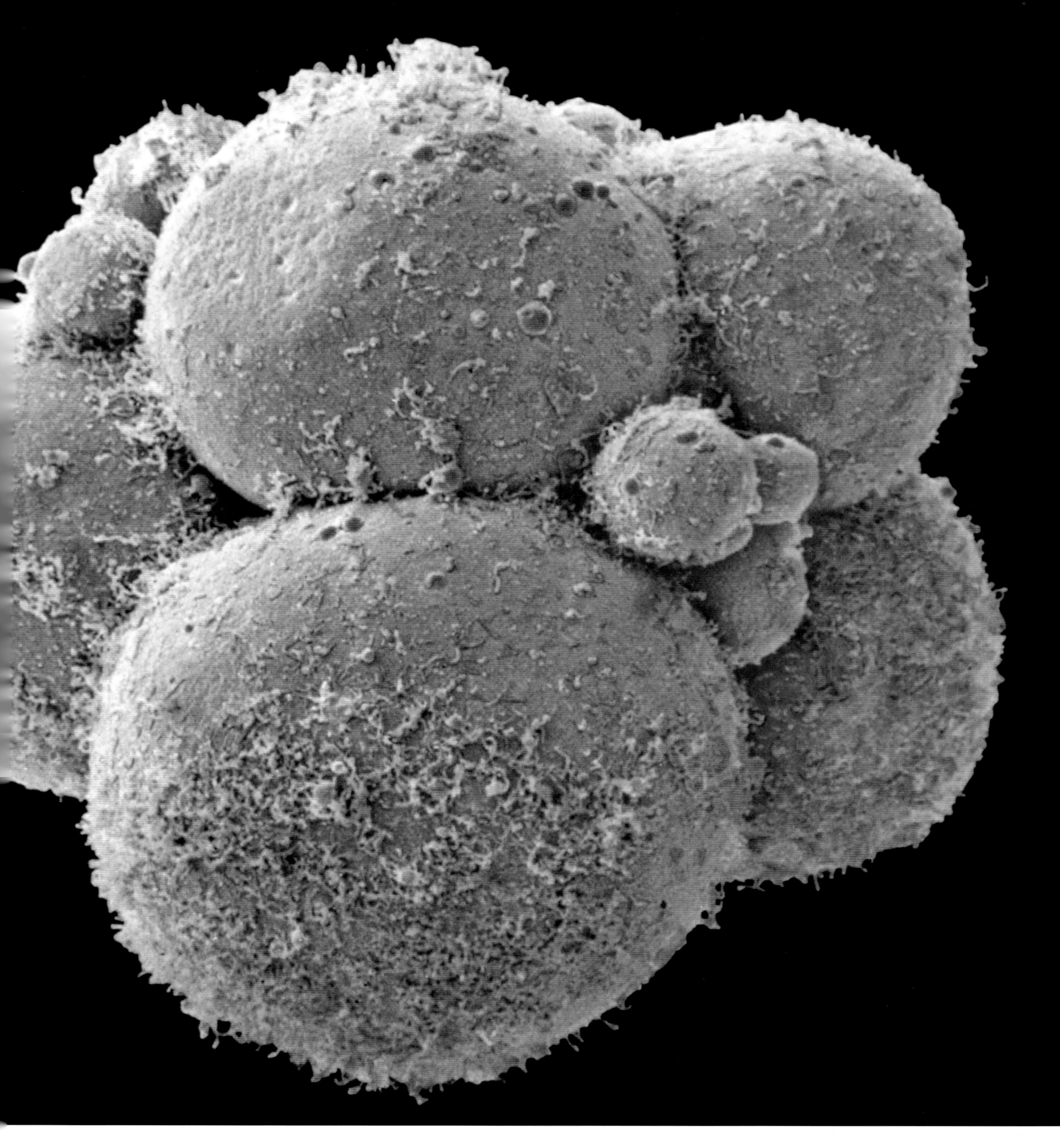

Human embryo at eight-cell stage [SEM]

THREE DAYS AFTER FERTILIZATION, a fertilized egg has divided to form a cluster of eight large rounded cells called a morula (Latin for blackberry). The smaller spherical structures (centre right and left) will degenerate. At this eight-cell stage the morula is still in the Fallopian tube, making its way towards the uterus (womb), where, if conditions are right, it will be implanted in the uterus wall and continue its transformation into an individual composed of billions of cells.

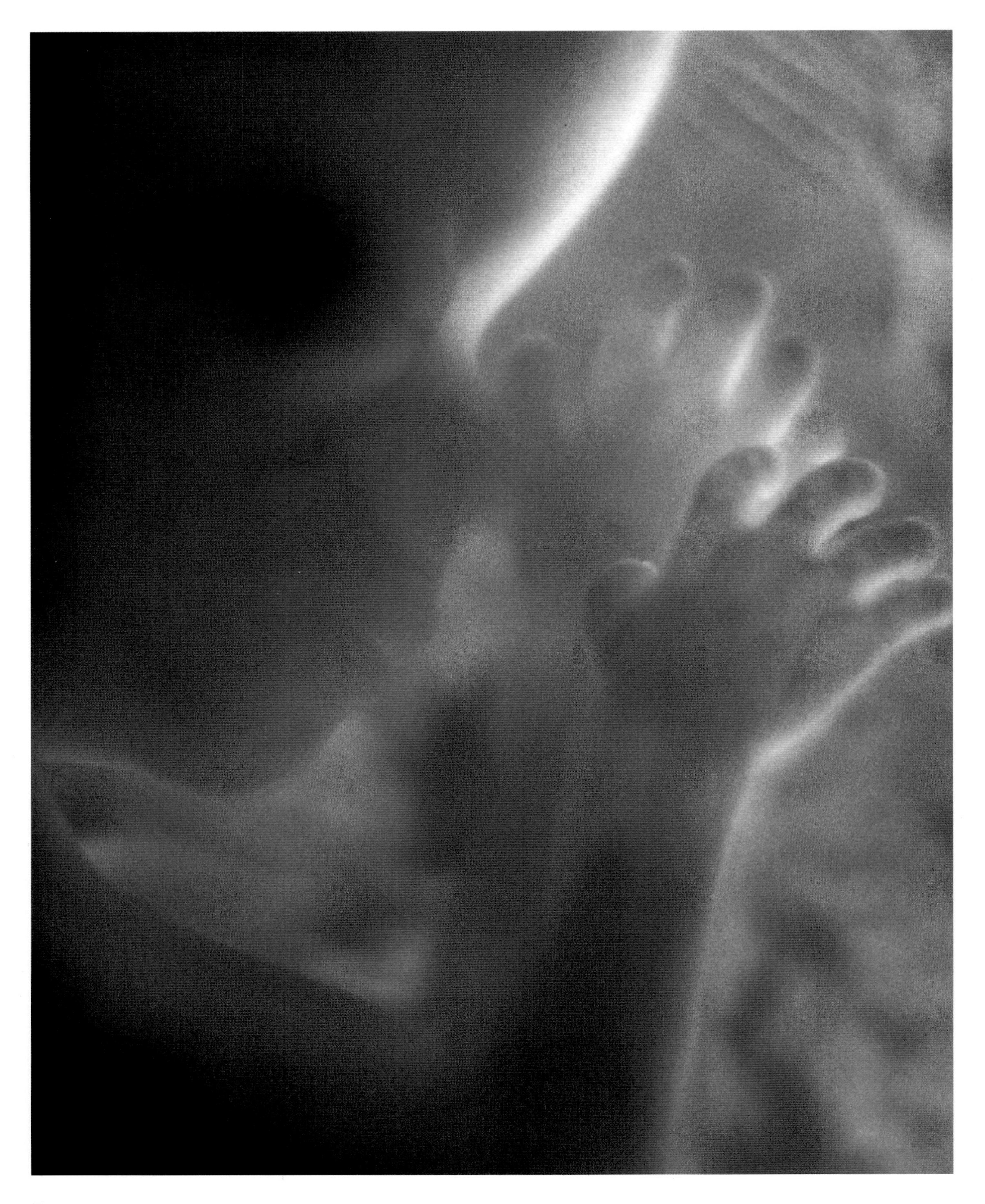

Fetus [Clinical photography]

A FETUS BECOMES RECOGNIZABLY human at about twelve weeks. This fetus – at fourteen weeks old – has a length of about 10cm/4in and weighs around 150g/5oz. An eye is visible as a dark patch, and the fingers and thumbs are well developed. By the age of six months, the fetus may suck its thumb.

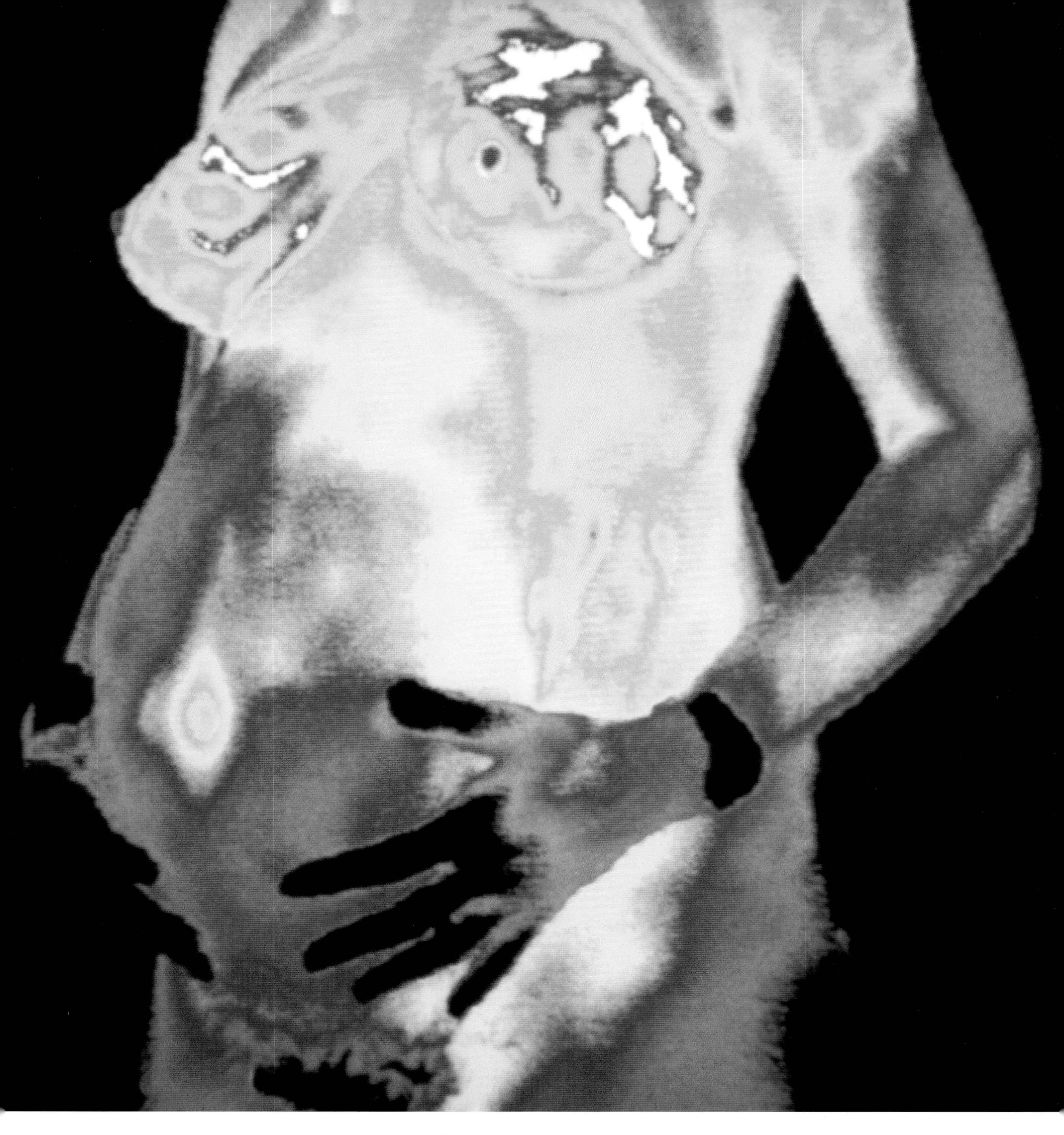

Pregnant woman [Thermogram]

THE DIFFERENT COLOURS in this image represent different skin temperatures, from black (coolest) to white (warmest). The image was taken with a thermographic camera, which records the long wavelength infrared radiation emitted by the surfaces of the body. Skin temperature may reflect the volume of blood supply across the body, and thermography can highlight a variety of circulatory disfunctions.

Each day the body's nervous system makes more connections than all the world's telephone systems put together. It is the system that co-ordinates all other systems, imposing order on what would otherwise be chaos.

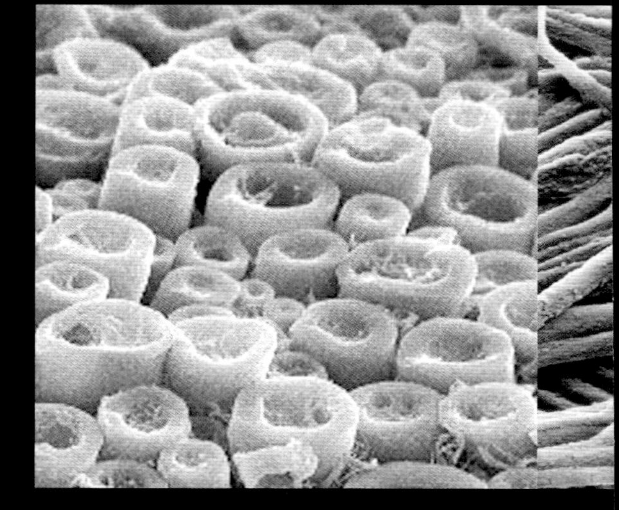

The nervous system is divided into the central nervous system, consisting of the spinal cord and brain, and the peripheral nervous system, which relays information from sense organs to the central nervous system.

The cells that make up the nervous system are called neurons. There are two main kinds of neuron: sensory neurons and motor neurons. Sensory neurons are activated by sensory cells that respond to stimuli – changes in the internal or external environment. Different sensory cells are specialized for detecting different kinds of stimuli, such as pressure, heat or light. The sensory neurons carry information about the stimuli to the central nervous system, which processes the information. If a response is necessary, a motor neuron carries instructions from the central nervous system to an effector organ, such as a muscle or gland. The process happens very fast: nerve impulses travel to and from the central nervous system at up to 290kph/180mph.

Reflex actions are rapid responses that do not involve the brain. The knee-jerk reflex is a classic example. If the right leg is crossed over the left and struck just below the kneecap, the lower part of the right leg jerks by reflex action. The path taken by nerve impulses during this action is called a reflex arc. Receptor neurons carry information straight to the spinal cord, and motor neurones carry instructions back to the muscle that jerks the leg. By the time the brain has registered the response,

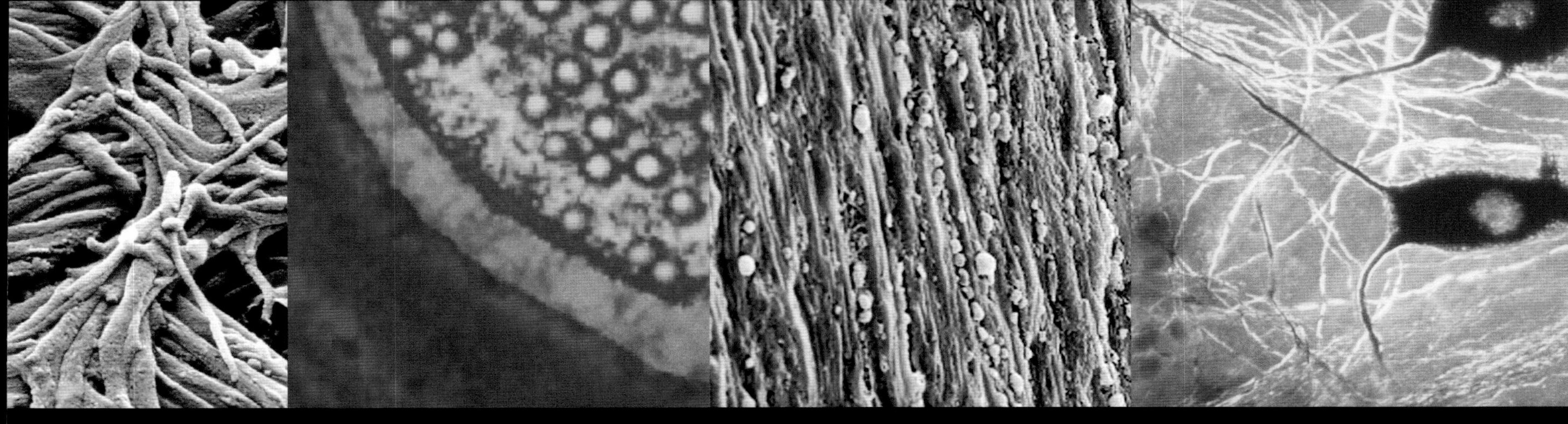

it has already happened. Reflex arcs cut down reaction time and in many cases – blinking, for example – reduce the risk of damage to the body.

Neurons vary considerably in shape and size, but their basic plan is a small, nucleated body sprouting long, thin extensions. Branching extensions called dendrites bring information from other neurons or sensory cells, while a single extension called an axon carries information away from it. Some axons run from the spinal cord to the toes.

Information is transmitted along an axon in the form of electrical impulses. Between the axon of one neuron and the dendrites of another is a microscopic gap, the synapse, which is bridged by a chemical transmitter. Certain neurotoxins can block the chemical transmitters, preventing nerve impulses reaching muscles and causing paralysis. One of these toxins, curare, is used by Amazonian hunters as an arrow poison. Synthetic curare is an effective muscle relaxant and is an important tool for treating tetanus and rabies, which are characterized by muscle spasms.

Spinal cord [SEM]

A CROSS SECTION through a spinal cord shows the central region of grey matter (brown), which is composed of nerve cells. The outer zone of white matter (yellow-brown) consists of tracts of nerve fibres. The white outer covering is made up of tough membranes called meninges. Inflammation of this tissue, which surrounds the brain as well as the spine, causes meningitis, a potentially fatal disease.

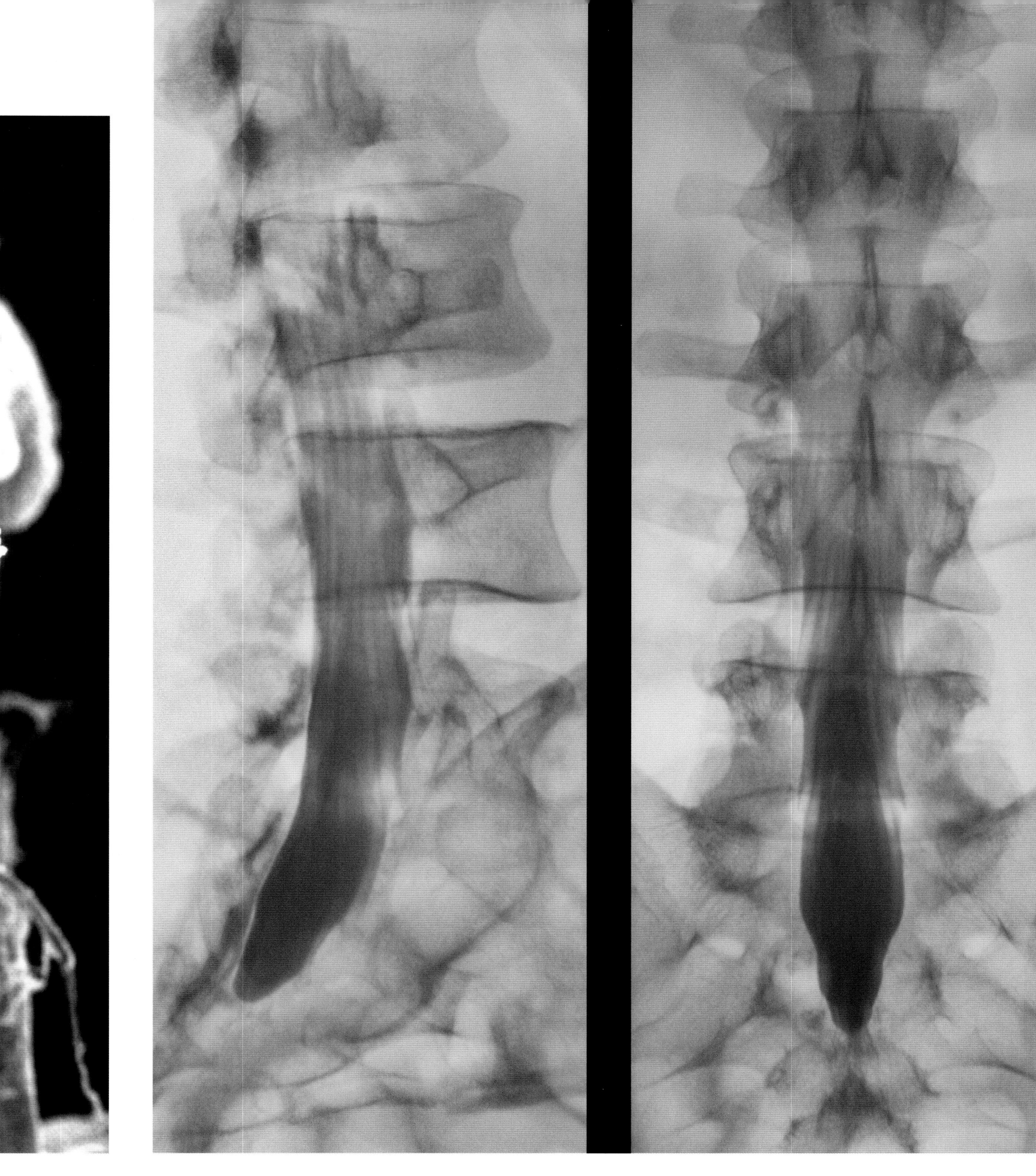

Spinal cord in the lower back [X-ray]

SEEN FROM THE SIDE (left) and front (right), the lower part of the spinal cord (blue) is shown enclosed within the protective vertebral column (yellow-brown). Nerves (not seen) branch from the cord, carrying information between the brain and the body. The spinal cord can also control simple actions without conscious instructions from the brain. These are the reflex actions, such as blinking when dust enters your eye, sneezing, or pulling your hand away from a sharp object or a flame.

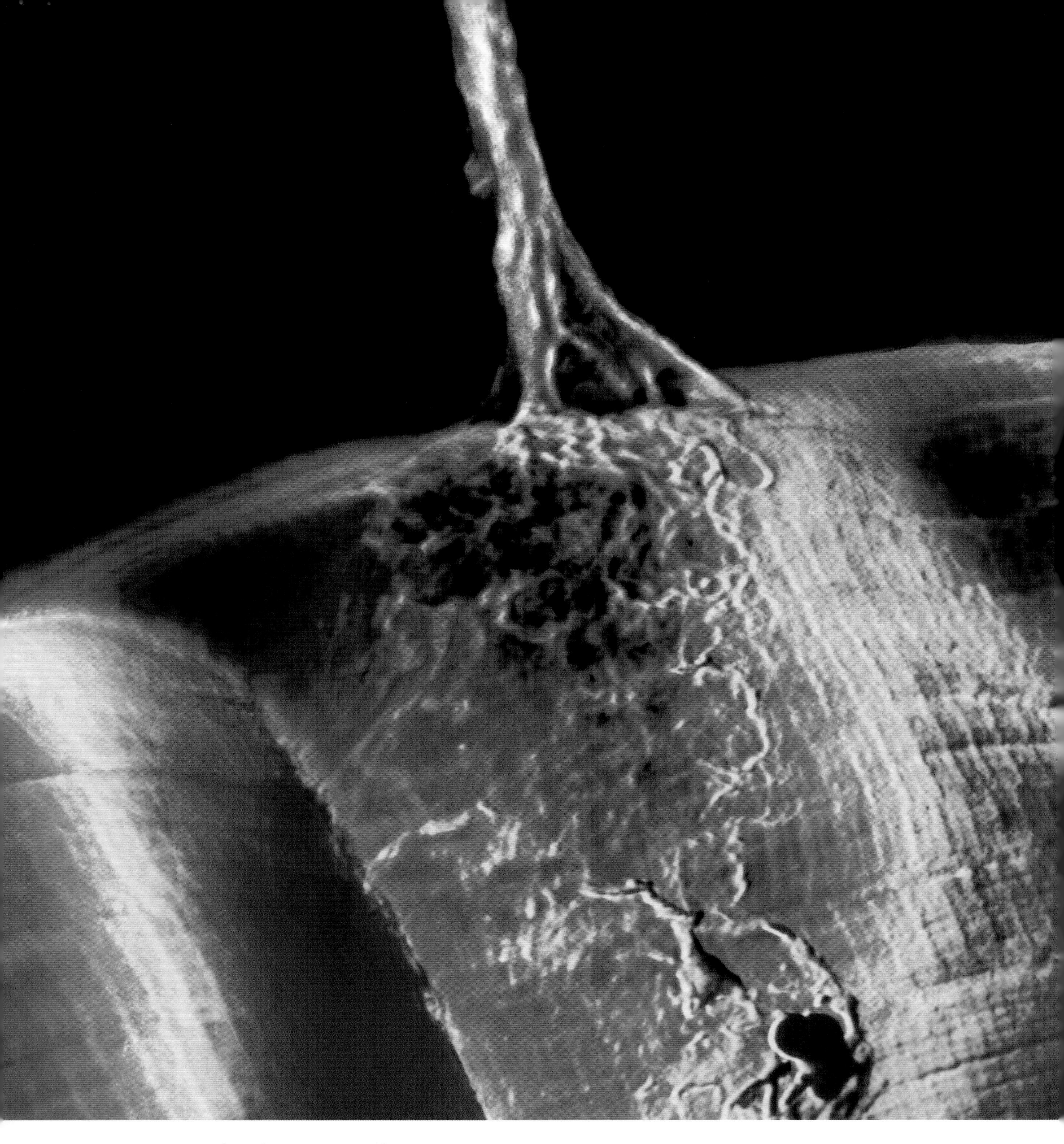

Nerve terminating on muscle [SEM]

IMPULSES FROM THE CENTRAL NERVOUS SYSTEM are transmitted to a muscle or gland by motor neurons, or nerve cells, which consist of a cell body and thread-like fibres called axons (pink). The branches at the end of the axon terminate at a junction called a synapse. When activated, the neuron releases chemical transmitters that cross the synapses to interact with receptors in the muscle cell.

Motor end plates [LM]

A NERVE CELL (line starting at bottom right) forms branches leading to individual muscle cells (bands running diagonally across frame). At its junctions with the muscle cells, the nerve cell divides further to form a cluster of small terminals, called boutons (dots), on the muscle surface. A synapse occurs where each bouton meets the muscle cell. Here, messages to contract the muscle are passed on by chemical neurotransmitters.

Synaptic junction [TEM]

A SECTION THROUGH THE JUNCTION between a nerve cell and muscle shows the end of the nerve (blue) separated by a narrow gap – the synapse – from the red muscle fibre at left. The small yellow spheres clustered near the junction contain the neurotransmitter chemicals, which pass across the synapse to stimulate receptors in the muscle. The green structure partly enclosing the nerve ending is a Schwann cell, which supports and insulates the nerve.

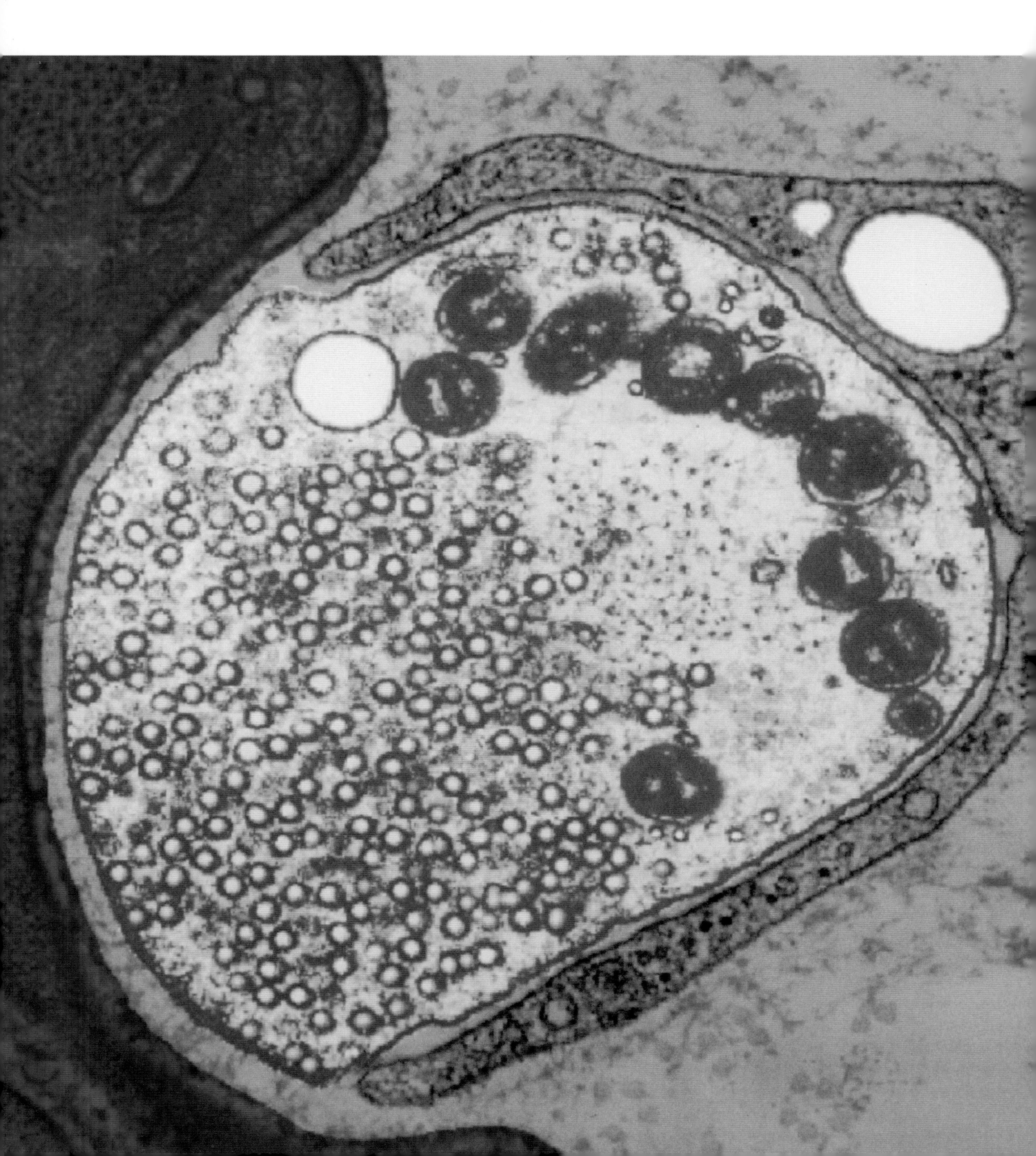

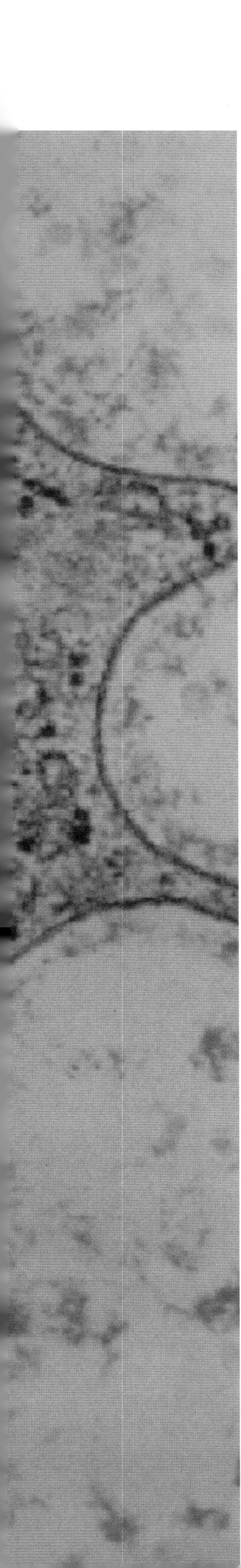

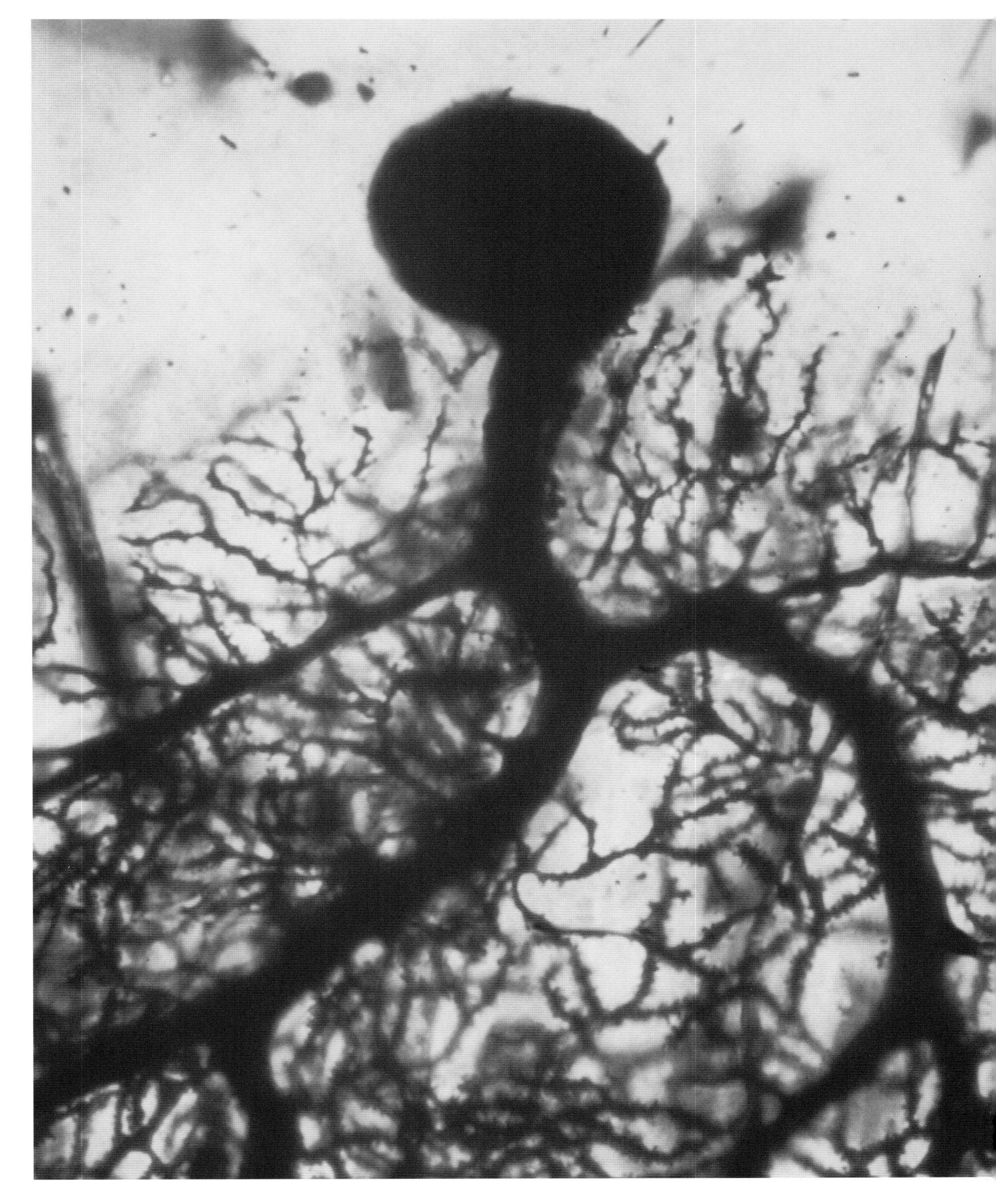

Purkinje cell [LM]

THIS DEVELOPING NERVE CELL in a one-year-old infant is from the cerebellum, a part of the brain involved in maintenance of muscle tone and balance. The numerous branches extending from the spherical cell body are dendrites, which transmit information from neighbouring nerve cells. The cell is known as a Purkinje cell, after its discoverer, a nineteenth-century Czech physiologist.

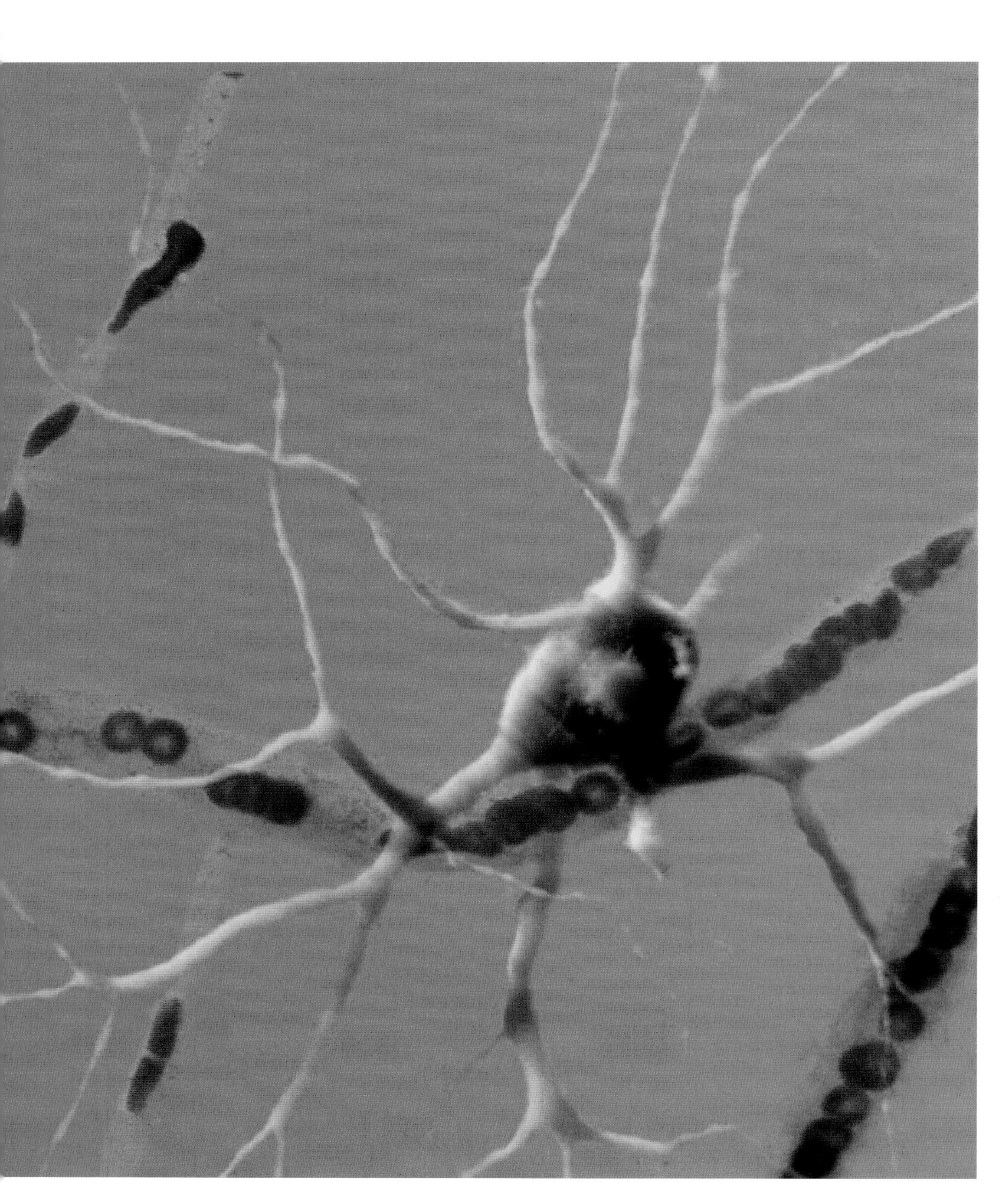

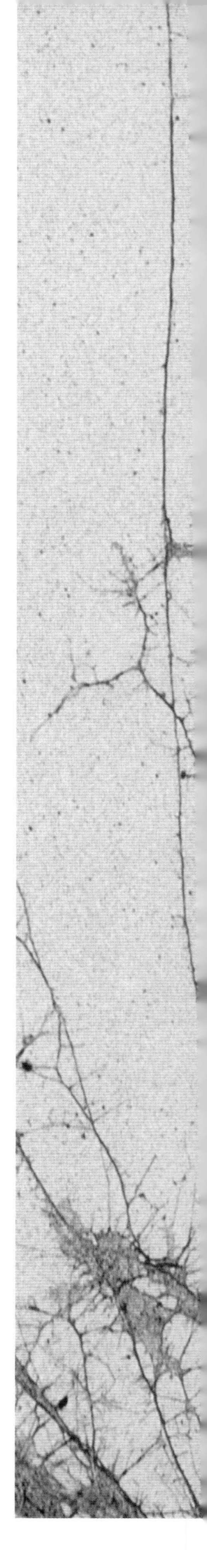

Nerve cell and red blood cells [SEM]

SHOWN IN GREEN, A NERVE CELL from the eye has long branches called dendrites, which transmit impulses from other nerve cells to the cell body. The pink tubes are capillaries, the smallest of the blood vessels, only just wider than the red blood cells that carry oxygen to the body tissues.

Nerve cell culture [SEM]

RESEARCHERS USE NERVE CELL CULTURE to investigate methods of regenerating nervous tissue as a cure for spinal paralysis. This cultured sample from a spinal cord shows the numerous branching strands, or neurites, growing out of a nerve cell body (not seen). These will connect with other nerve cells, forming a network of nervous tissue.

Nerve cells from small intestine [LM]

MOVEMENT OF THE WALLS of the intestine is controlled by a network of nerve cells (dark brown), connected by long projections (narrow brown lines). When the intestinal wall is stretched by its contents, the nerve cells cause the muscles above the dilation to contract, and those below it to relax, resulting in a wave-like movement that propels the contents of the intestine down the gut.

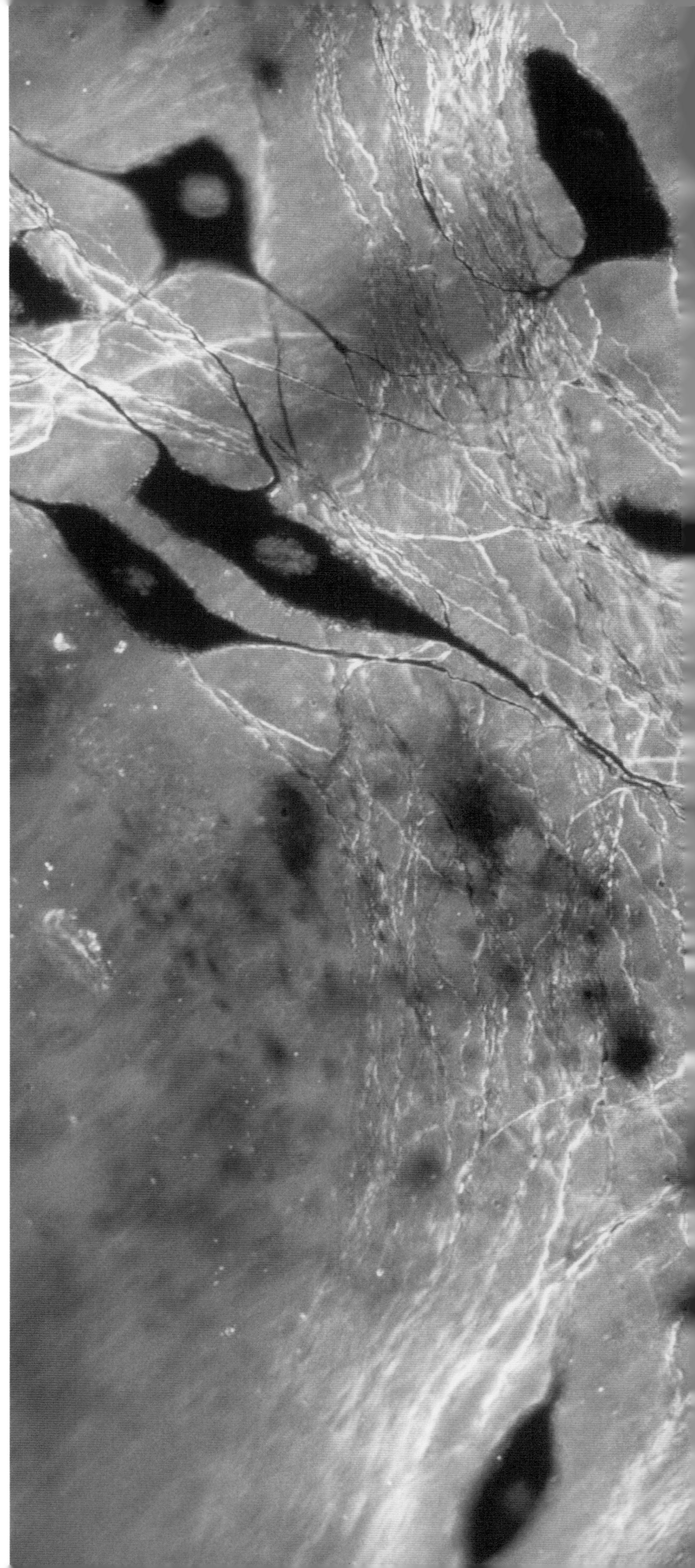

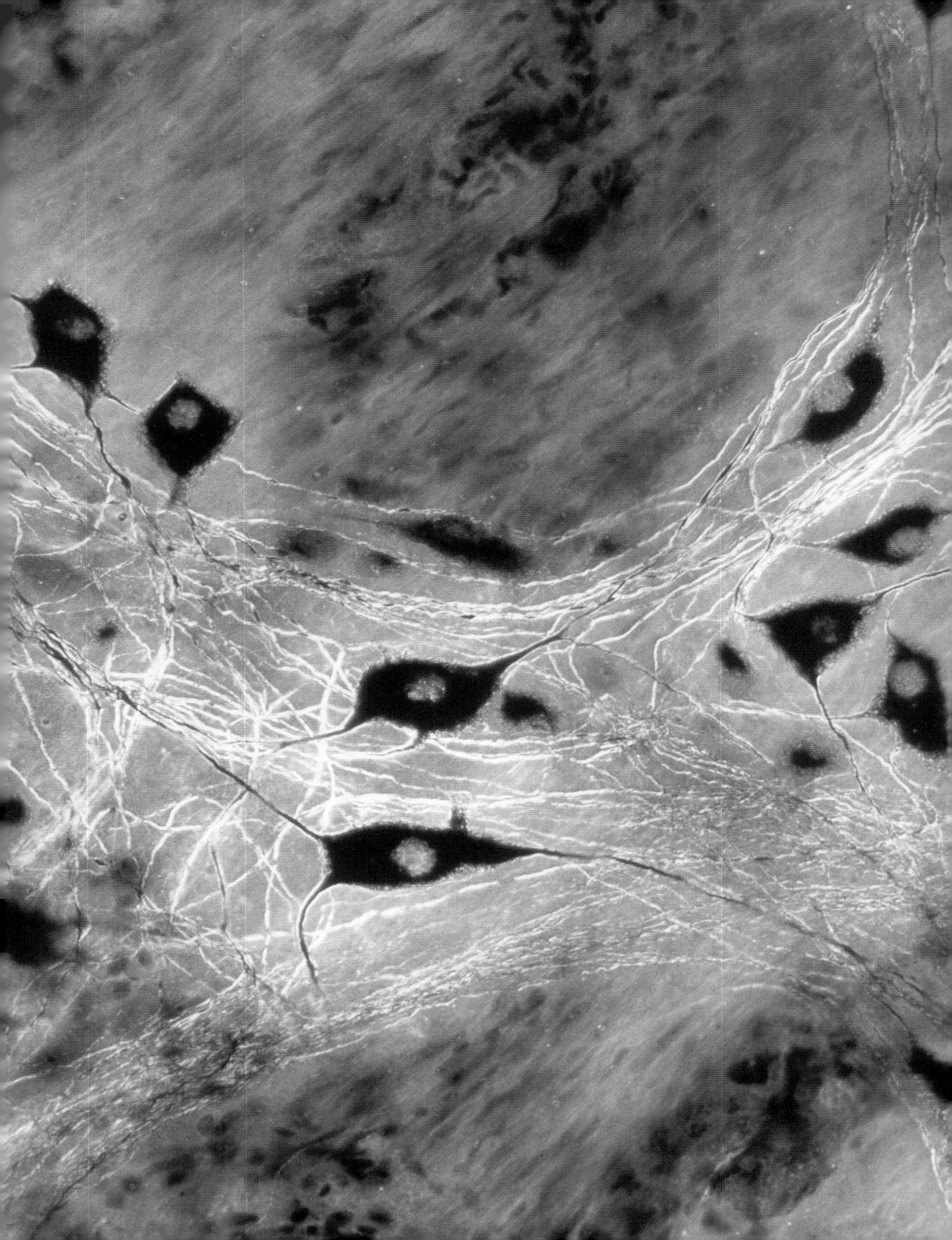

Nerve cells from spinal cord [LM]

THE LONG STRANDS EXTENDING from the bodies of these two nerve cells are dendrites, which receive information from other nerve cells. Each nerve cell is linked to its neighbours by up to 50,000 dendrites. The information they carry is processed in the nerve cell body, and passed on through output nerve fibres called axons.

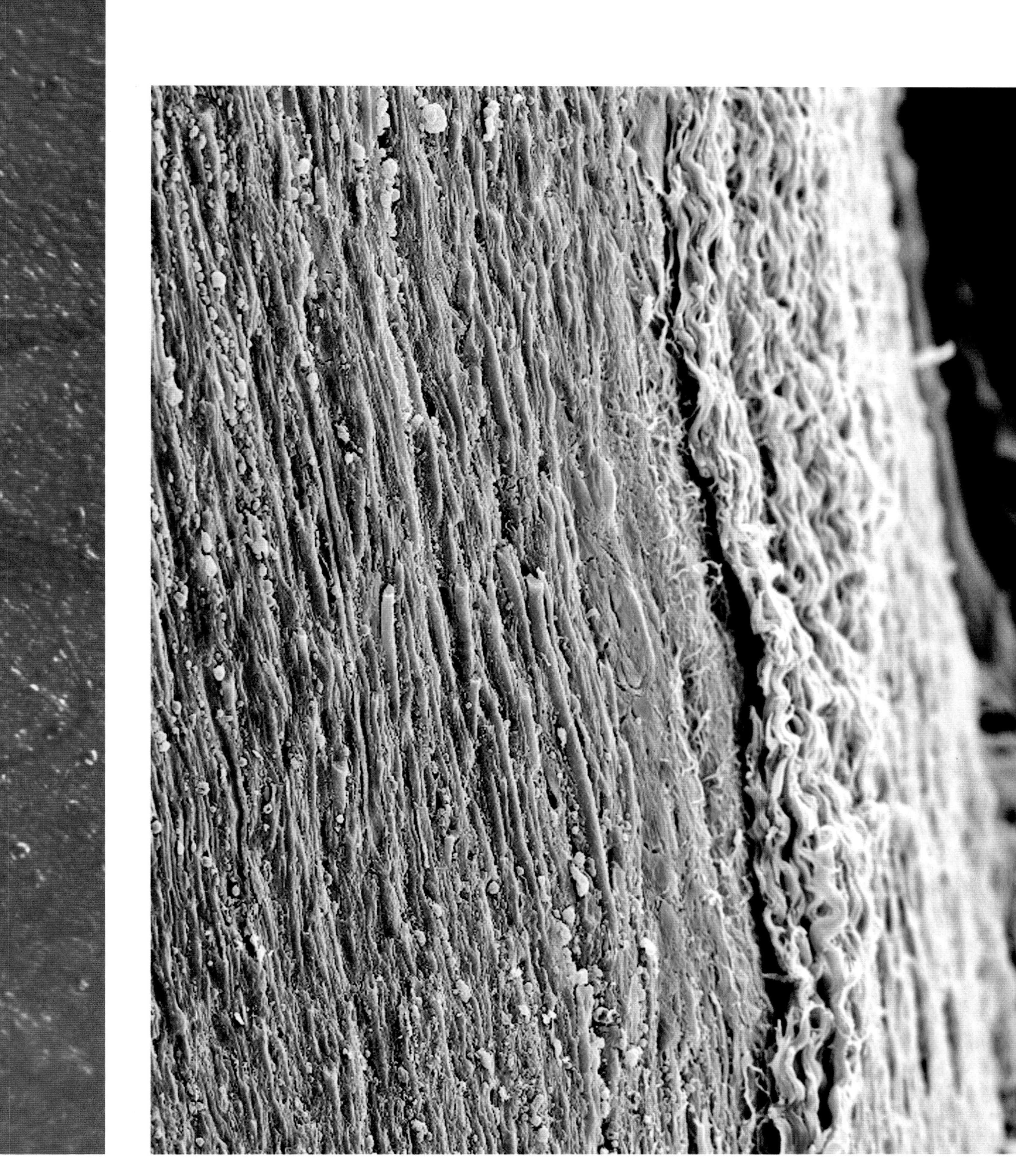

Spinal cord white matter [SEM]

KNOWN AS WHITE MATTER, the outer tract of the spinal cord consists of parallel nerve fibres each sheathed in a layer called myelin. This insulates the nerve fibres, rather like the plastic sheath around an electrical cable. Nerve fibres sheathed in myelin can transmit signals up to forty times faster than nerve fibres without myelin sheaths.

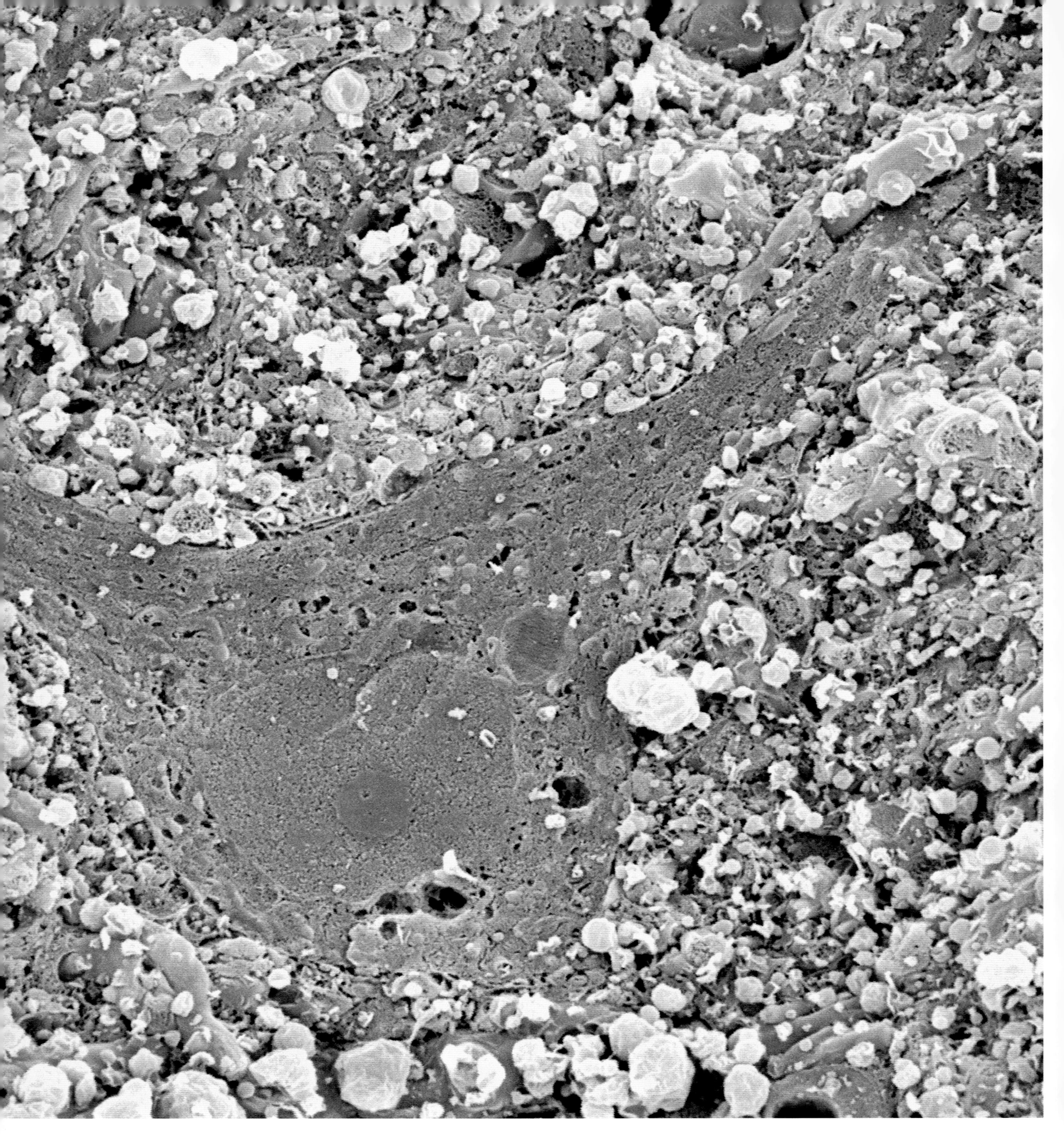

Nerve cell from spinal cord [SEM]

THIS IMAGE OF A NERVE CELL from the spinal cord shows the cell nucleus (dark grey), the 'control centre' containing the cell's genetic material. Unlike most other nucleated cells, nerve cells do not usually reproduce once humans reach adulthood. Although there is a steady decline in number, humans have so many nerve cells – 100 billion in the brain alone – that the loss of the odd million or so has no apparent effect.

Nerve fibres [SEM]

SEEN IN CROSS SECTION, a bundle of nerve fibres consists of nerve cell axons (small black cores), the output processes of nerve cells, surrounded by insulating myelin sheaths (pink). Myelin sheaths increase the transmission speed of electrical nerve signals. The disease multiple sclerosis occurs when areas of myelin degenerate, disrupting the transmission of nerve signals.

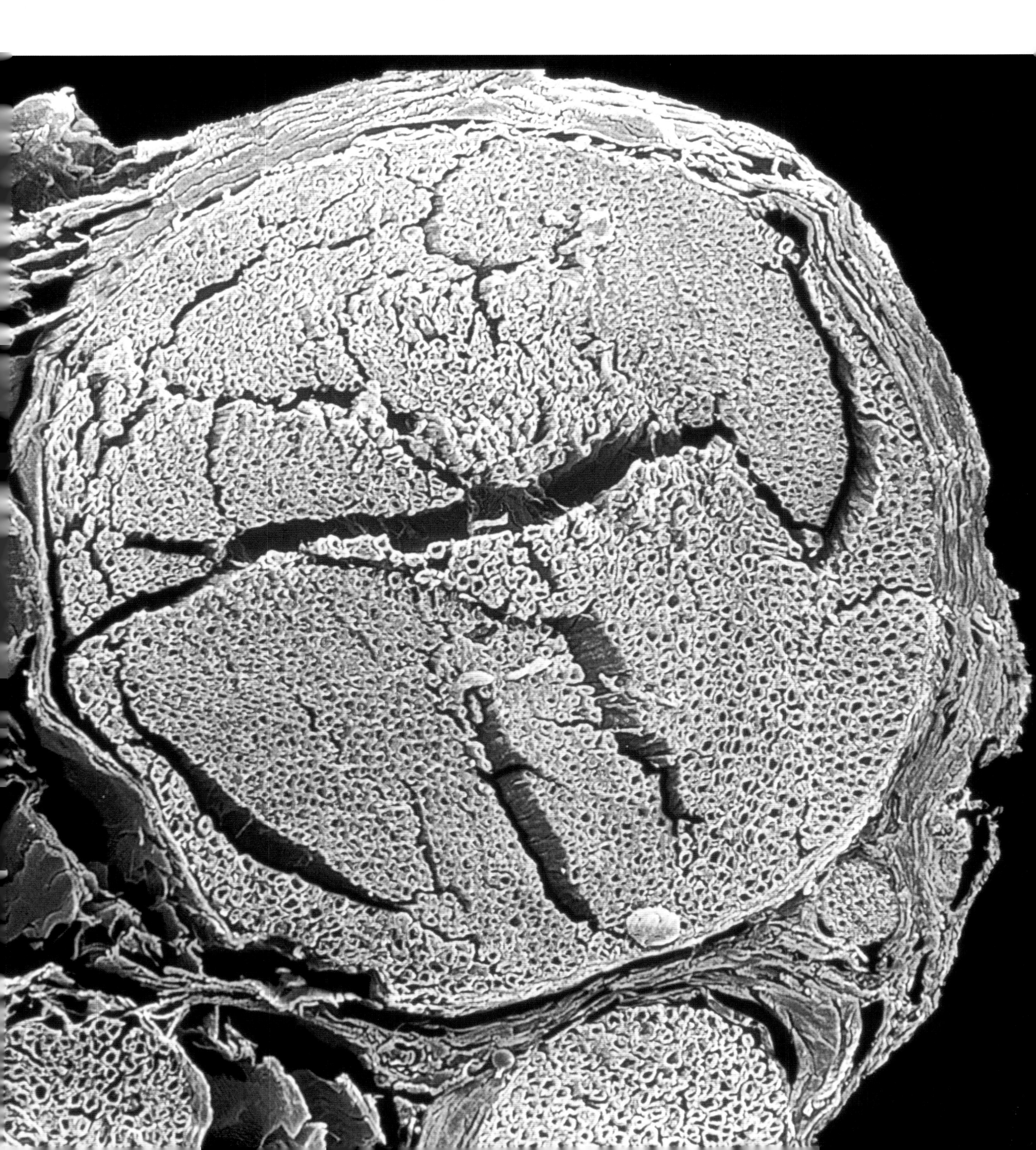

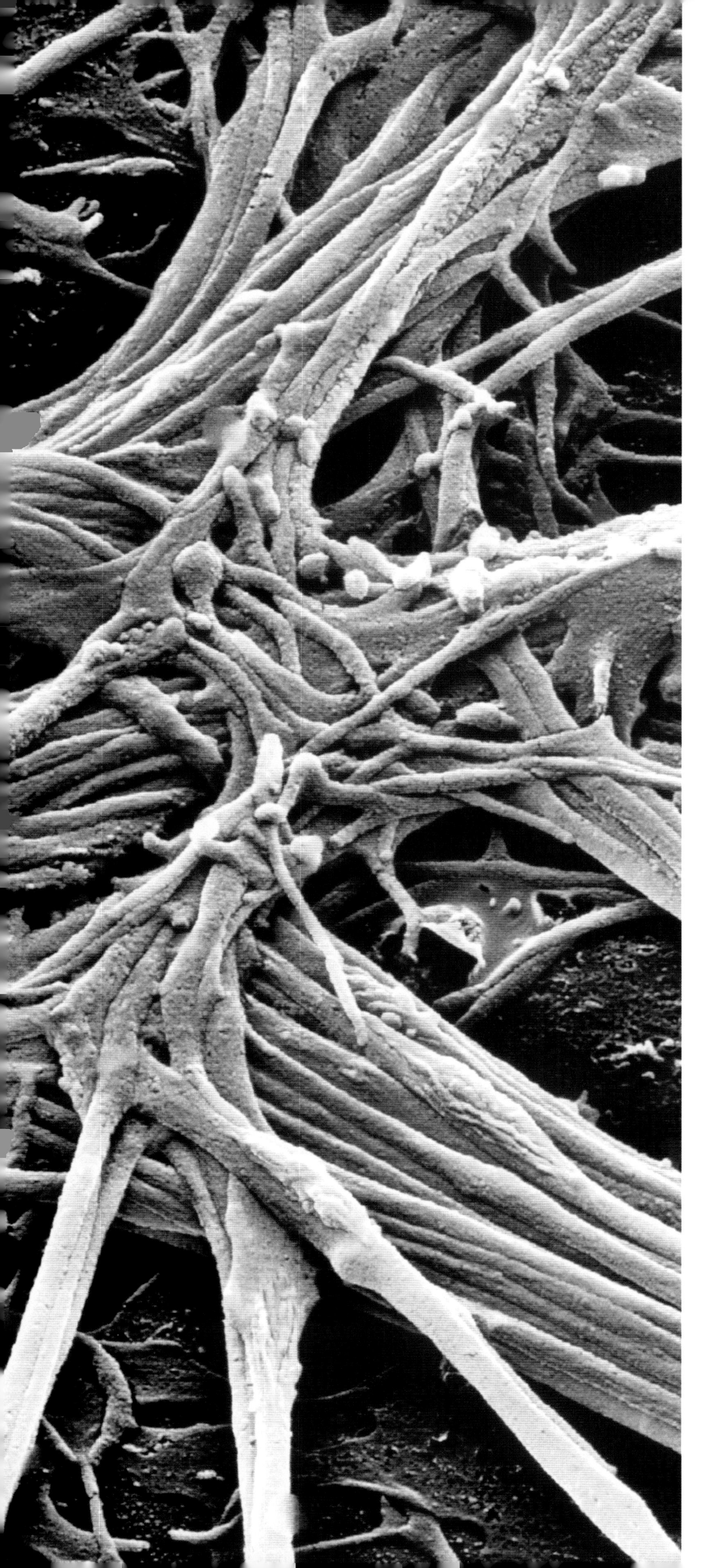

Overlapping nerve fibres [SEM]

EACH FIBRE IS MADE up of several axons – long extensions that carry signals from one nerve cell body to another, or to an effector cell such as a muscle or gland. Some axons stretch the length of the spinal cord, making them the longest cells in the body.

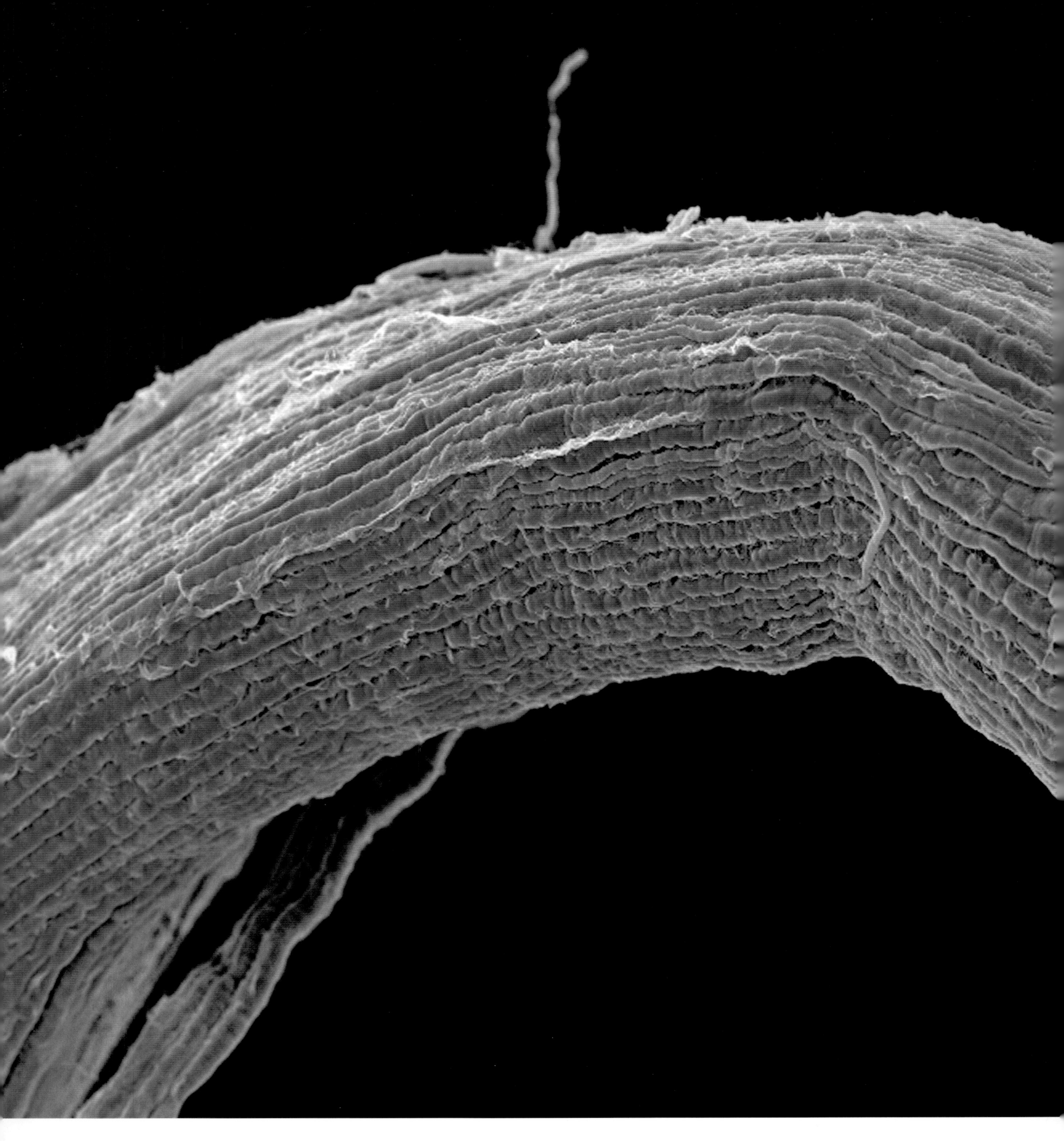

Nerve fibre bundle [SEM]

A GROUP OF NERVE FIBRES is bundled together in a structure called a fasciculus. The bundle is held together by collagen, a protein that gives strength to connective tissue. Each nerve fibre has a central strand (see right), the axon, which carries nerve signals and connects nerve cells together, allowing the transmission of electrical impulses around the body. These fibres are covered with a fatty insulating sheath of myelin, which increases the speed of nerve transmission.

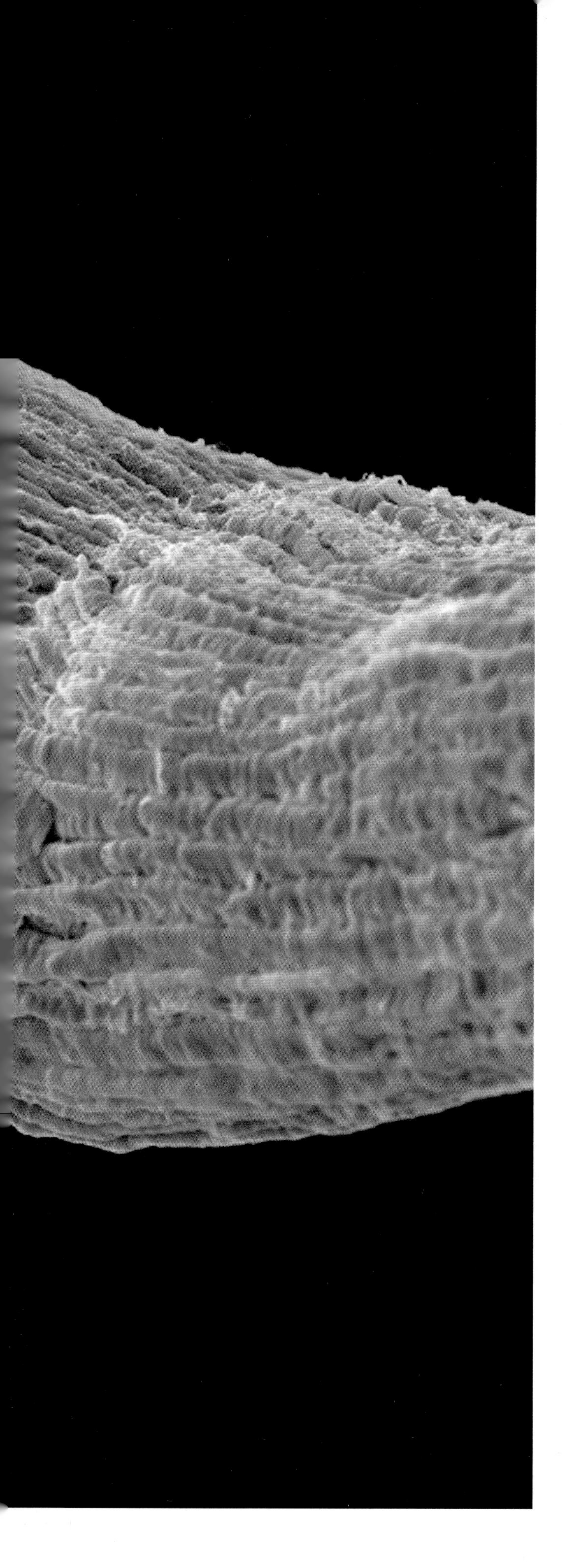

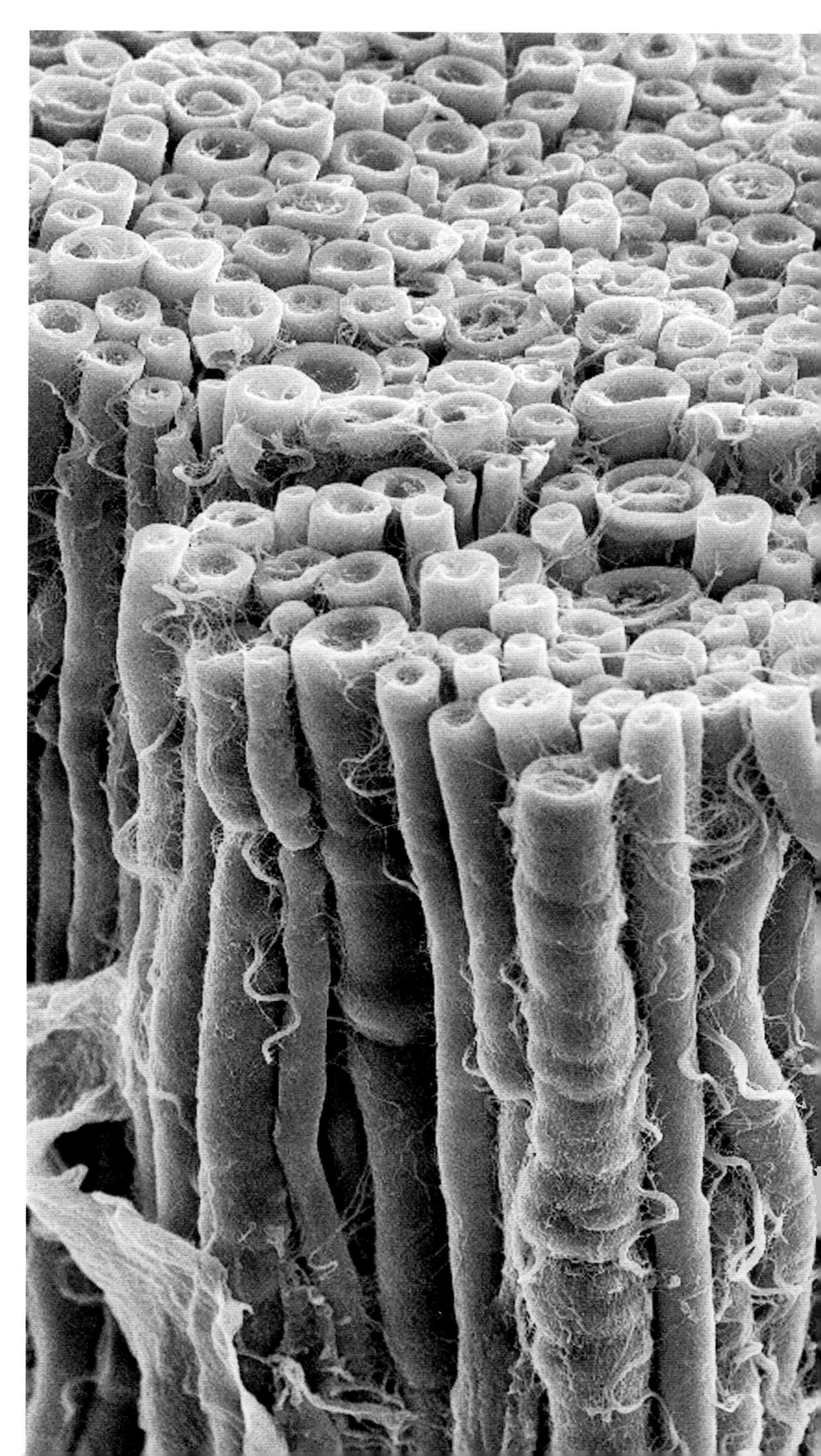

Our bodies are under constant siege by potentially harmful organisms such as bacteria, viruses and fungi. Our skin keeps most of these pathogens at bay, but if they manage to breach this barrier, they find themselves opposed by the many and varied defences that make up the body's immune system. In essence, the immune system works on the principle of 'us' against 'them'; any protein that the body does not recognize as being part of itself is treated as an enemy.

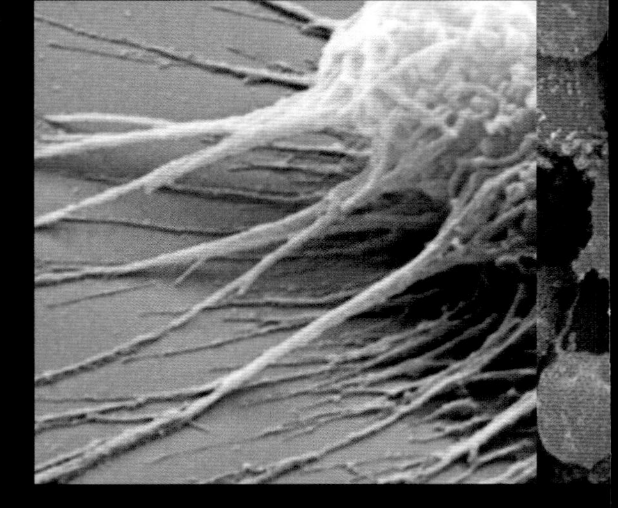

The main defences against invading pathogens are white blood cells produced in the bone marrow. Some types of white blood cell, called phagocytes, engulf and destroy pathogens. Some phagocytes roam through the body and can slip through the walls of blood vessels into infected tissue. Others are based in the lymph nodes – tissues in the body that act as screening centres for pathogens.

Mast cells, another type of white blood cell, release the chemical histamine, which attracts phagocytes and causes capillaries to dilate, increasing blood supply to infected tissue. This produces the familiar symptoms of inflammation: redness, swelling and itching or pain. Inflammation associated with allergies is caused by mast cells producing histamine in response to normally harmless substances such as dust or pollen. Hay-fever sufferers and other victims of mild allergies can usually find relief by taking antihistamine medication.

If the pathogens have not been wiped out by the phagocytes, they are targeted by another group of white blood cells, the lymphocytes. Found mainly in the lymph nodes, lymphocytes are stimulated

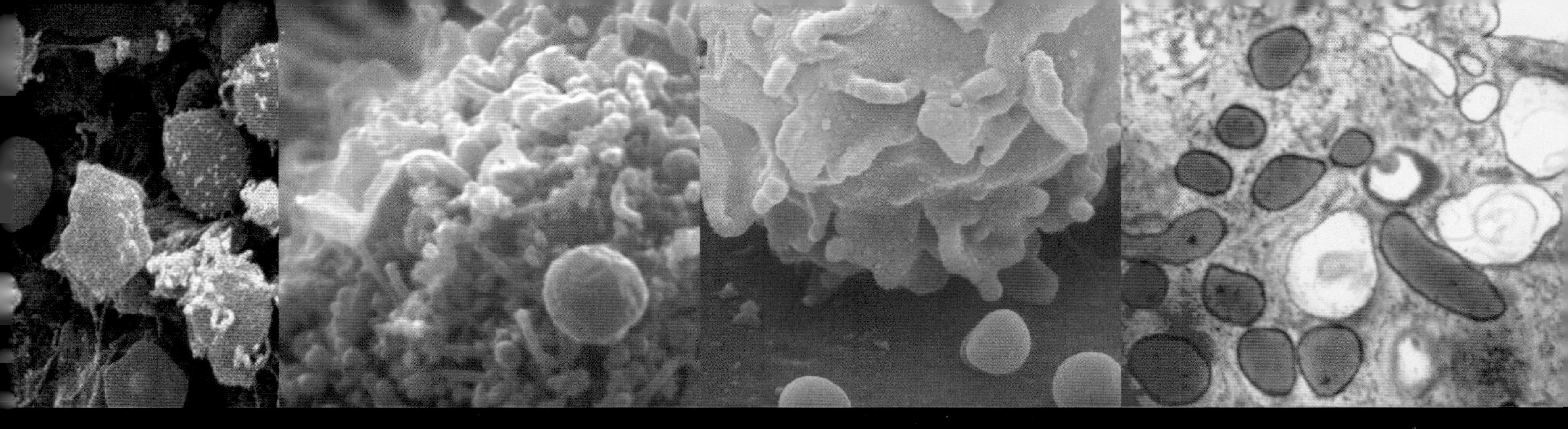

into action by foreign molecules (antigens) on the pathogen. One type of lymphocyte attacks antigens directly; the other type releases antibodies into the blood and kills pathogens at a distance.

After being exposed to a particular antigen, the immune system 'remembers' it. The next time it encounters that antigen the immune response is more rapid and effective. Immunity acquired in this way can last for years. However, some infectious diseases such as polio, tetanus and diphtheria are so virulent that they can overwhelm the natural immune response and prove fatal on first exposure. Protection against such diseases may be provided by a vaccine, a less virulent form of the pathogen, which is introduced into the body to induce the specific antibody reaction that produces immunity against that particular disease.

Bone marrow tissue [SEM]

FOUND MAINLY IN THE SKELETON'S flat bones and in the spinal column, bone marrow produces all the blood cells, including the white blood cells that are an essential part of the body's immune system. There are several types of white blood cell, each with a specific function. Some are attracted to bacteria and engulf and destroy them, while others produce antibodies that render foreign proteins (antigens) harmless.

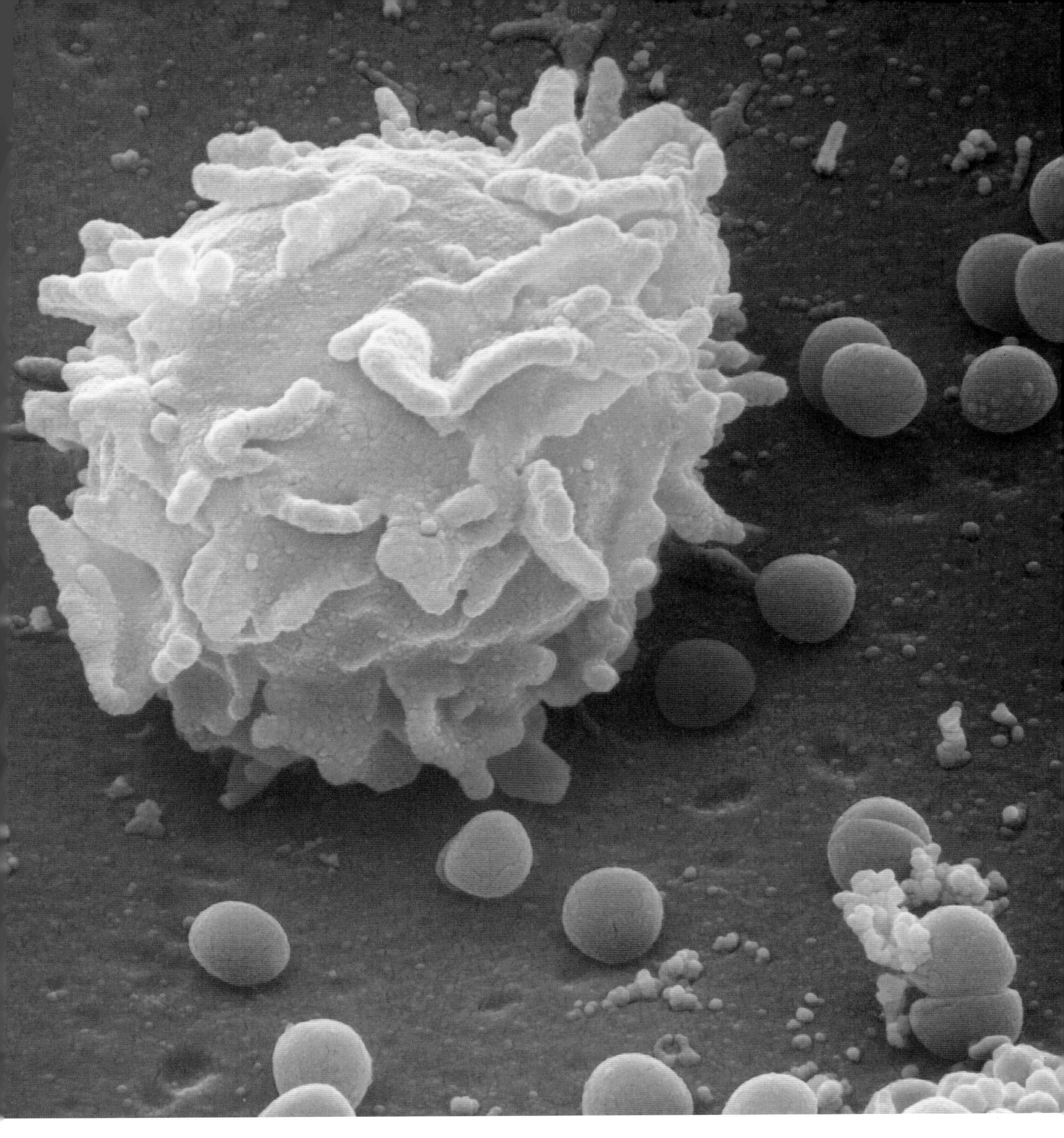

White blood cell [SEM]

A WHITE BLOOD CELL, or leucocyte, is shown surrounded by bacteria that can cause boils and food poisoning. This type of white blood cell occurs throughout body tissues, not just in the blood. It destroys pathogens such as bacteria by engulfing them and then digesting them with enzymes.

Dendritic immune cell [SEM]

LOOKING RATHER LIKE AN INSECT, this dendritic immune cell has long projections that help it to move to sites of infection. When dendritic cells engulf foreign bodies, they may then display them to other cells as a warning of the infection.

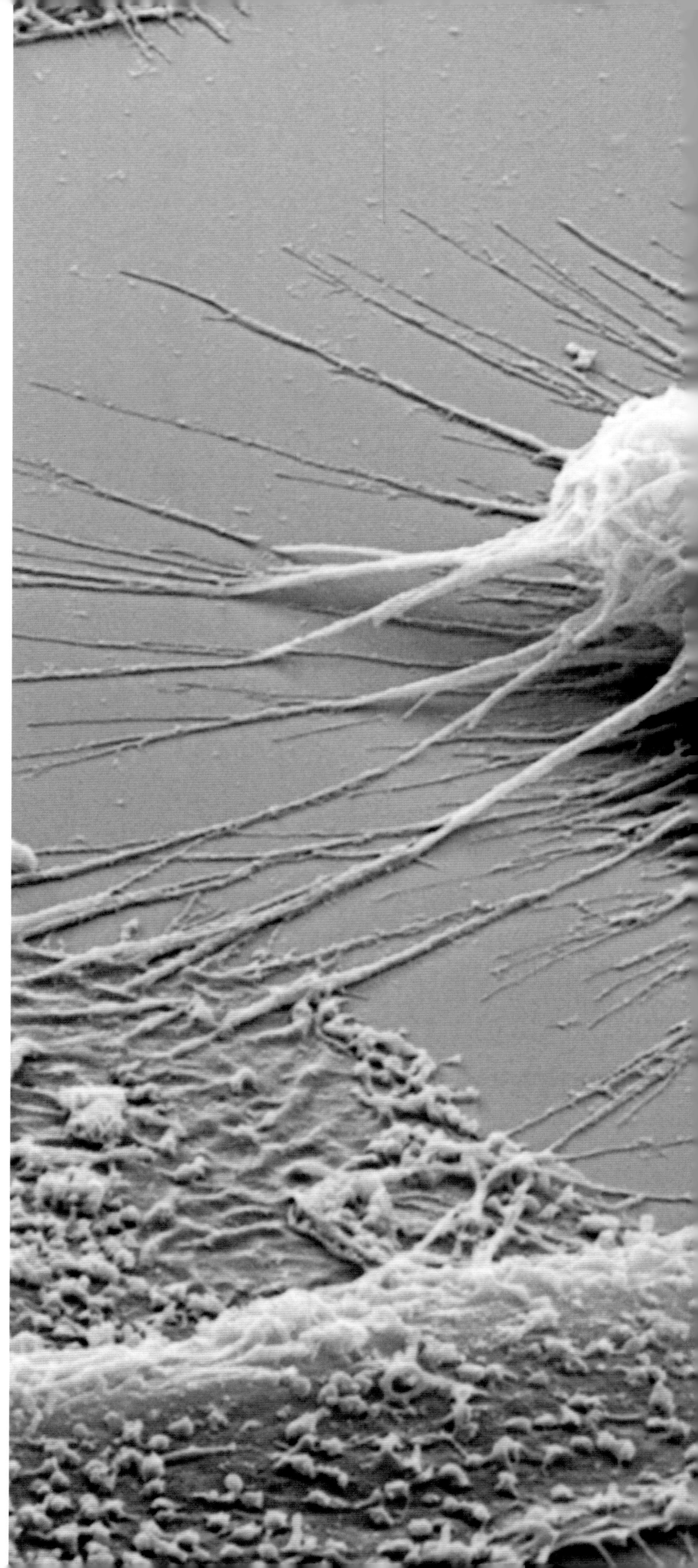

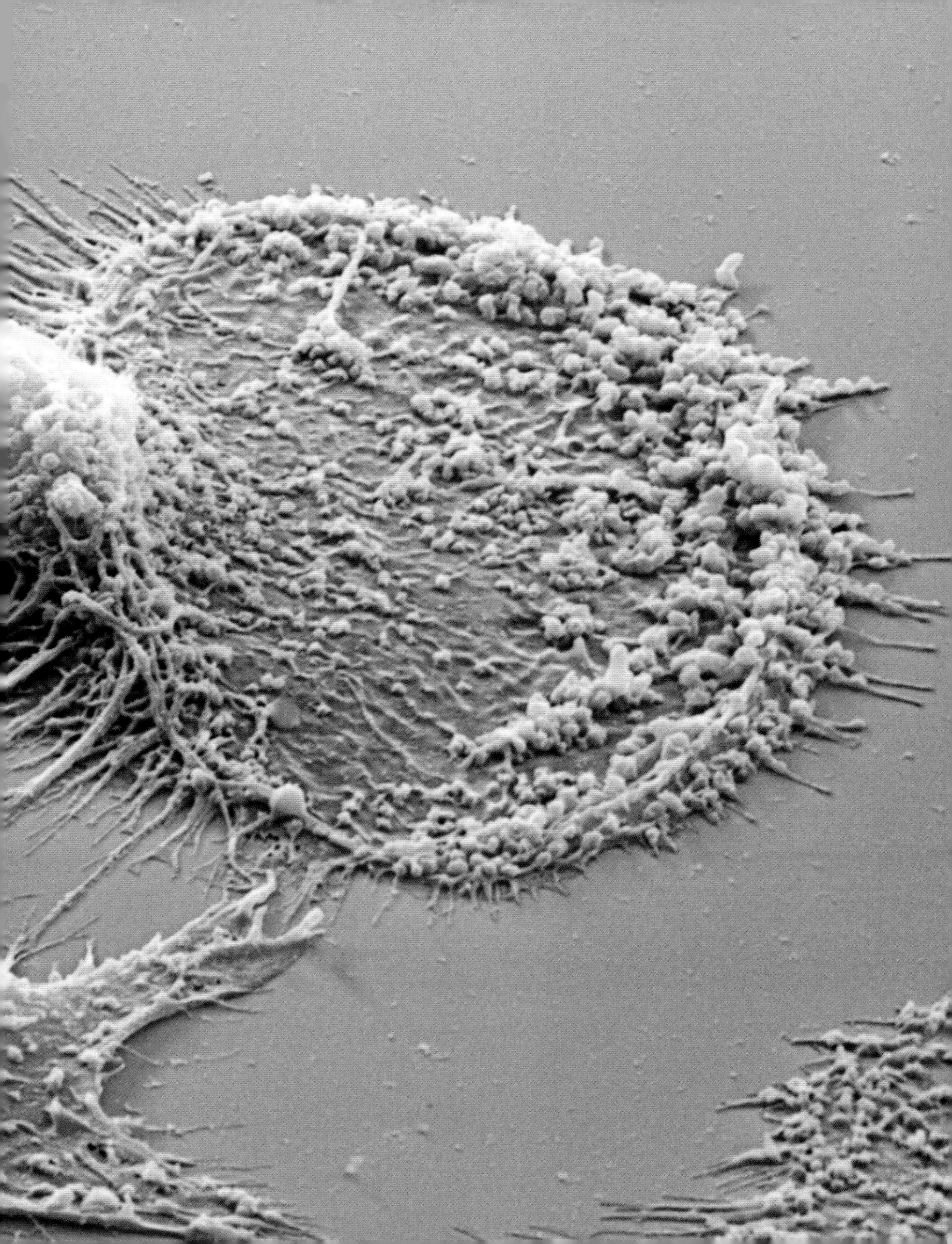

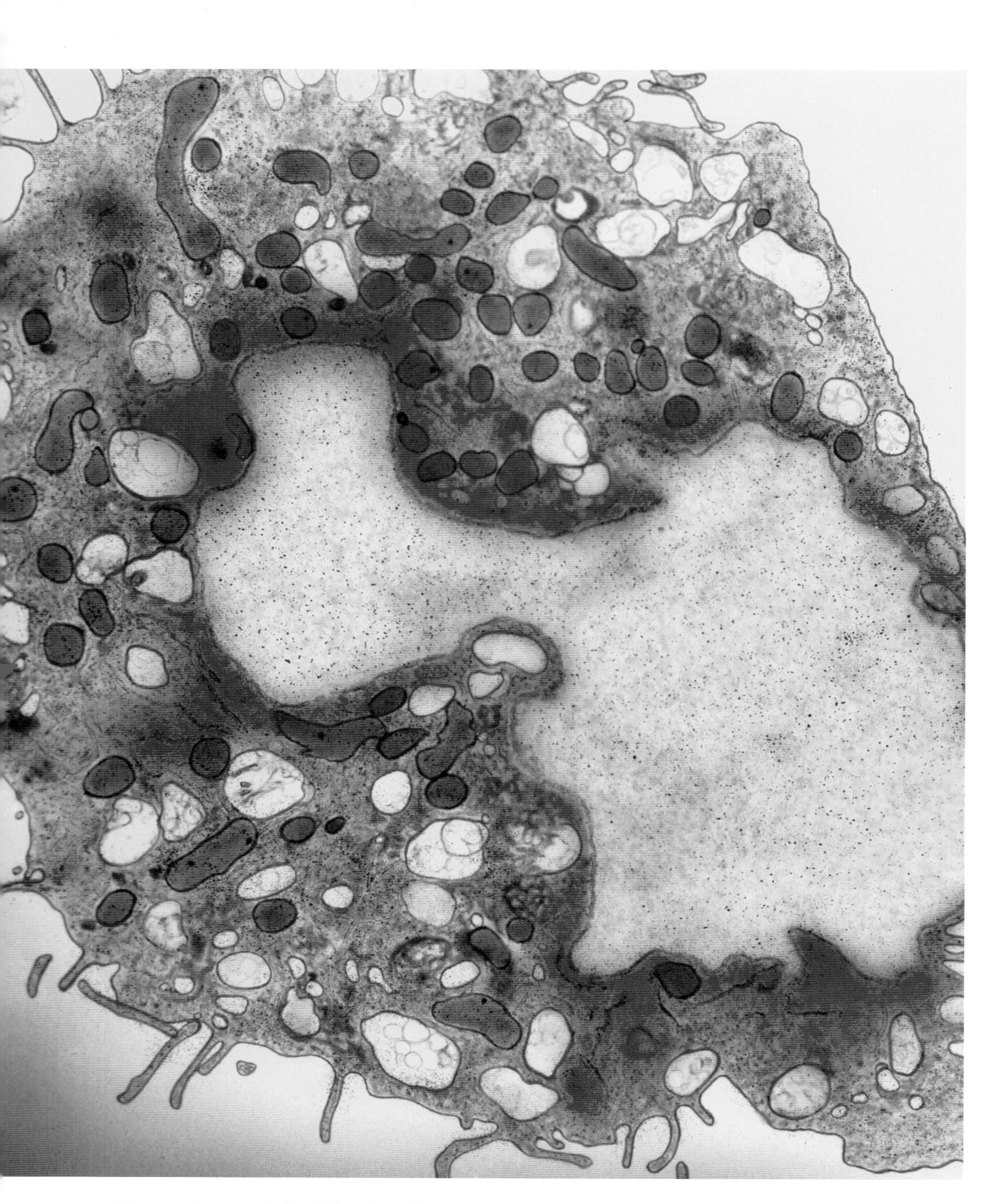

Macrophage white blood cell [TEM]

A SECTION THROUGH A MACROPHAGE white blood cell shows its large nucleus (green). The oval, yellow objects in the cytoplasm around the nucleus are lyosomes, small vesicles that discharge the enzymes that digest organisms engulfed by the cell. A macrophage can 'eat' up to twenty-five bacteria before the toxic products of breakdown kill the cell.

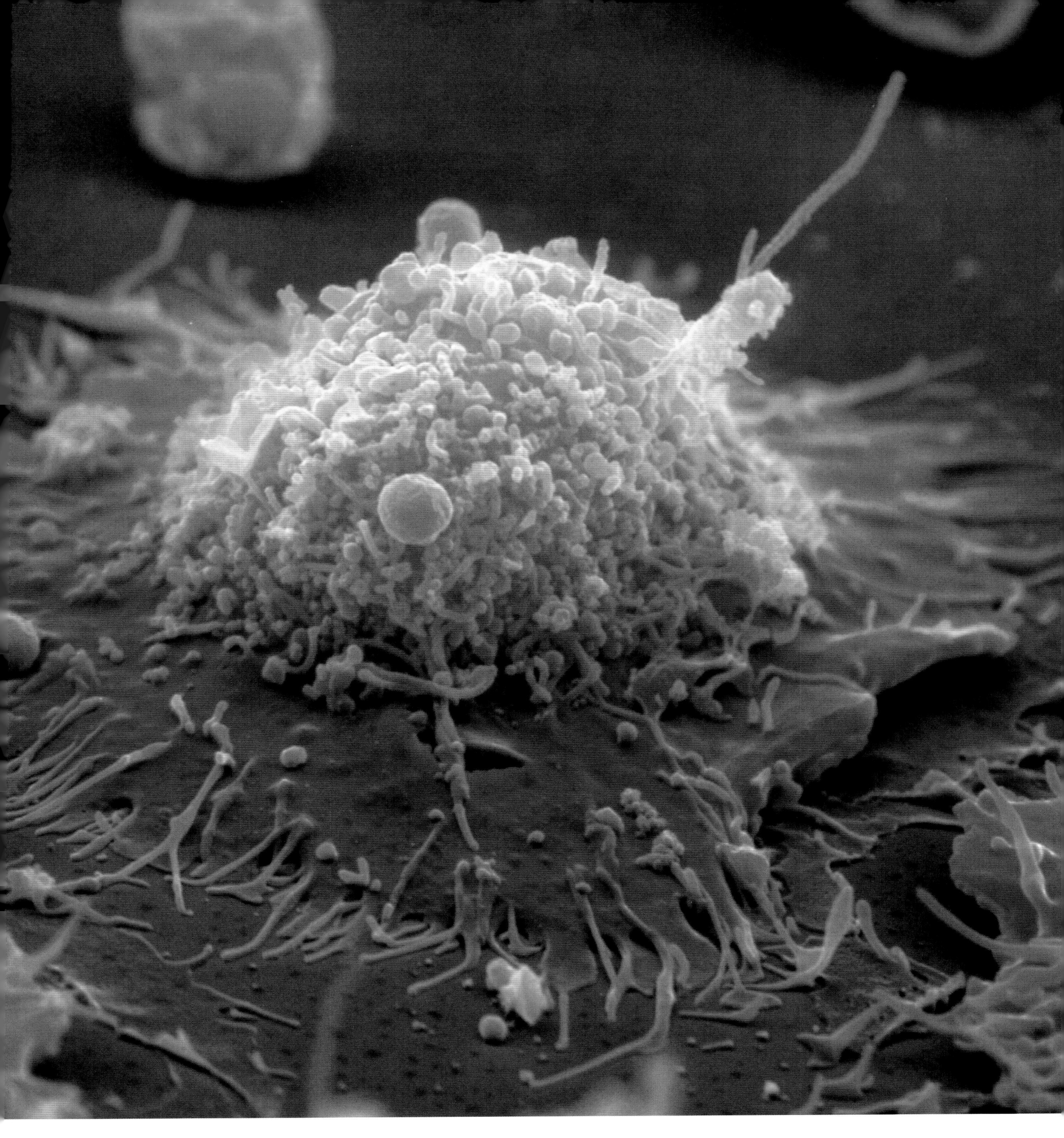

B-lymphocyte white blood cell [SEM]

WHEN A B-LYMPHOCYTE ENCOUNTERS a pathogen for the first time, it divides rapidly to produce a large population of cells that produce an antibody which specifically targets the invading pathogen and renders it harmless. When the infection is over, most of the cells die, but some B-lymphocytes remain in the body for years. These memory cells 'remember' the pathogen, and if they encounter it again they respond by dividing and producing the specific antibody.

Macrophage engulfing bacteria [SEM]

A MACROPHAGE WHITE CELL (pale brown) is shown engulfing bacteria (blue). Once the macrophage has taken the bacteria inside its cytoplasm, it will digest it with enzymes. Macrophages accumulate at the sites of bacterial infection. In some cases toxins secreted by the bacteria kill the white blood cells, which liquefy and, together with dead bacteria, form pus.

Antibody molecule [Computer graphic]

ANTIBODIES ARE Y-SHAPED MOLECULES composed of a vertical fragment (blue) and two oblique branches (green and yellow). Although the body produces a huge range of antibodies with this basic shape, the tips of the branches of each type are differentiated to match precisely a specific antigen (red).

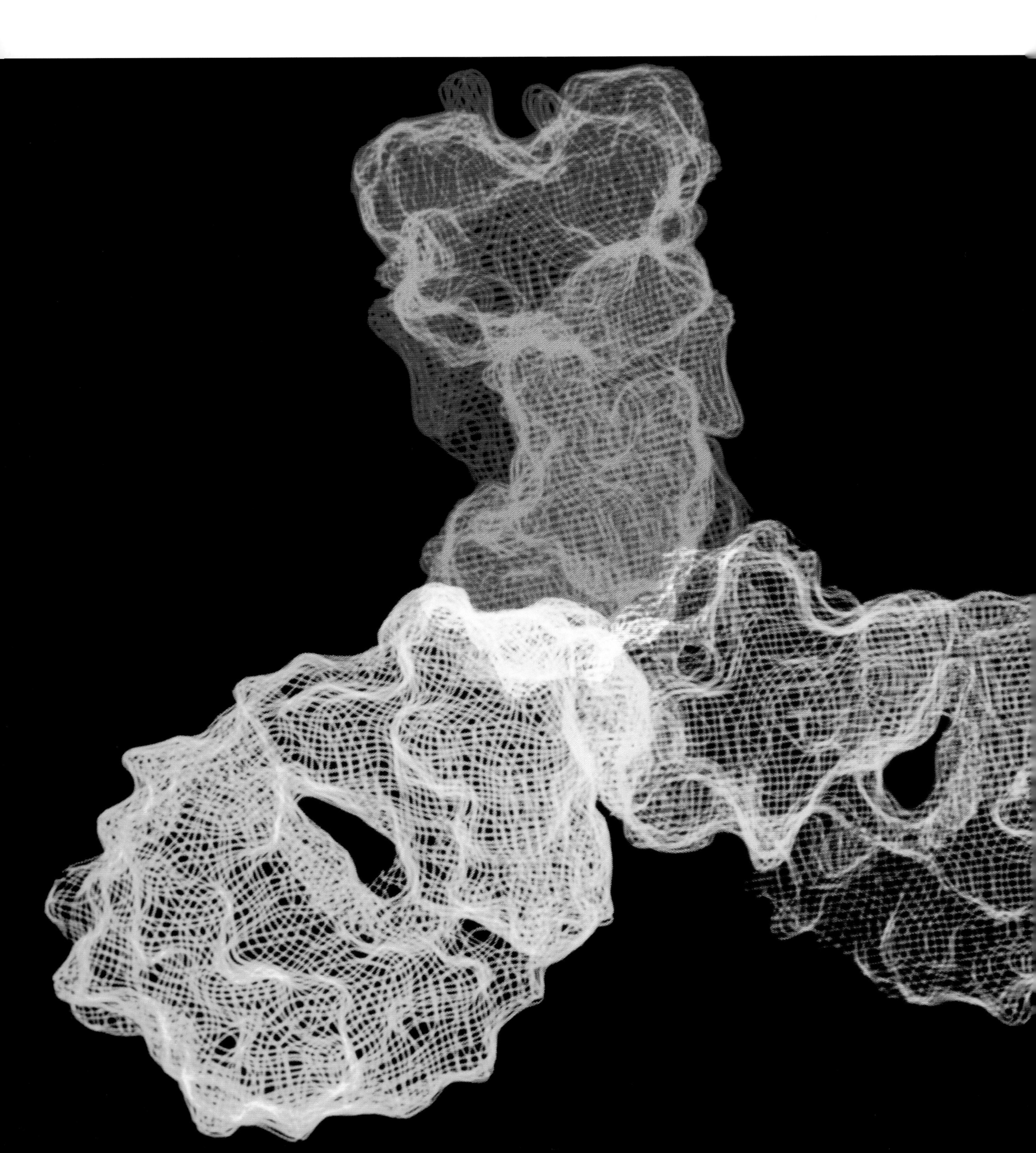

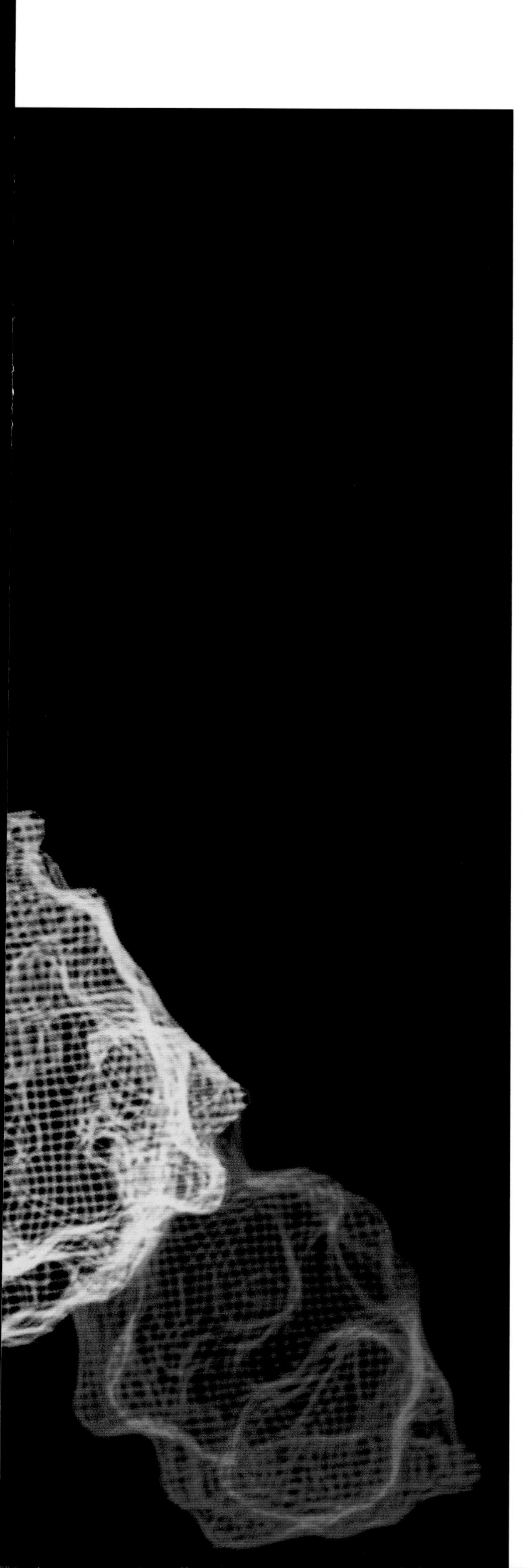

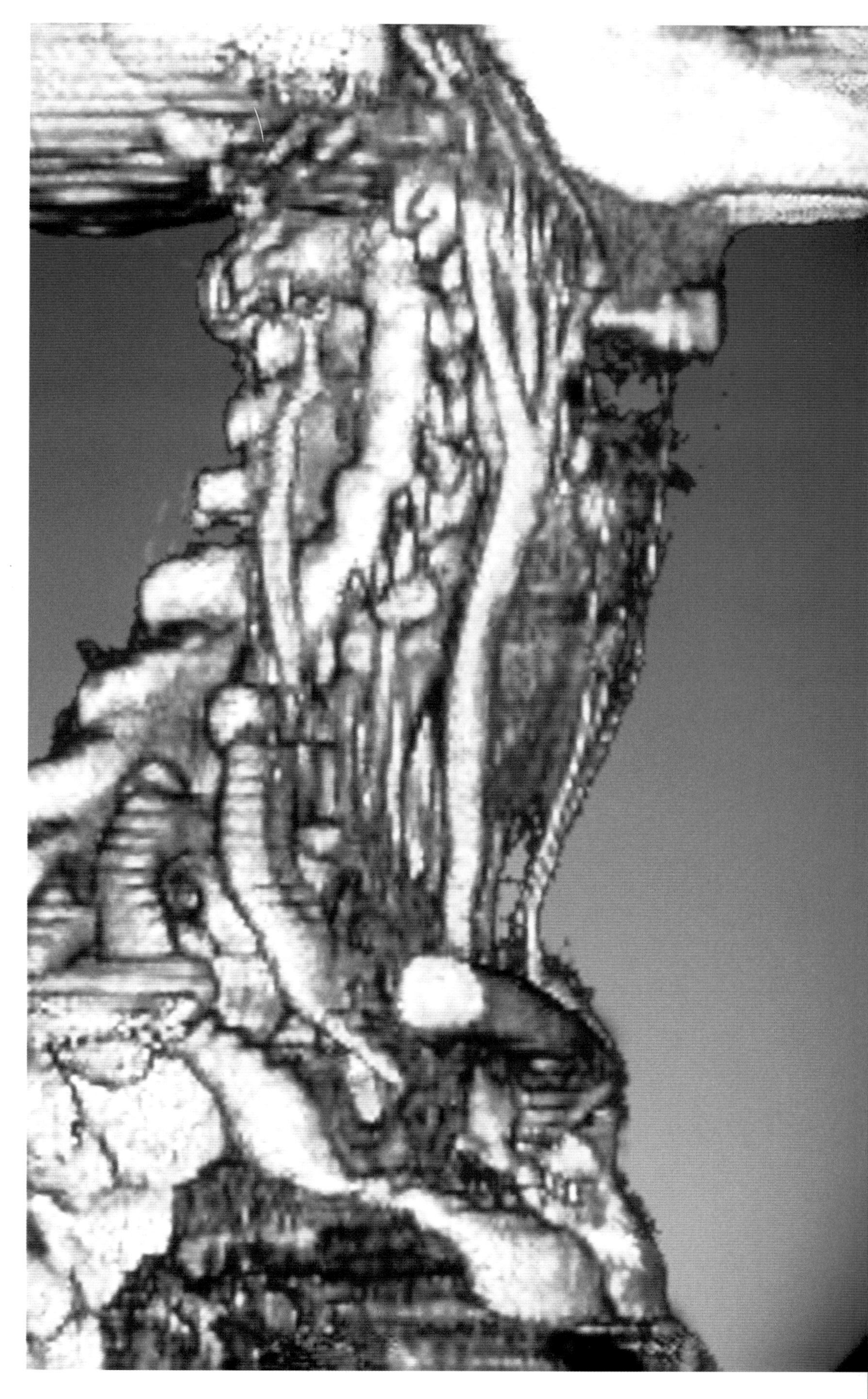

Lymph node in neck [CT]

THE ELONGATED RED OBJECT inside the front of the neck is one of the body's lymph nodes, which play an important part in immune defence. The nodes store and release the B- and T-lymphocytes responsible for detecting and destroying invading pathogens. They also filter out debris or bacteria for digestion by macrophagous cells.

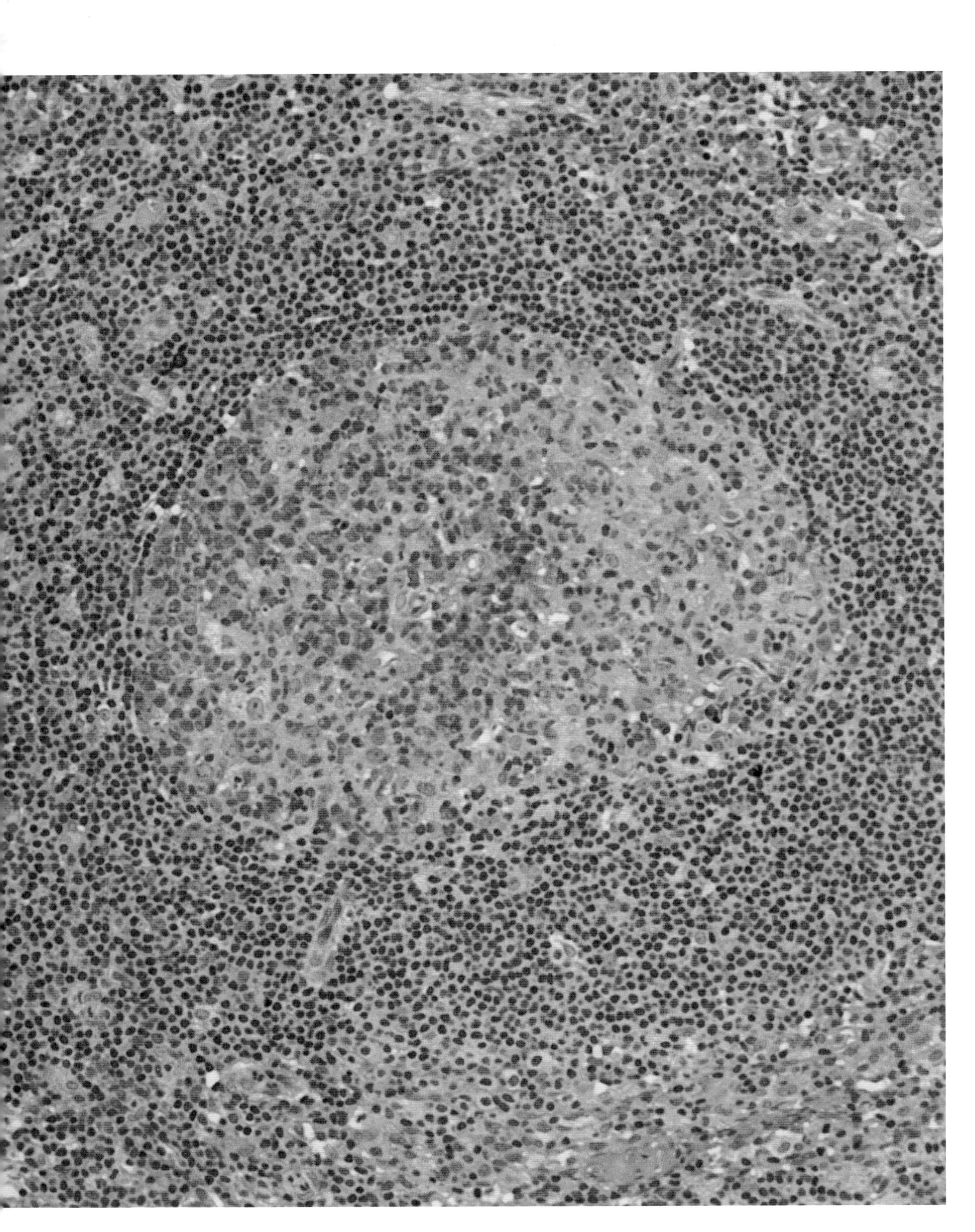

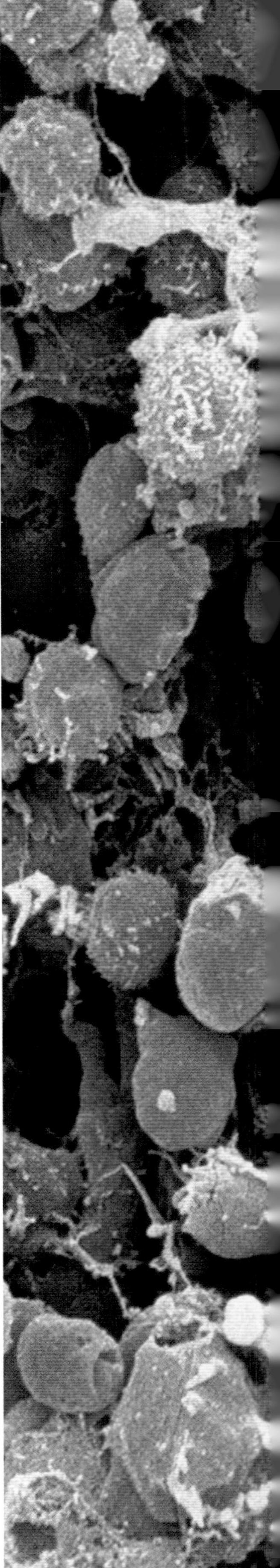

Lymphoid follicle in lymph node [LM]

THE FOLLICLE IN A LYMPH NODE is rich in B-lymphocytes, white blood cells that protect the body against disease organisms. Each B-cell responds to a different foreign molecule, or antigen. Specialized cells in the lymph node present antigens to the B-cells, and when a B-cell is exposed to its specific antigen it begins to divide in the germinal centre (pale pink, at centre).

Lymph node [SEM]

A SECTION THROUGH A LYMPH NODE shows some of the weapons of the immune system. The pink cells are lymphocytes, which either attack foreign bodies directly or destroy them with antibodies. The pale brown cells are macrophages, which engulf bacteria and debris, and then digest them with enzymes. The red cells are red blood ones, and the midbrown cells are reticular ones, which make up a mesh of connective tissue.

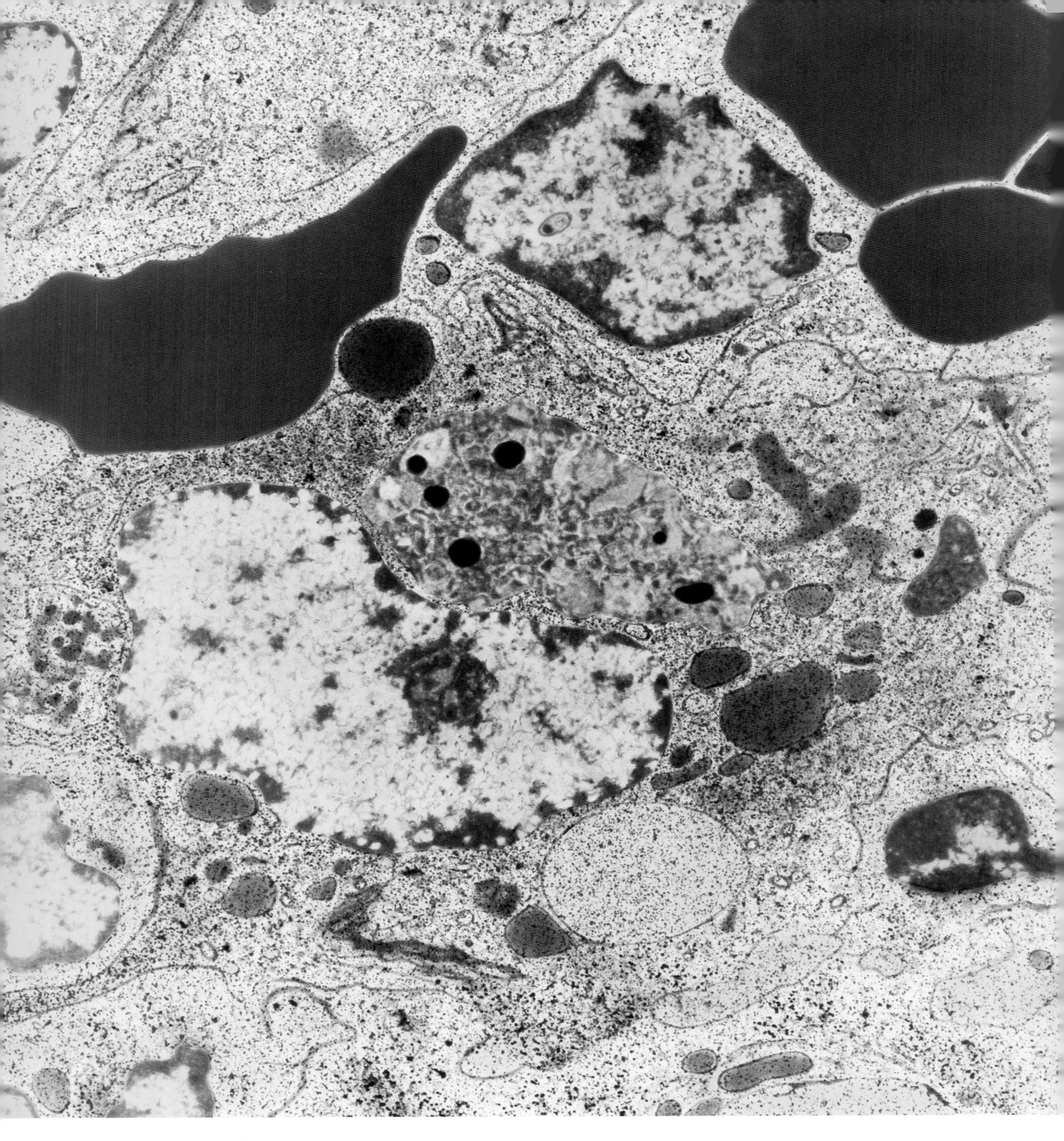

Macrophage in the spleen [TEM]

SEEN HERE SURROUNDED BY LYMPHOCYTES (green), the monocyte white blood cell at centre is recognizable by its large, kidney-shaped nucleus (blue). Monocytes are macrophages – rubbish-collecting cells that 'eat' foreign bodies. In the spleen, macrophages ingest blood-borne foreign bodies and aged red blood cells. Some healthy red blood cells can be seen at upper left and top right.

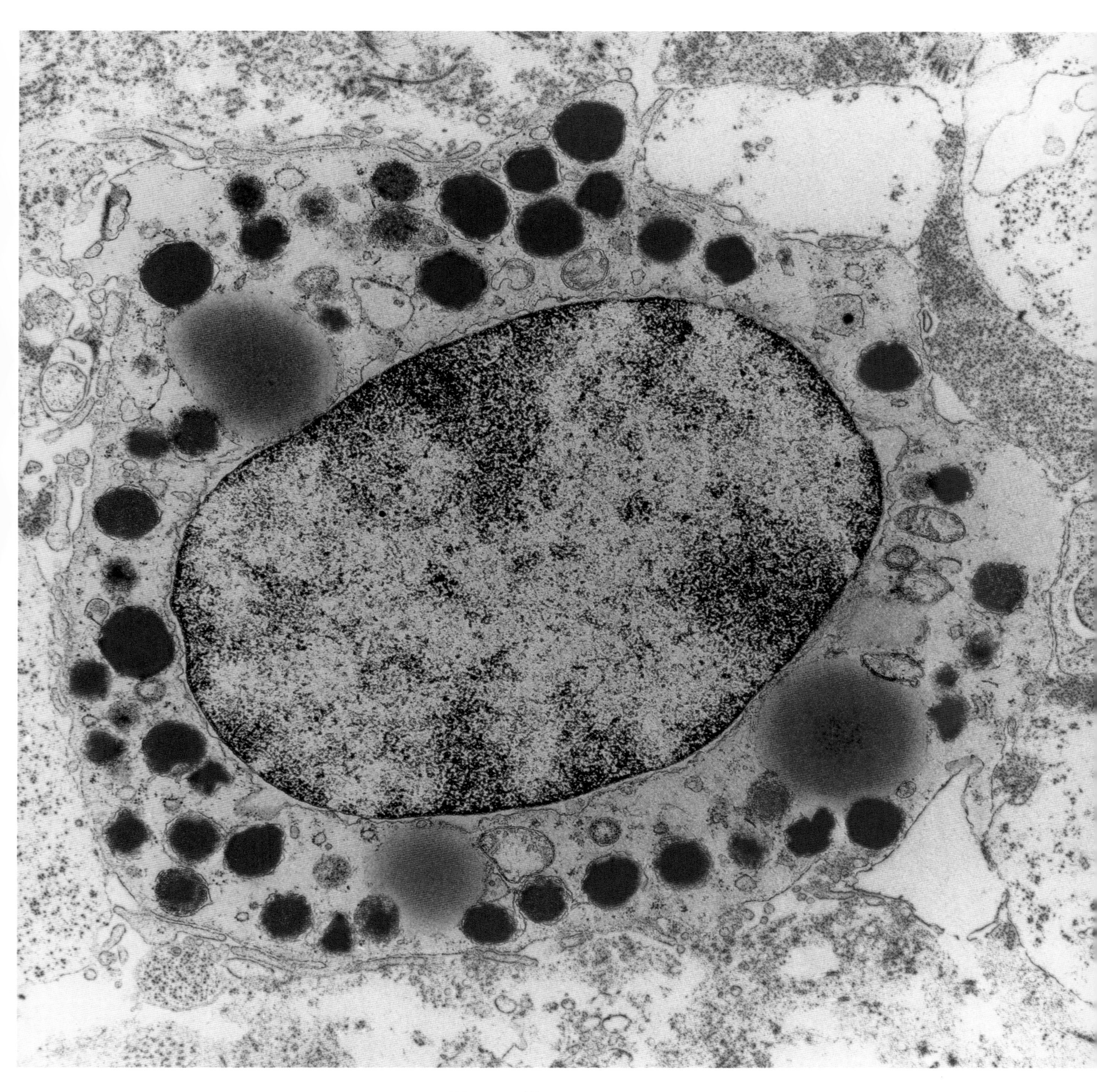

Mast cell [TEM]

THE PURPLE GRANULES in the mauve cytoplasm of this mast cell contain chemicals that are involved in inflammation or allergy. When tissue is damaged, mast cells release heparin, an anticoagulant, and histamine, which dilates blood vessels, increasing blood flow. It is the extra blood flow to the damaged tissue that causes the familiar symptoms of redness, heat, swelling and pain.

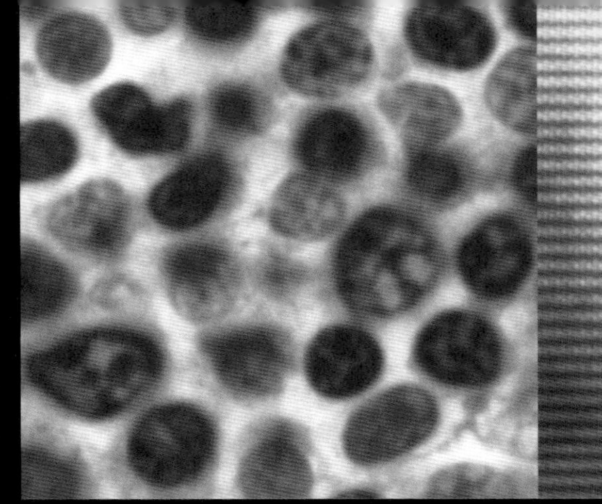

Most of our basic body functions are regulated with no conscious input from ourselves. Many are controlled automatically by the endocrine system, a series of glands that communicate with organs and tissues through chemical messengers called hormones. Working singly or in combination, hormones control and co-ordinate everything from our mood and metabolism to growth rate and sexual development.

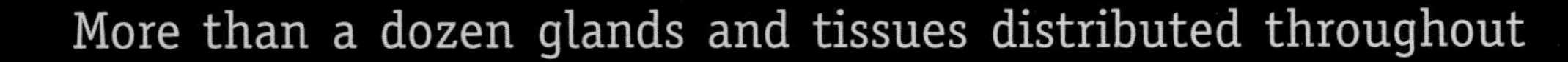

More than a dozen glands and tissues distributed throughout the upper body make up the endocrine system. Some parts, including the pituitary gland, thyroid gland and adrenal glands, are endocrine specialists whose only task is to produce one or more hormones. Other parts of the system secrete hormones in addition to their other functions. The ovaries and testes, besides forming eggs and sperm, produce the male and female hormones oestrogen and testosterone, which influence sexual form and function. The pancreas, as well as producing digestive juices, secretes the hormones insulin and glucagon, which regulate the level of energy-giving glucose in the bloodstream.

Endocrine glands secrete their hormones directly into the blood for transport to target organs whose cells are genetically programmed to respond to specific hormones. For example, the liver is a target organ for insulin secreted by the pancreas. When insulin locks on to its target cells, it transmits chemical instructions that stimulate the liver to decrease the rate at which glucose is released. When blood sugar falls to a certain level, the pancreas senses the change, stops producing insulin and secretes glucagon, a hormone that instructs the liver to release some of its stored supply of glucose. This balancing mechanism, known as 'negative feedback', can be compared to the way a thermostat switches a central heating system on and off to maintain a constant temperature. Some endocrine

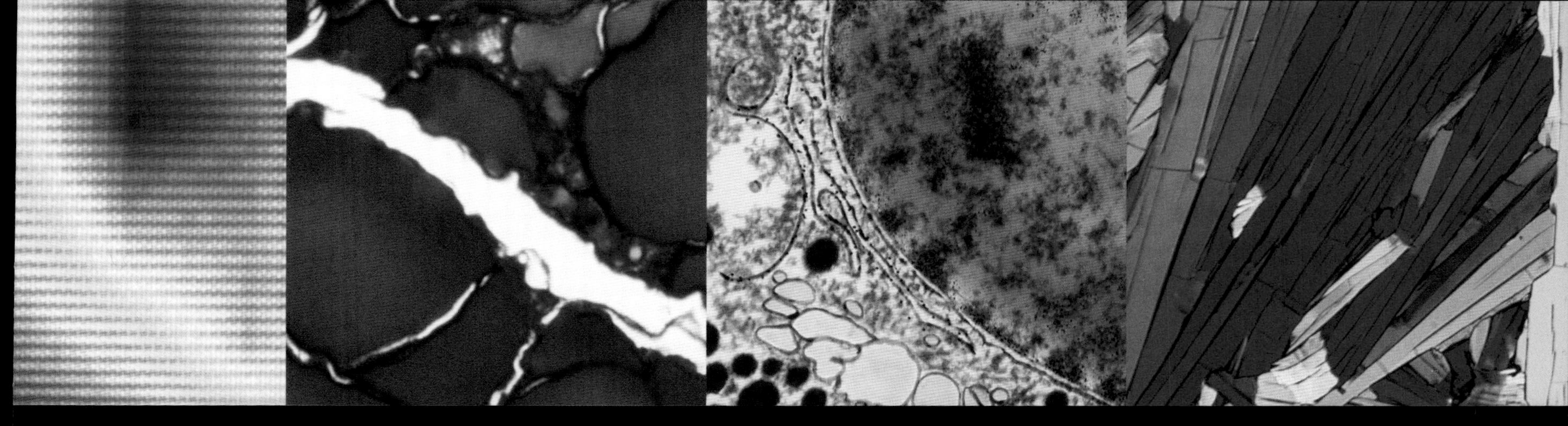

glands, such as the pancreas, are self-regulating, but most are orchestrated by the pituitary gland, the so-called 'master gland'.

Tiny amounts of hormone can produce rapid and dramatic effects. We have all experienced the surge of energy that flows through us when we are confronted by danger. No sooner do we realize the threat than our heartbeat and breathing rates speed up and our blood pressure rises, priming our bodies for action. This fight-or-flight reaction is caused by the secretion of adrenaline from our adrenal glands. One molecule of adrenaline can release millions of glucose molecules. Other hormonal changes – the onset of puberty, menstruation – are more profound but slower acting, taking place over weeks, months or even years.

Hormone levels can fluctuate, especially during periods of increased hormonal activity such as puberty and pregnancy. Stress, infection and diet also influence hormone levels. Serious malfunctions of the endocrine system can produce a variety of disorders. If the pancreas fails to produce enough insulin to regulate blood sugar, diabetes occurs. Too much growth hormone in childhood causes excessive growth; too little results in restricted growth. In most cases, treatment involving synthetic hormones or surgery is highly successful in restoring the body's chemical equilibrium.

Endocrine system

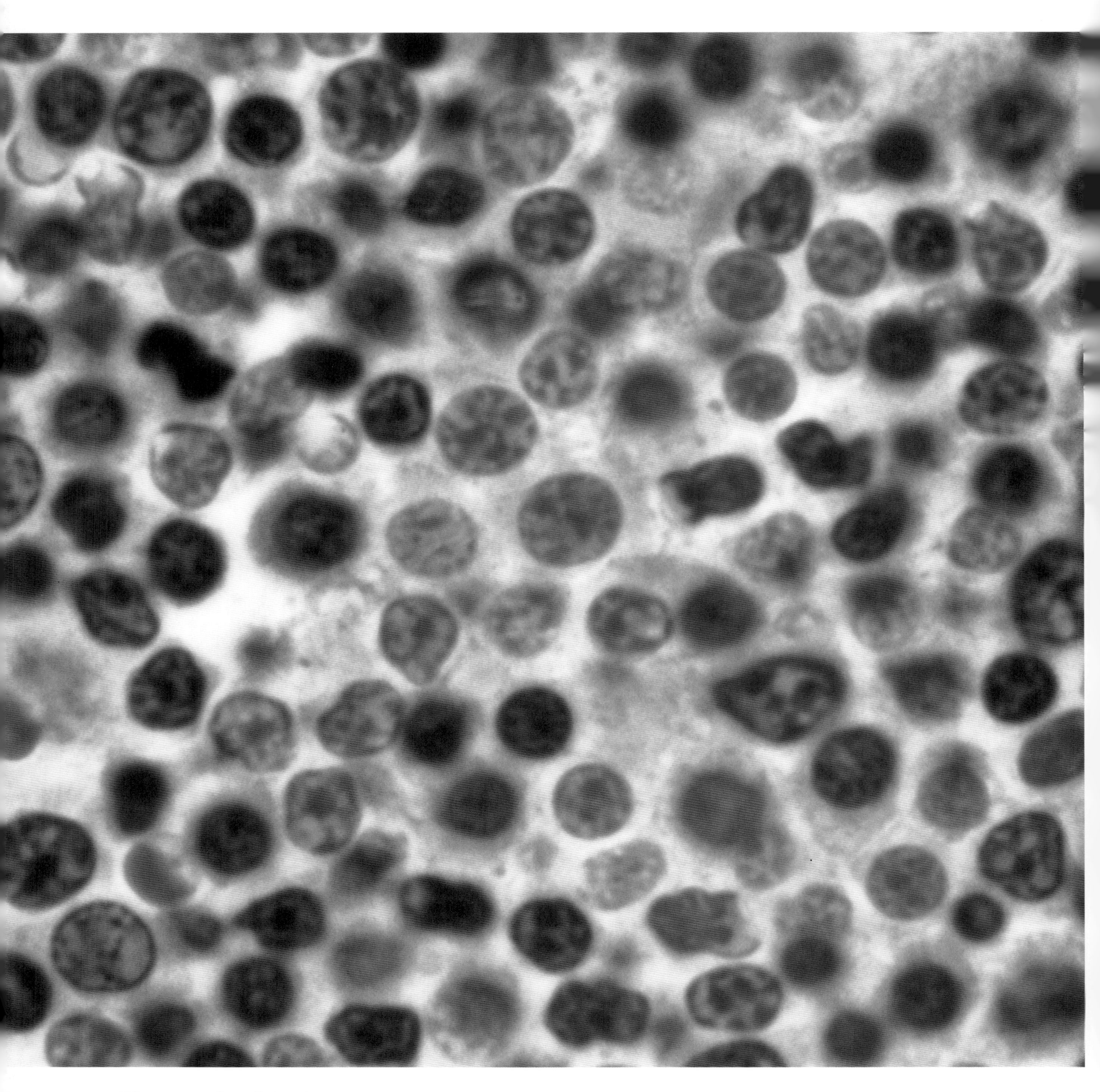

Thymus gland [LM]

LOCATED BENEATH THE UPPER PART of the sternum, the thymus gland is considered part of the endocrine system because it secretes a substance essential for the development of immune responsiveness in very young children. If it is absent, few lymphocytes (dark nuclei) are produced, immune processes do not function, and death results from infection.

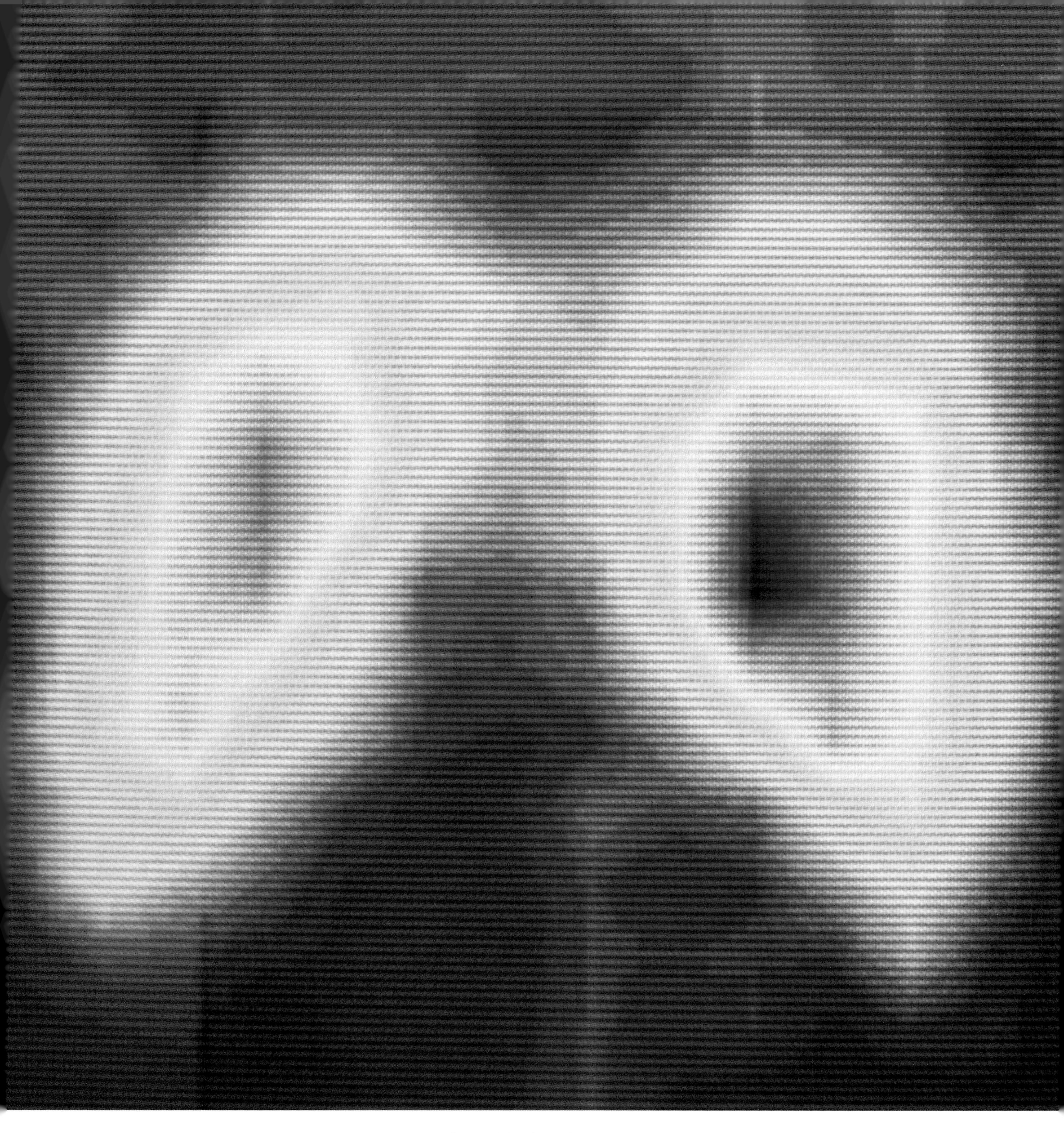

Thyroid gland [Gamma scan]

THIS IMAGE OF THE TWO LOBES of a thyroid gland highlights areas of high activity (blue) and low activity (green). The thyroid produces thyroxin, a hormone containing iodine. If iodine is deficient in the diet, the thyroid enlarges and forms a swelling, or goitre. The condition is prevented by iodizing drinking water or table salt.

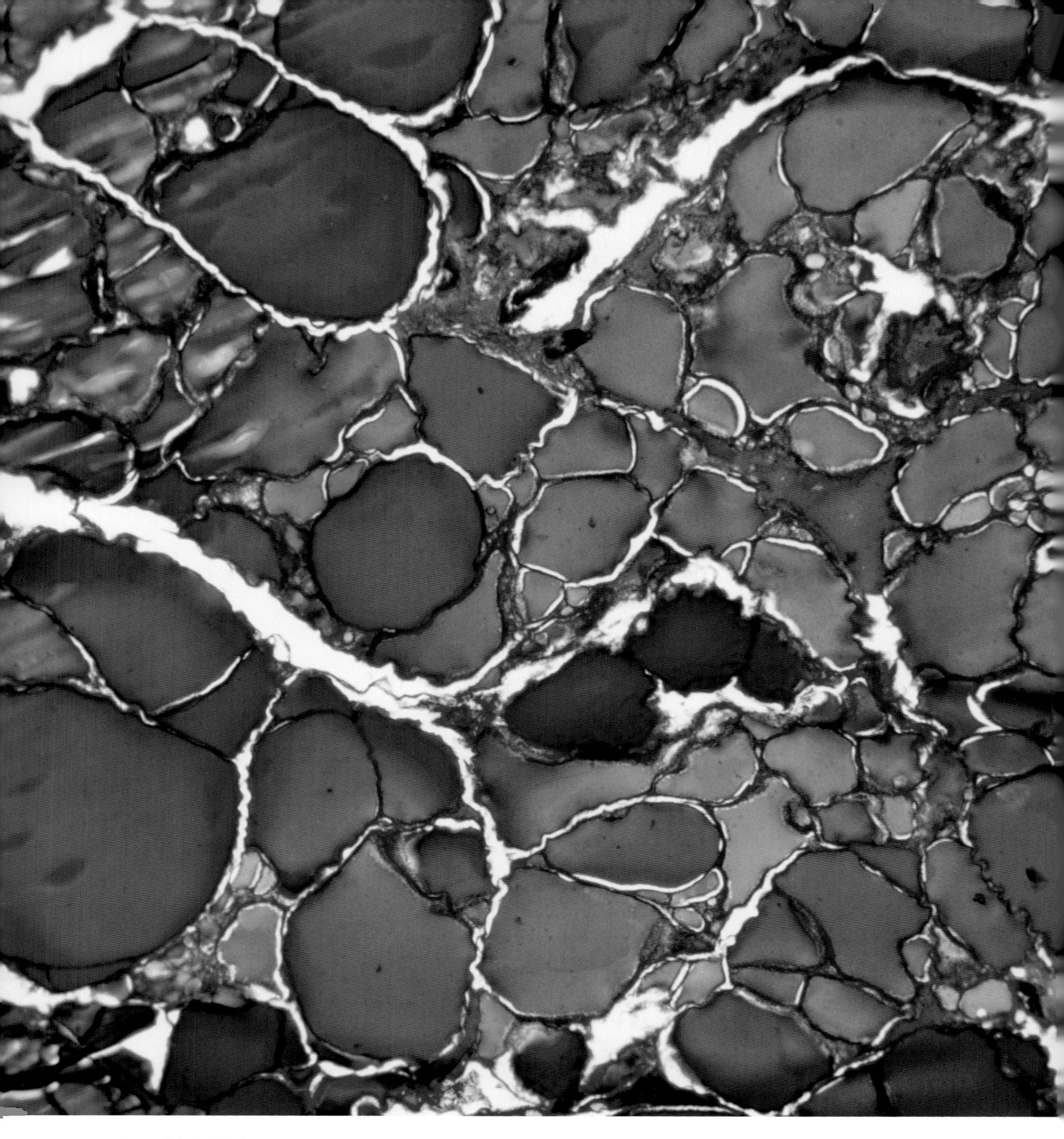

Thyroid follicles [LM]

A SECTION THROUGH a human thyroid gland shows the follicles in which hormones regulating growth and development are produced. The follicles store a form of the hormones (orange) that has been synthesized and secreted by the single layer of blue lining cells. In less active mode, as here, the lining cells are flattened; when actively secreting hormones, they are larger and cube shaped.

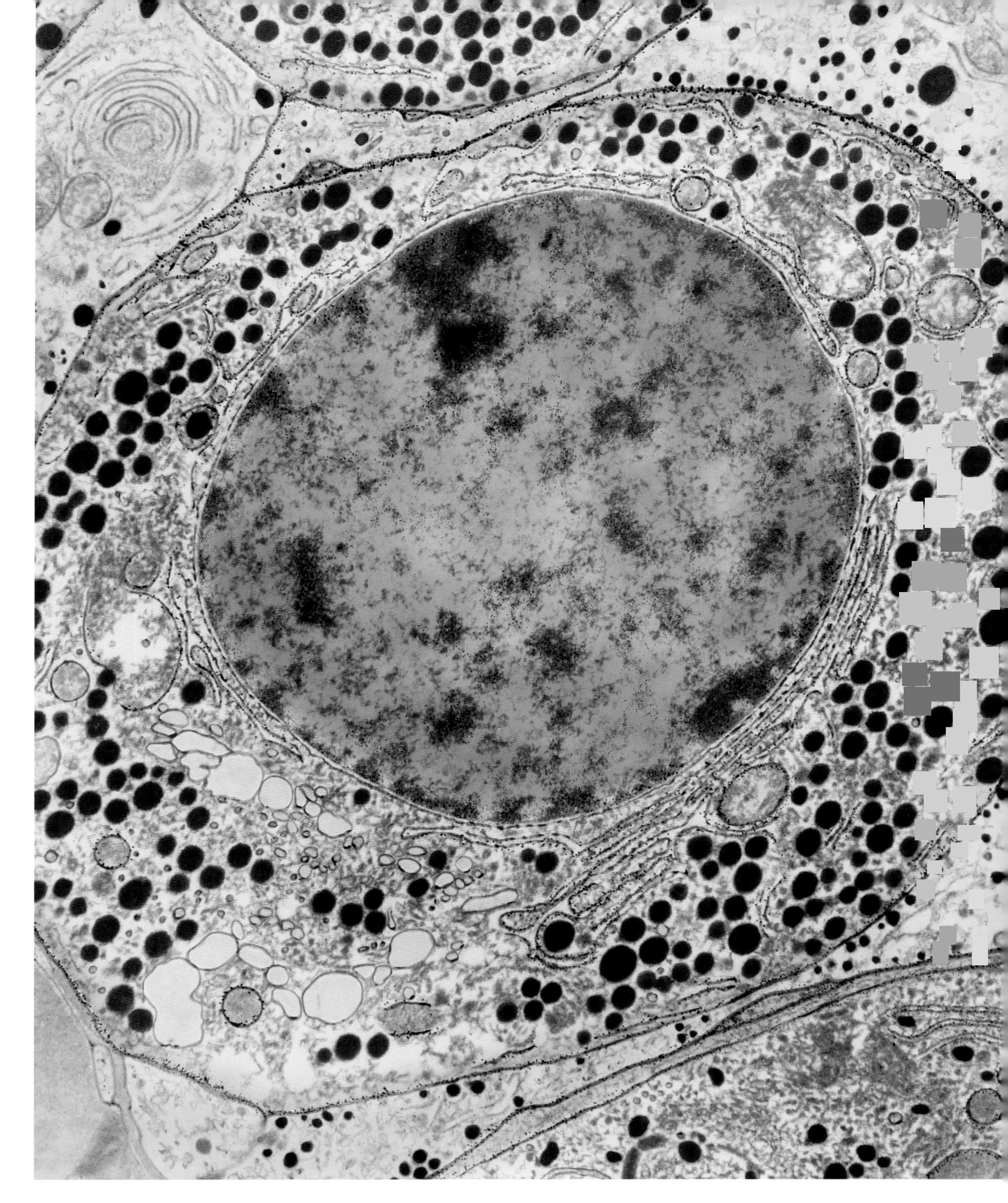

Hormone-producing cell in pituitary gland [TEM]

THE BROWN GRANULES in the yellow cytoplasm of this pituitary gland cell are hormones that promote growth, especially of bone and muscle. A pea-size gland located at the base of the skull, the pituitary also produces other hormones that stimulate milk production, regulate body water balance and control the activities of other glands.

Pituitary gland [SEM]

HORMONE-SECRETING CELLS (pale brown) surround a capillary in the pituitary gland. The pituitary is known as the master gland because it regulates the activities of other endocrine glands. It is influenced by the brain, which communicates with the gland by means of nerve impulses and chemicals. The yellow object in the capillary is a macrophage white blood cell, which digests bacteria and other foreign bodies.

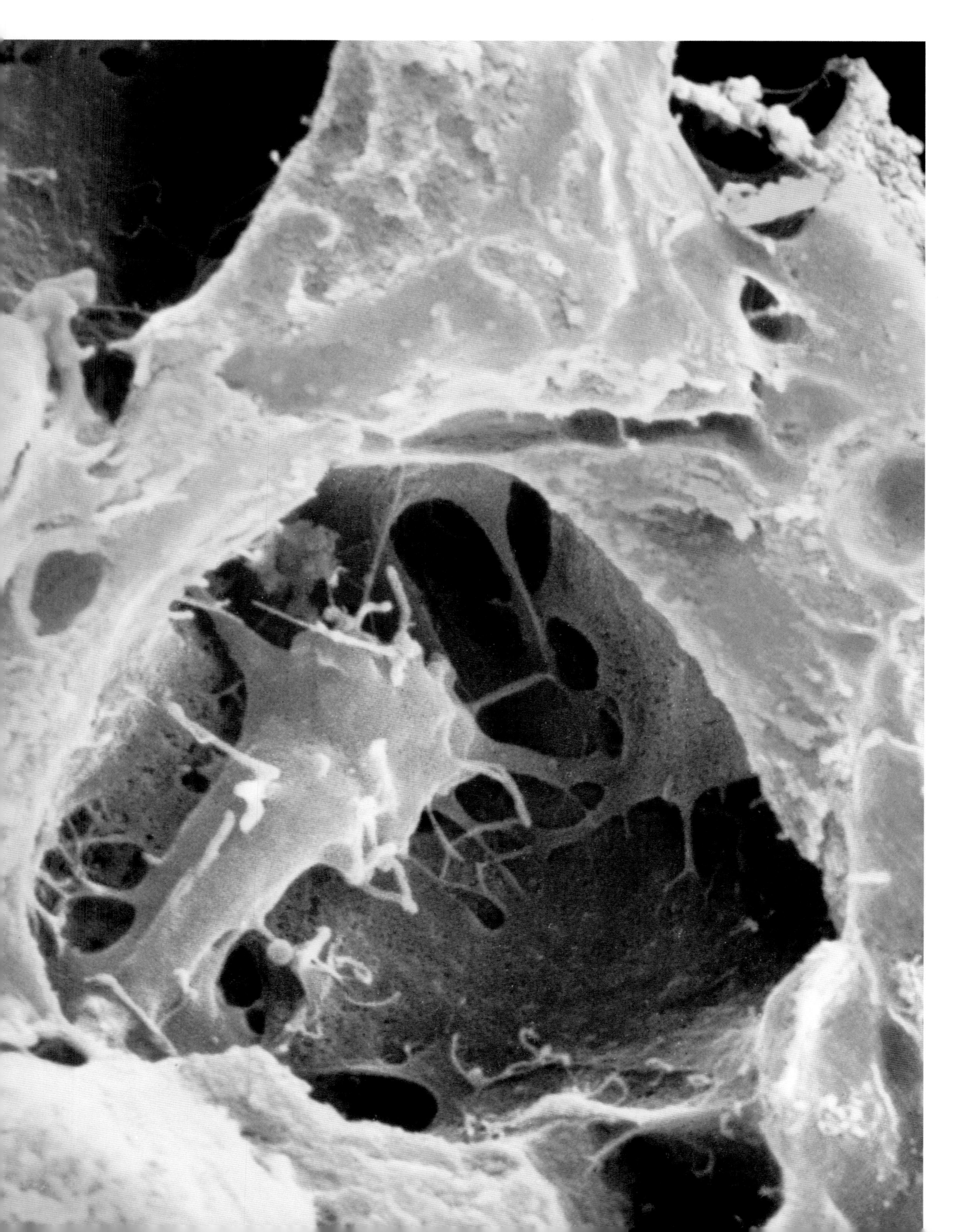

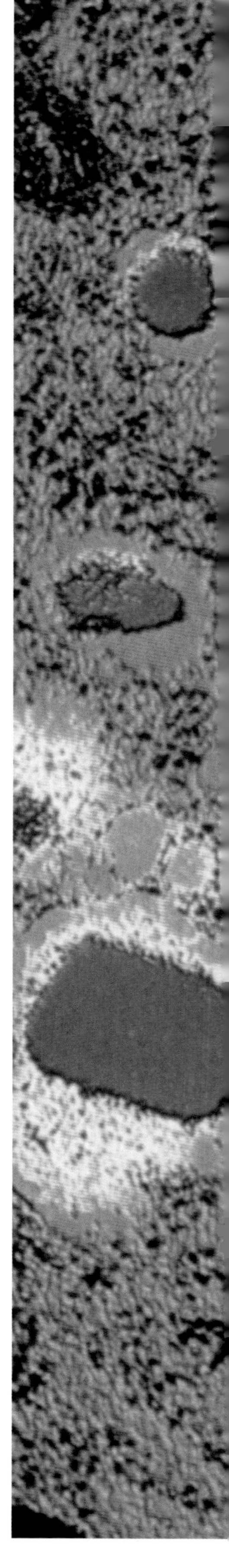

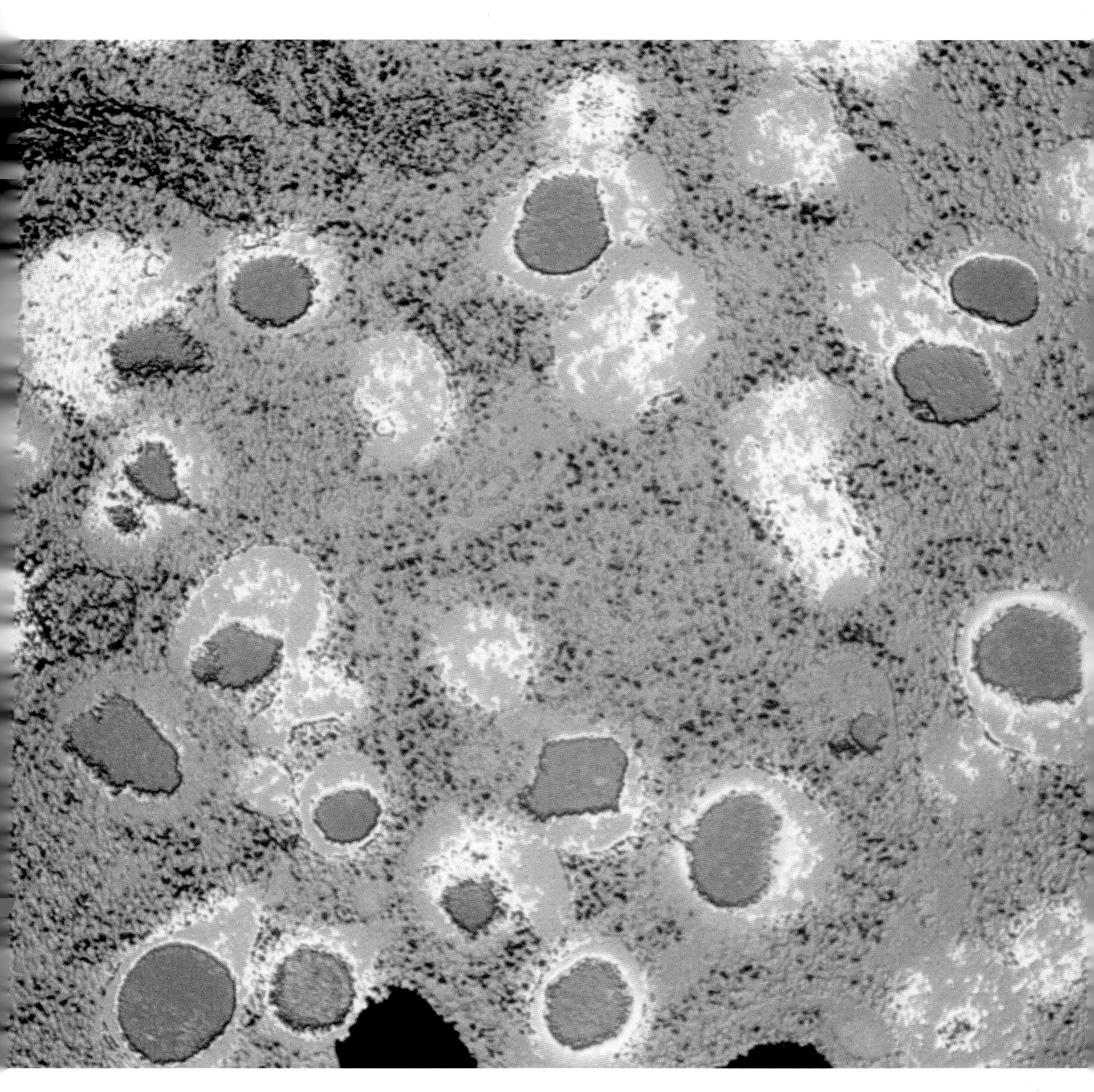

Insulin granules [TEM]

PRODUCED IN THE PANCREAS, insulin is a hormone that increases the rate at which blood sugar is converted to glycogen in the liver. It also stimulates fat cells to convert blood sugar into stored fat. Insulin is usually secreted in response to rising blood-sugar levels – after a meal, for example.

Adrenal gland cells [TEM]

AS WELL AS PRODUCING the hormone adrenaline, the adrenal glands secrete several other hormones. These three cells (two with large, round nuclei) from the outer part of an adrenal gland produce the hormone cortisol, which regulates the body's immune system, reducing inflammation and combating the stressful effects of infection or shock.

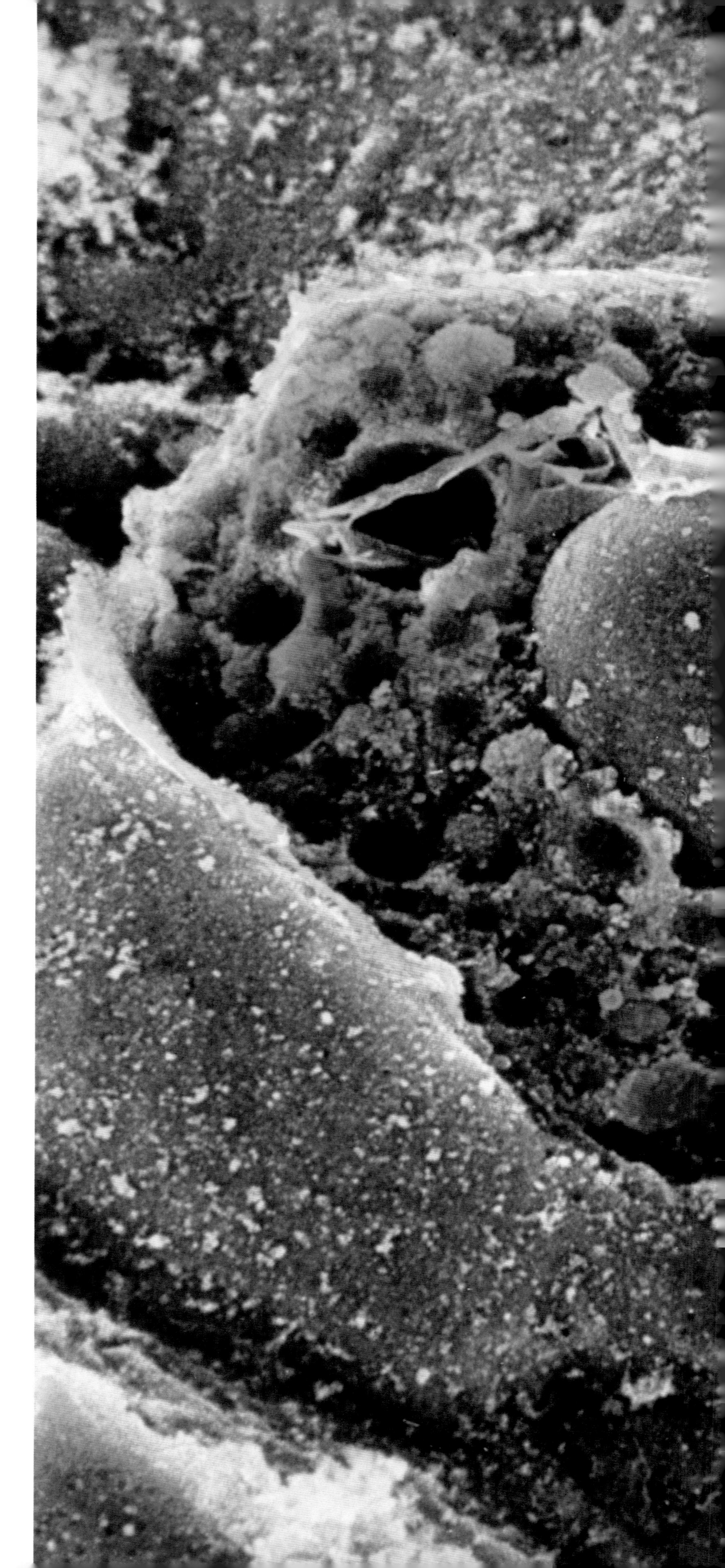

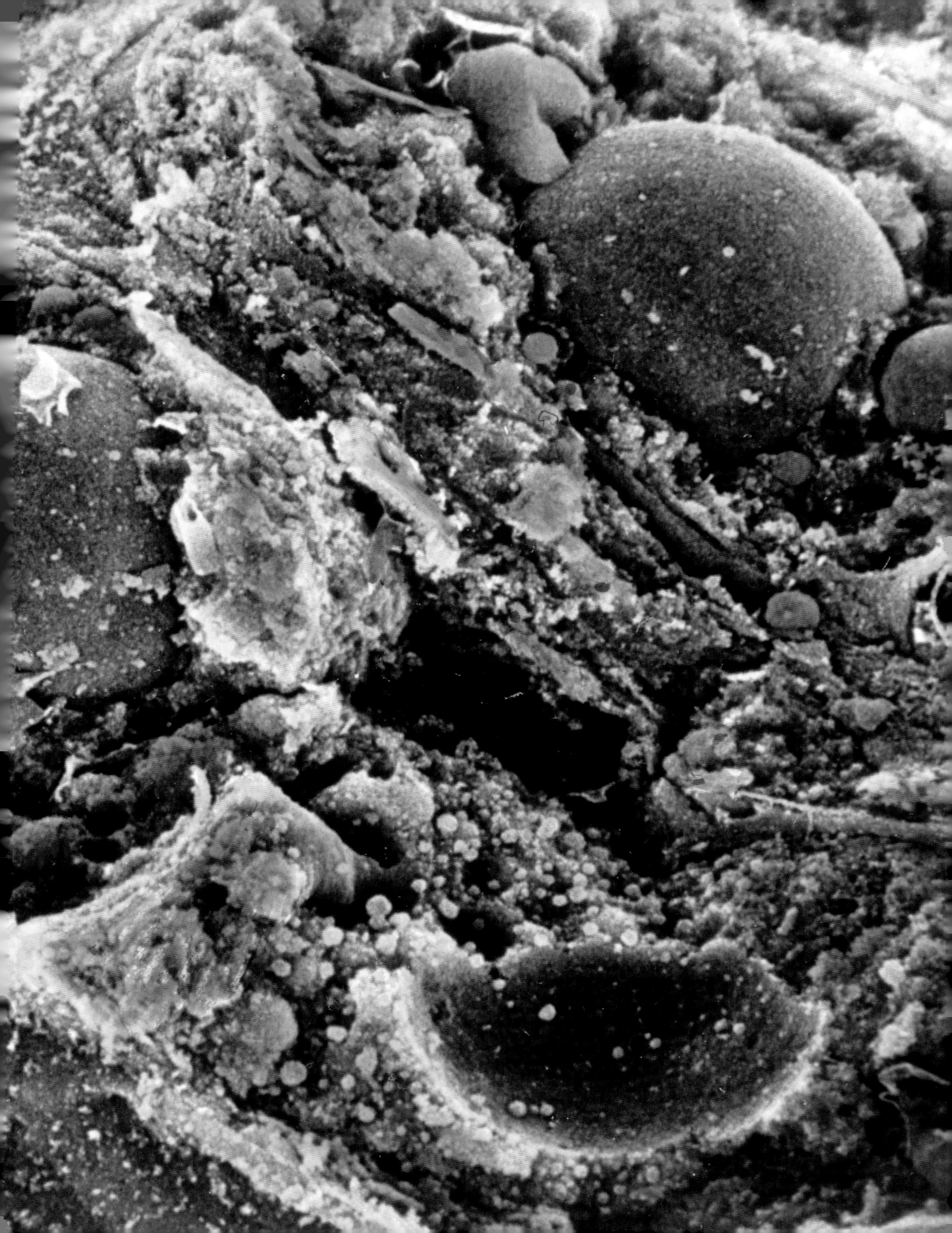

Oestrogen crystals [LM]

THE BLUE STRUCTURES ARE CRYSTALS of oestradiol, the most potent of the six naturally occurring oestrogen hormones that stimulate the development of the female reproductive system and secondary sex characteristics, such as breast formation. Oestradiol is secreted by the ovaries and is the major controlling hormone in the regulation of the menstrual cycle.

Testosterone crystals [LM]

TESTOSTERONE IS THE MAIN ANDROGEN, the class of steroid hormones responsible for stimulating the development of the male secondary sex characteristics. Testosterone causes the growth of the beard and pubic hair, the deepening of voice and the development of the penis. It is produced mainly in certain cells of the testes by the action of enzymes on a steroid derived from cholesterol.

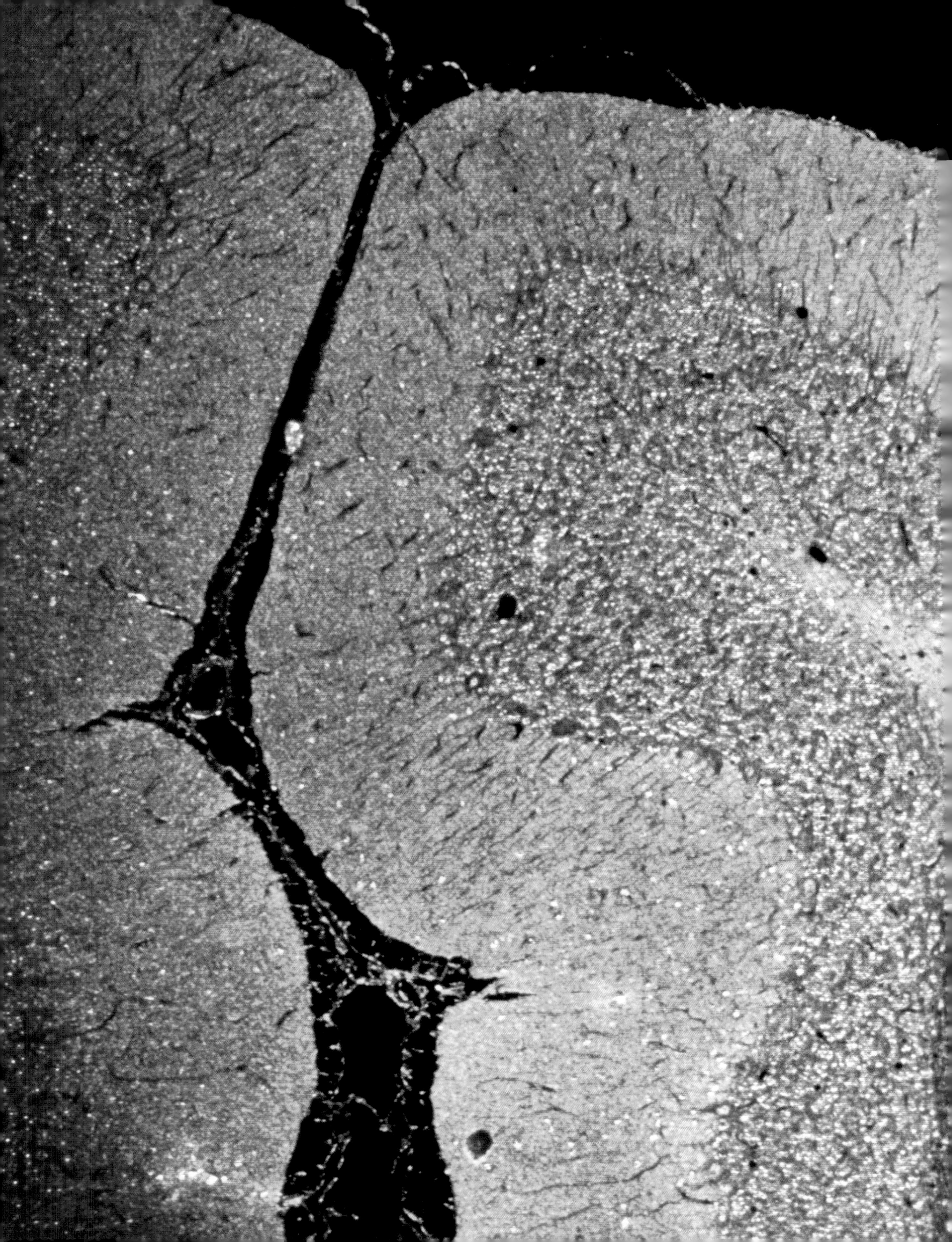

BRAIN AND SENSES

THE CEREBELLUM is involved in co-ordinating muscle activity and maintaining posture. This view shows the cortex, which is composed of an outer molecular layer (green) and an inner granular layer (brown). These layers, composing the so-called grey matter, are arranged around the cerebellum's central core, the white matter (dark green).

The human brain weighs about 1.5kg/3lb, has the consistency of a soft-boiled egg and possesses no obvious moving parts. Encased in the skull, it has no direct contact with the world. It cannot feel, see or hear. Yet this superficially rather uninteresting organ is the most complex structure known. It processes huge amounts of information from the world around it, deciding what to act on and what can be ignored. It stores thousands of memories and can recall them vividly decades later. It monitors basic body functions such

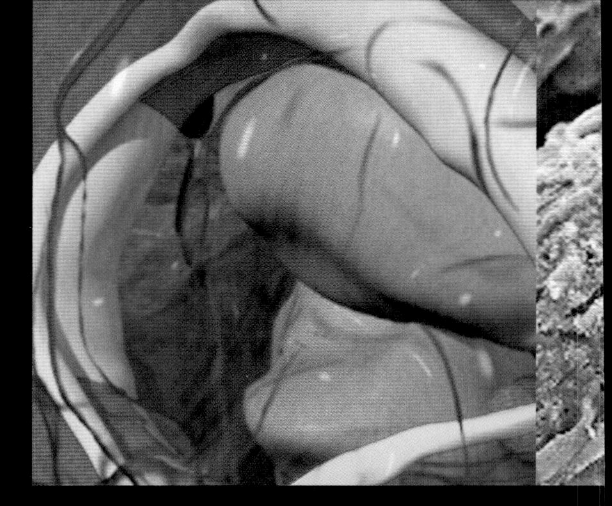

as breathing and heart rate without humans having to think about them, and it co-ordinates conscious activities such as running, driving a car or writing. The brain also lets humans think, dream, reason and experience emotions. An estimated ten thousand million nerve cells are packed into the brain. They are so active that, although the brain accounts for only 2 per cent of body weight, it uses about 20 per cent of the body's glucose-sugar energy supply.

The brain has three main structural parts, which some scientists place in an evolutionary hierarchy: the brainstem, the cerebellum and the cerebrum. The brainstem at the rear of the brain is concerned with life processes that take place automatically. Heartbeat, breathing and blood supply are controlled in the medulla, the lowest part of the brainstem, which merges with the spinal cord. Sleep and wakefulness are controlled by nerve cells in the centre of the brainstem. Just in front of the brainstem is the cerebellum. This greatly folded extension of the hindbrain is involved in balance and muscle co-ordination. Humans owe their manual dexterity in part to the functioning of the cerebellum.

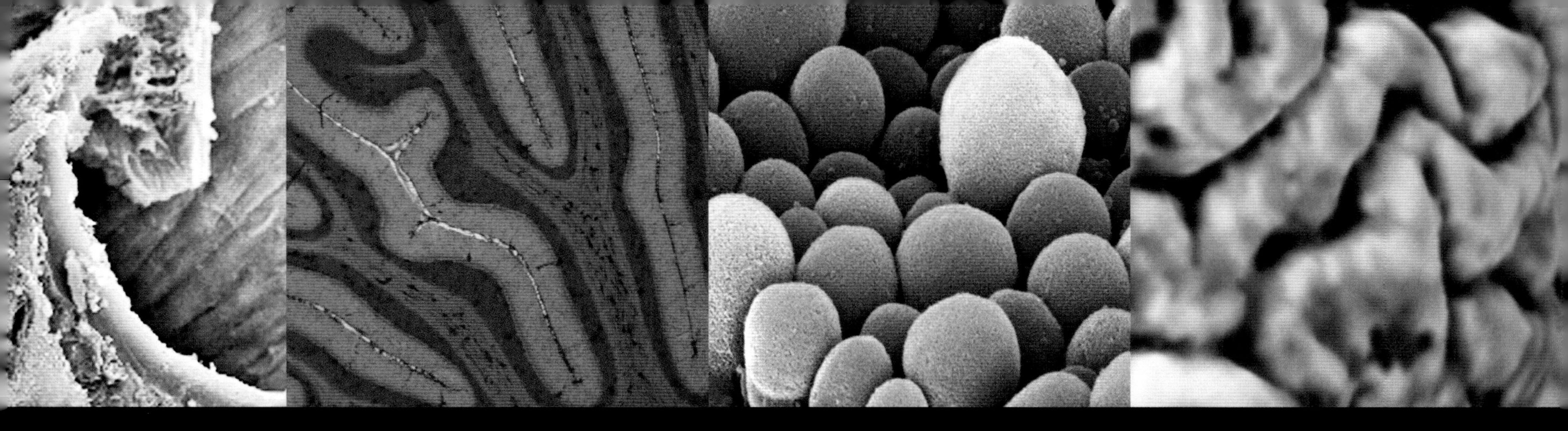

The cerebrum, divided into two halves, called hemispheres, is the largest and most highly developed part of the human brain. Each cerebral hemisphere has a thin but highly folded surface layer called the cerebral cortex. The cortex is the site of nerve centres for the senses and of conscious thought and memory. Different parts of the cerebral cortex have different functions – speech, vision, memory and so on. However, much of the cerebral cortex has no specific function, but is concerned with the interrelating of sensory input with the higher levels of consciousness and personality.

In humans, many of the nerve fibres from the two sides of the body cross over as they enter the brain, so that the left cerebral hemisphere is associated with the right side of the body, and vice versa. Experiments suggest that the right hemisphere is responsible for spatial and musical sense, while the left hemisphere is concerned with mathematical ability and deductive reasoning. For left-handed

Brain arteries [SEM]

SEVERAL CEREBRAL ARTERIES form part of the network of vessels that supply blood to the brain. Up to 20 per cent of the blood pumped by the heart is directed to the brain providing it with oxygen, which is consumed at the rate of about 46cc/2.8cu in per minute.

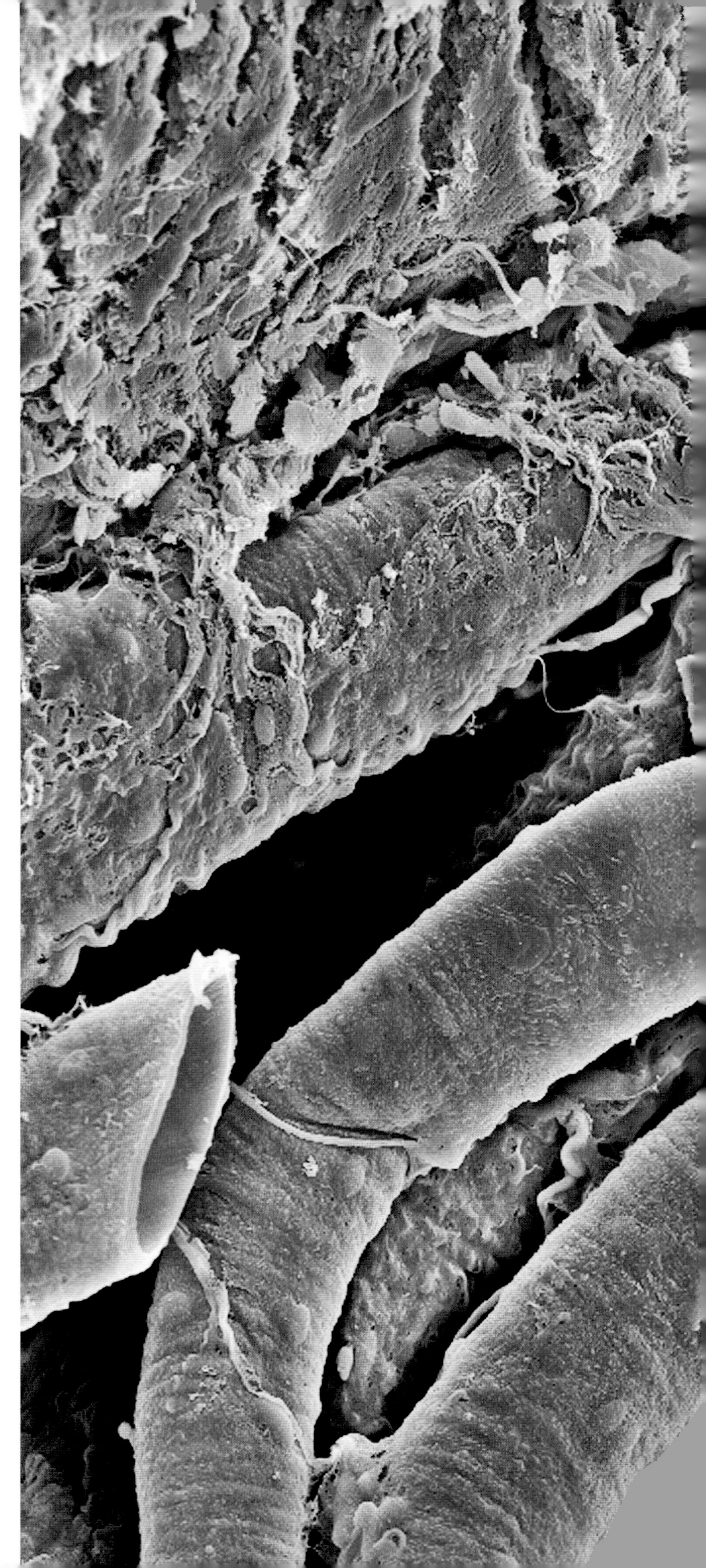

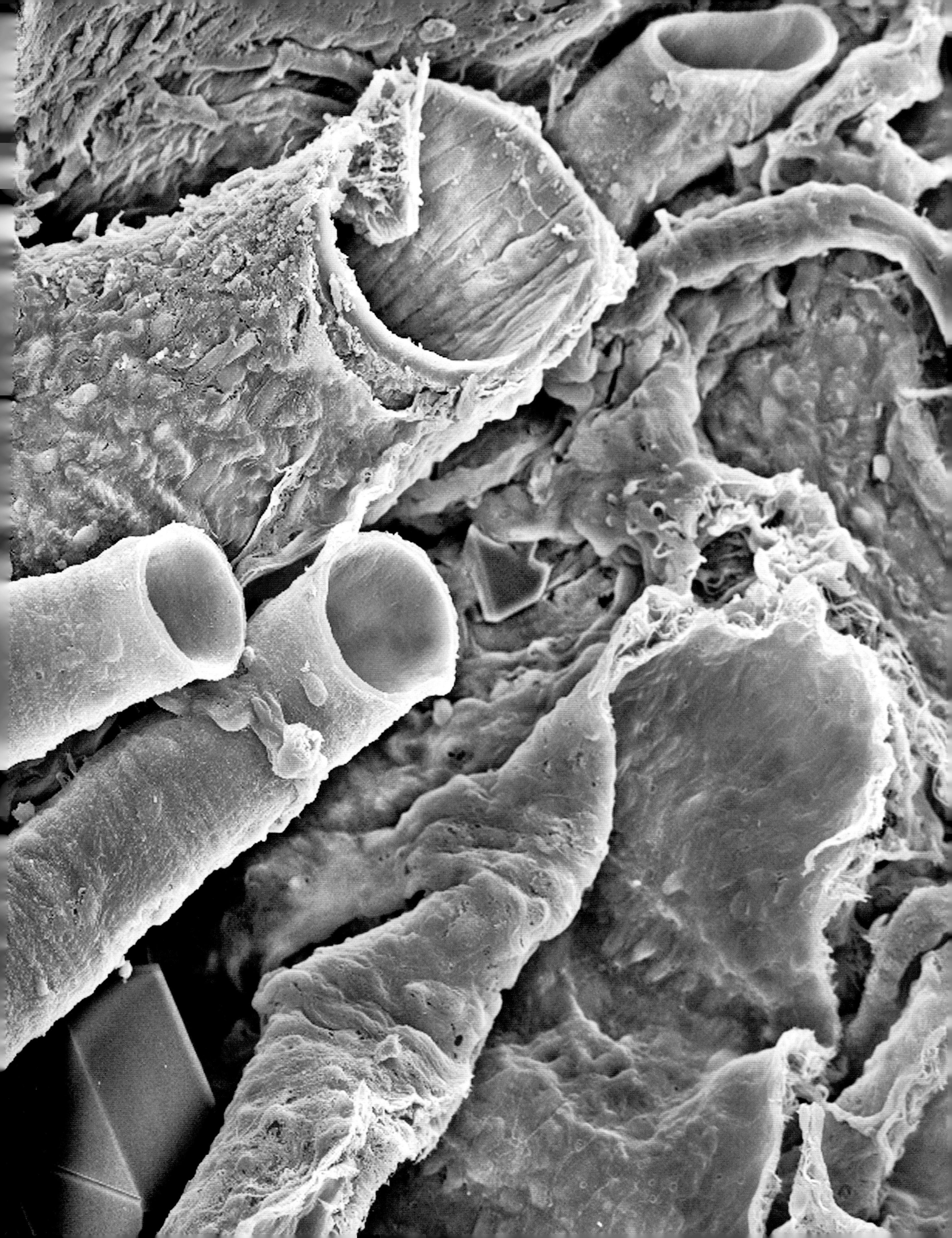

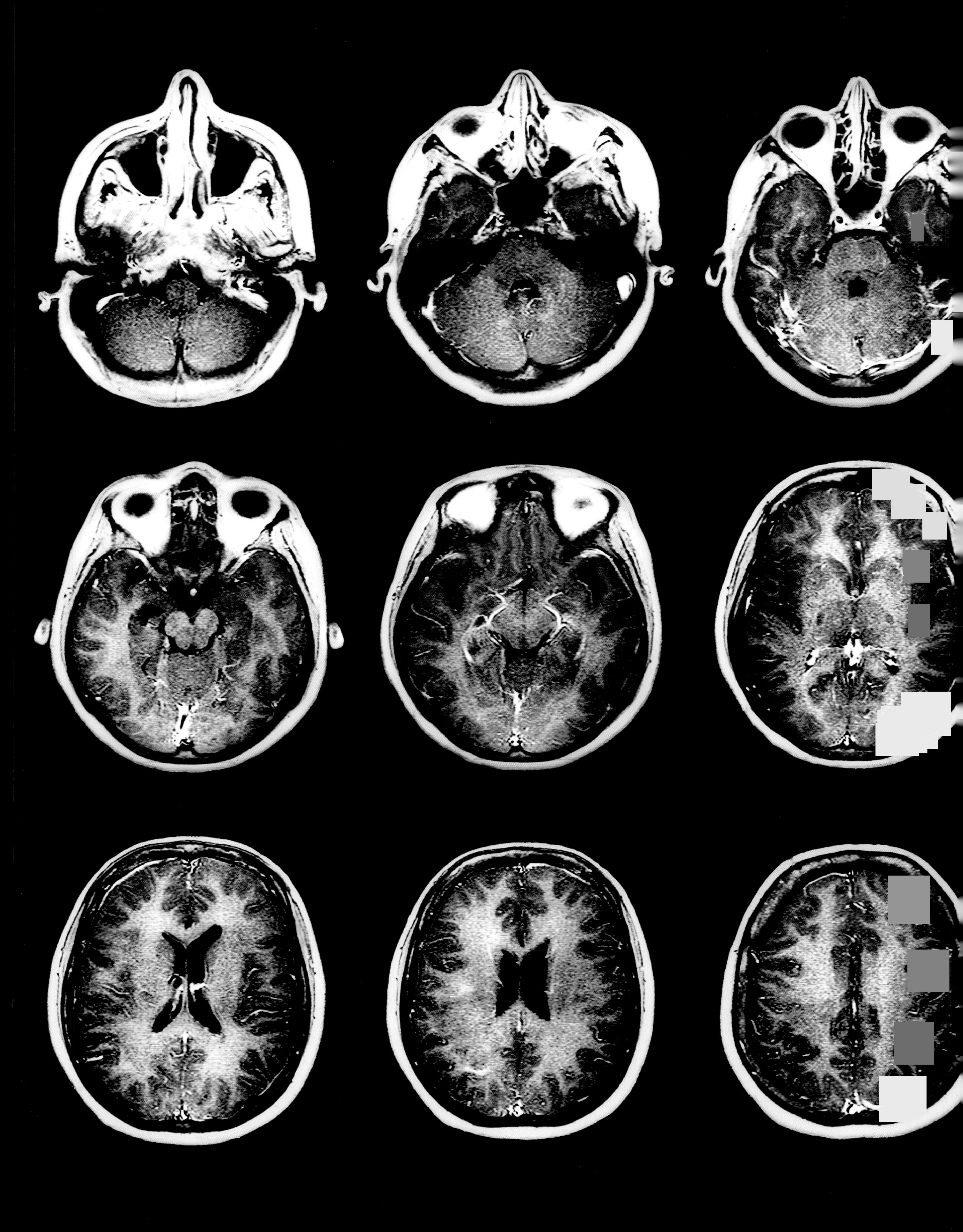

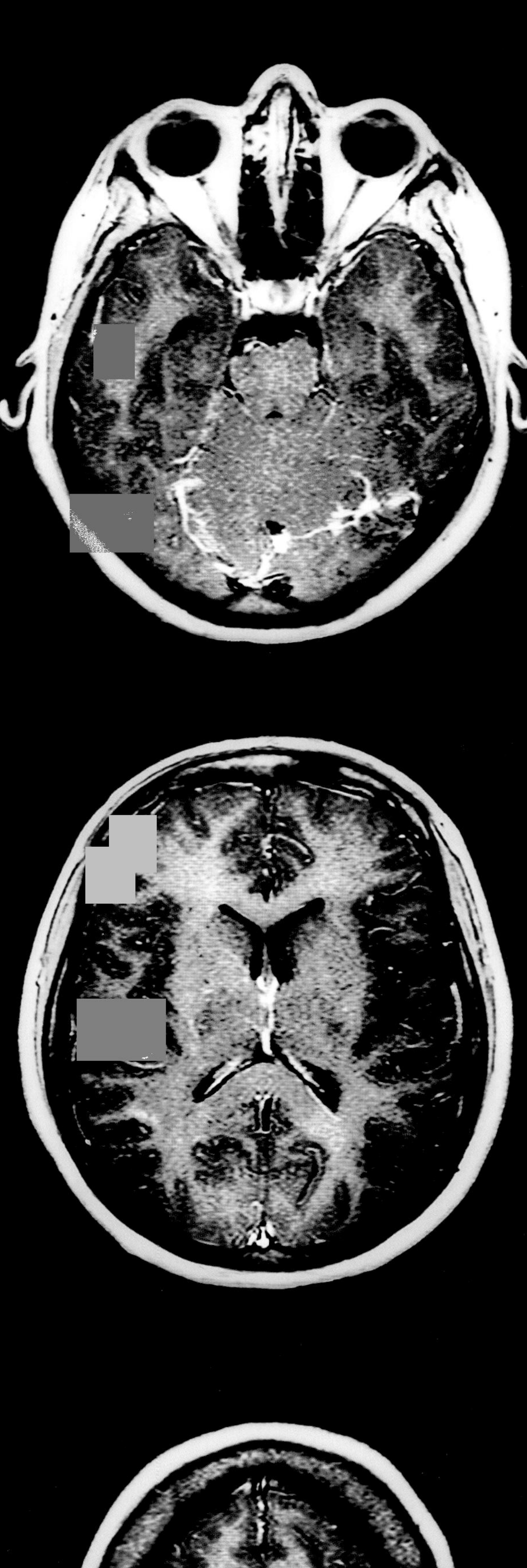

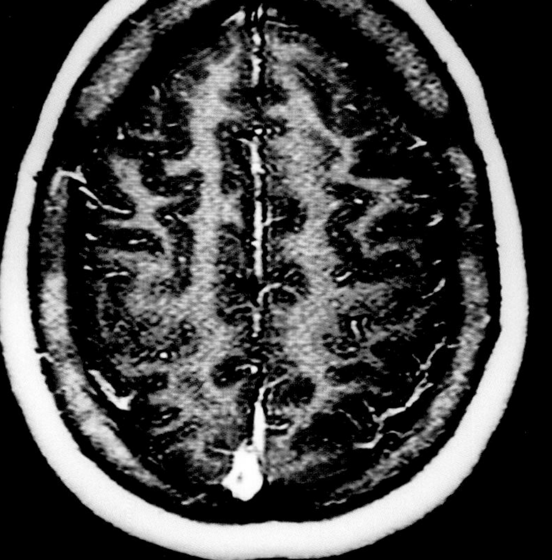

Brain anatomy 1 [MRI]

THIS SEQUENCE OF SCANS passes from the base of the brain (top left) up to a level near the top of the brain (bottom right). The eyes and nose appear in the first six scans. In the first three, the brain tissue consists mostly of the cerebellum and brainstem, regions that control subconscious functions such as balance and breathing. The cerebrum dominates the last two rows, with its folded outer cortex (best seen in the last scan) being responsible for conscious thought and movement. The ventricles or cavities dividing the brain hemispheres are seen in scans 8–10.

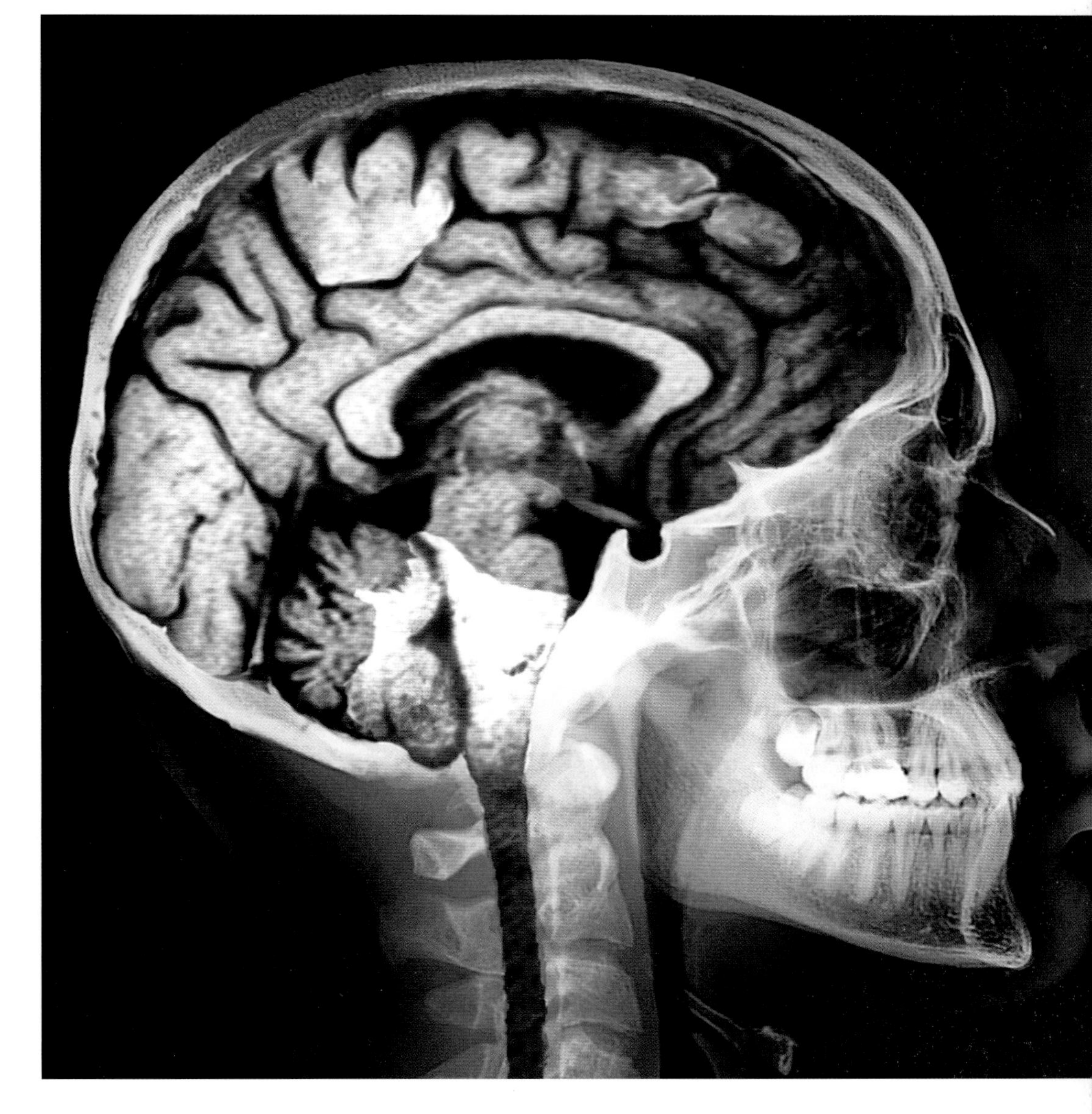

Brain anatomy 2 [MRI/X-ray]

CREATED BY COMBINING an MRI scan and an X-ray, this image reveals the internal anatomy of the brain in sagitall section. The brain's major area is the cerebrum, with its folded outer layer, the cerebral cortex, which is responsible for memory, language and conscious movement. The brainstem (centre), at the base of the brain, connects to the spinal cord in the neck. It controls subconscious functions such as breathing. The cerebellum (left of the brainstem) controls balance, posture and muscular co-ordination.

Cross section of cerebellum [LM]

THE CEREBELLUM IS LOCATED in the lower part of the hindbrain and is mainly concerned with learned skills, such as muscular co-ordination and balance. Like the cerebrum, its outer layer or cortex is deeply lobed. The cortex is divided into two zones, an outer molecular layer (orange) and an inner granular layer (red). These layers, composing the 'grey' matter, surround the central core of white matter, which consists of millions of densely packed nerve fibres.

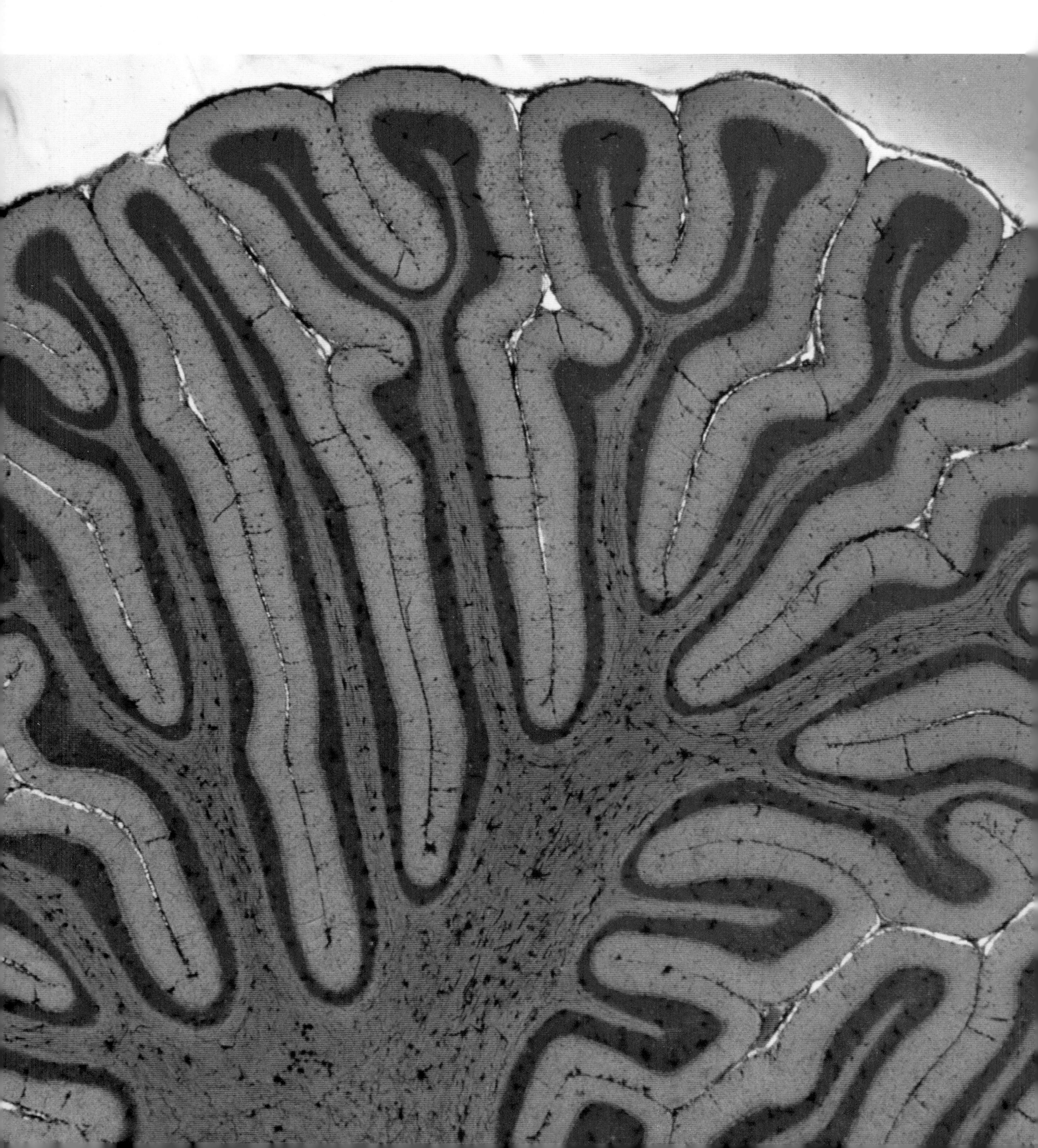

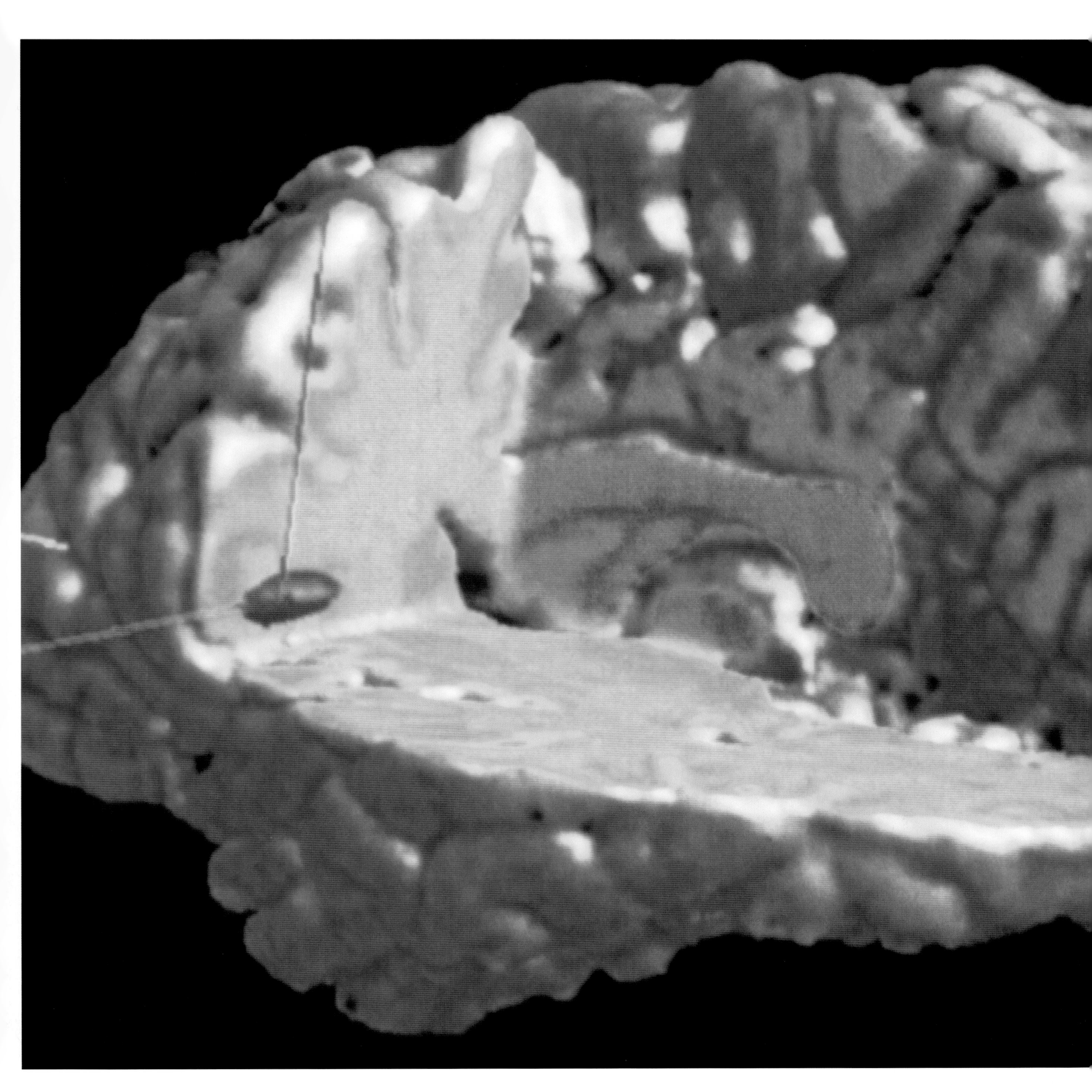

Brain responding to pain [PET/MRI]

THE RED AREA SHOWS cerebral blood flow in an area of the brain associated with pain perception. Our perception of pain is affected by our emotional state, which is controlled by the forebrain. While a particular stimulus may be perceived as painful in certain circumstances, it may be painless or even pleasurable in other situations. The brain itself is insensitive to pain.

Choroid plexus secretory cells [SEM]

THE SWOLLEN TIPS of these cells secrete the cerebrospinal fluid that cushions the brain and spinal cord against shocks. The fluid, which bathes the outside of the brain and fills the four brain chambers (ventricles), is produced in blood vessels called the choroid plexus. If the flow of fluid is blocked, the brain swells to produce the potentially serious condition of hydrocephalus, or 'water on the brain'.

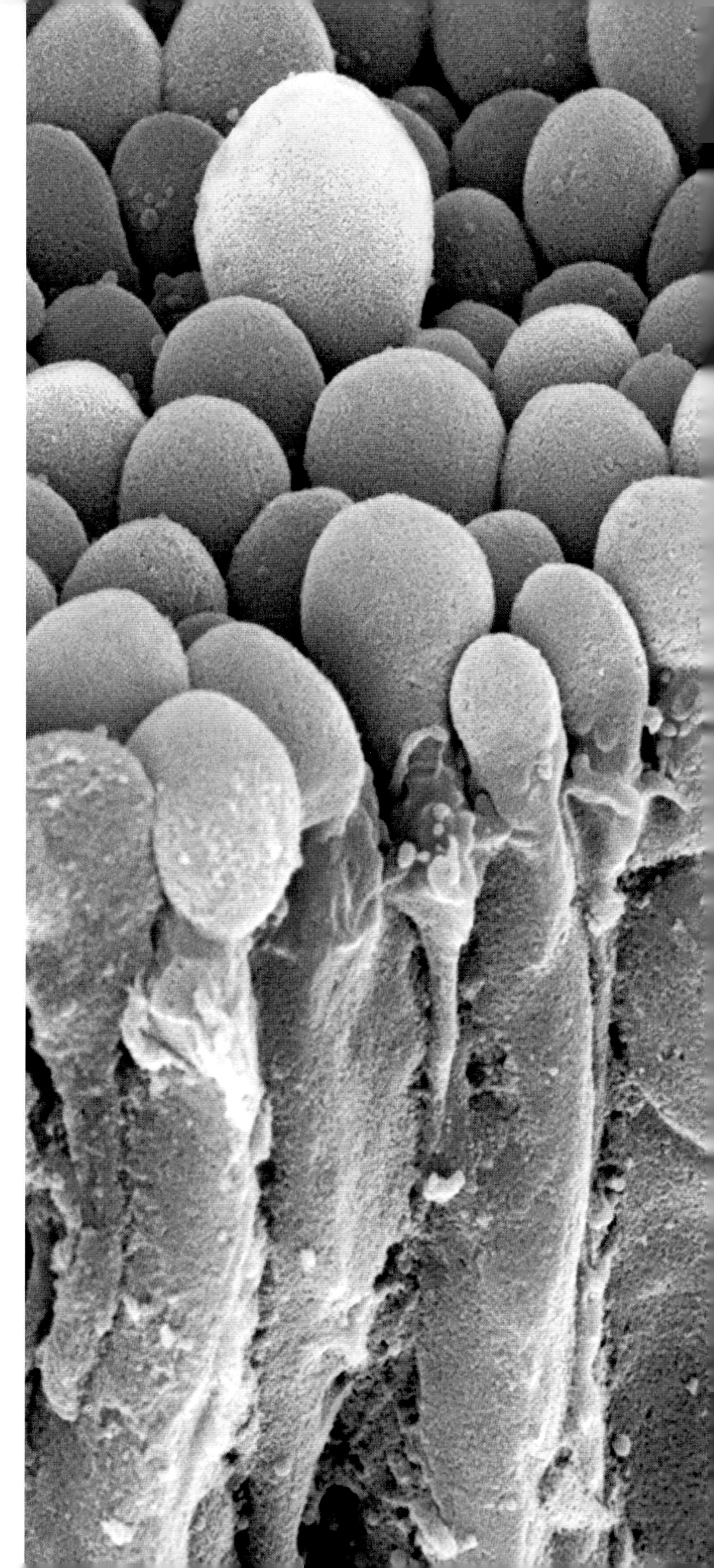

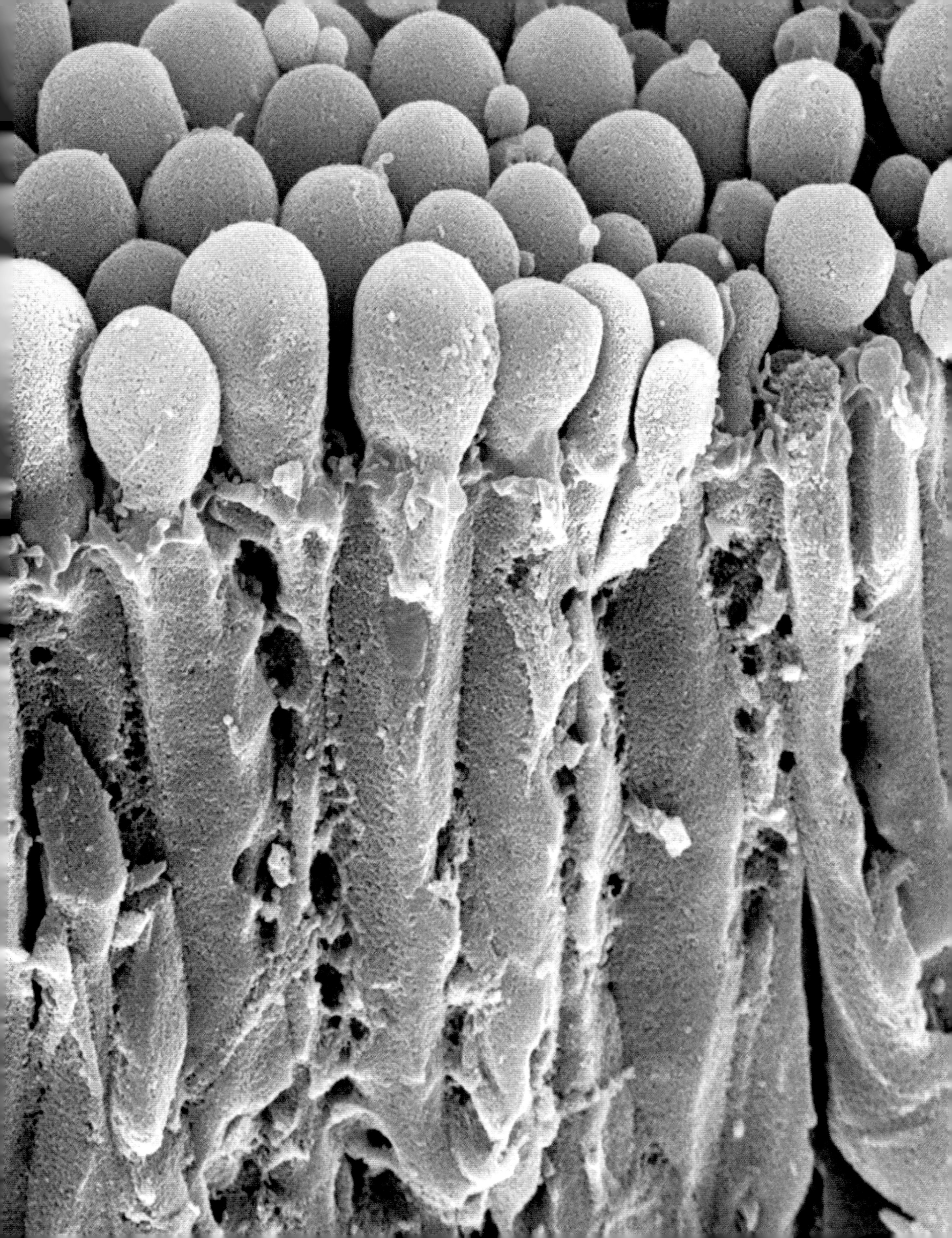

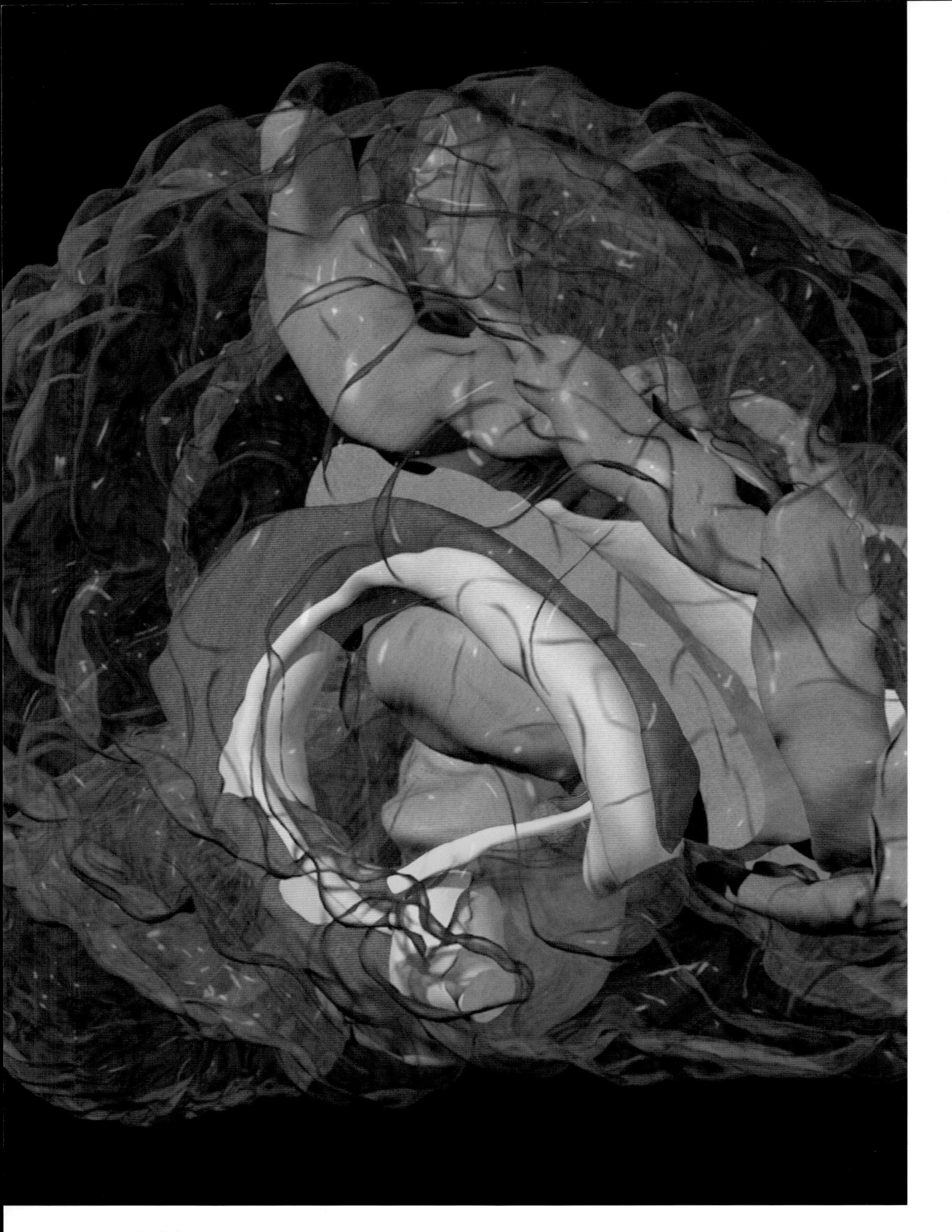

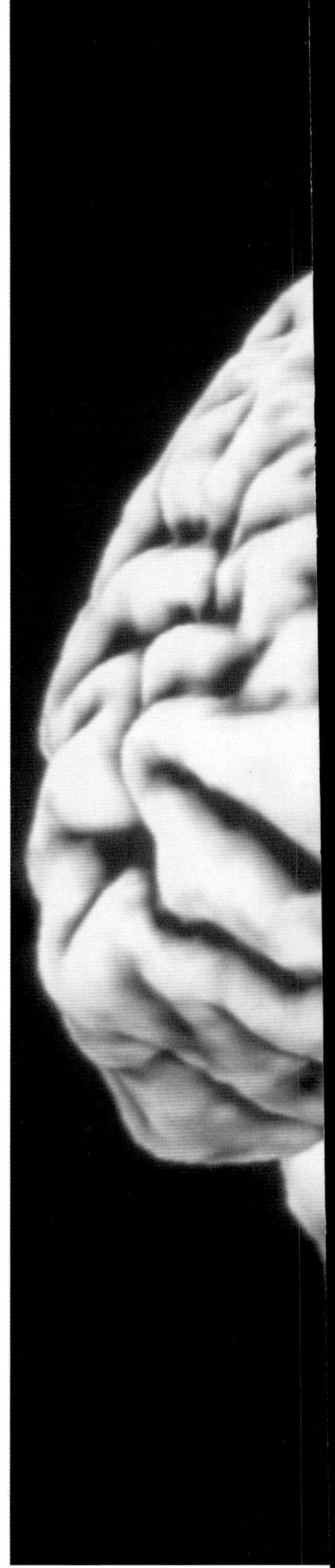

Limbic system [MRI]

COMPILED FROM SUCCESSIVE MRI scans, this image shows the limbic system deep within the cerebral hemispheres. The limbic system is responsible for basic physiological drives, instincts and emotions. Within the system are areas associated with intense sensations of pleasure, pain or anger.

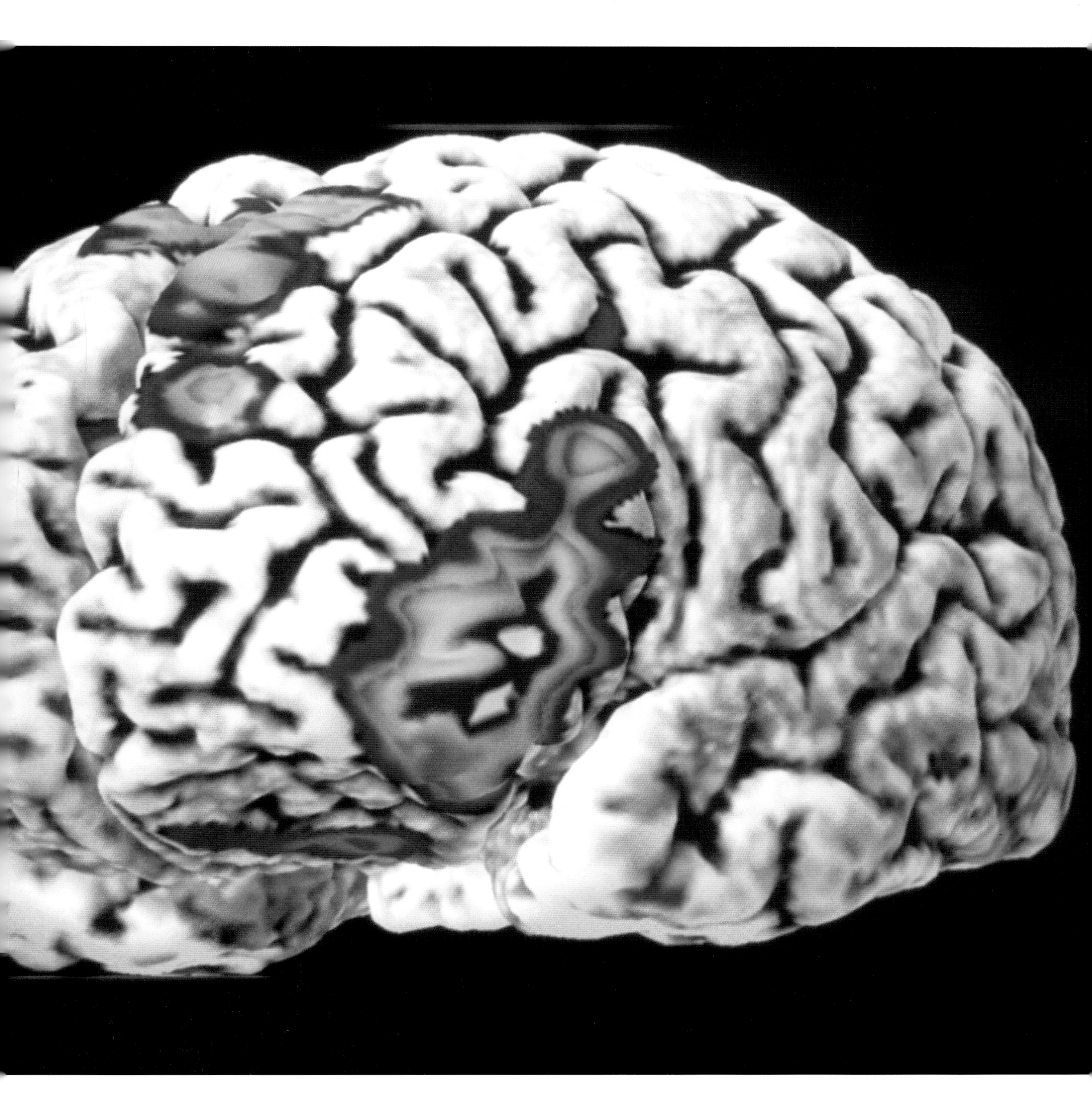

Brain responding to speech [PET/MRI]

THE AREAS OF THE BRAIN associated with speech (here coloured white, red, yellow, green and blue) are in the speech cortex of the brain's frontal lobe. The image was created by introducing radioactive substances into the subject's blood, and then recording metabolic activity in the brain during a speech exercise. To show the location of the speech cortex, the scan has been superimposed on an MRI image of the brain.

The ear is a miniature receiver, amplifier and signal-processing system. Sound waves reach the outer ear, which acts as a funnel, directing the sound down the ear canal. The sound waves reach one ear a split second before the other and at slightly different pressure. The brain detects these minute differences, so is able to judge the direction the sound comes from.

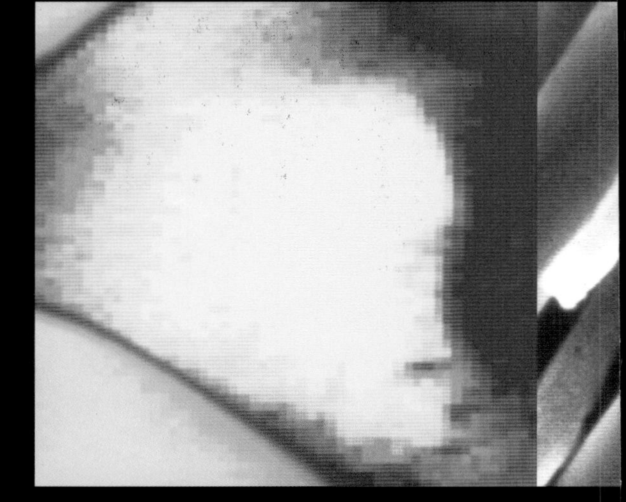

Sound is channelled to the eardrum, a fan-like membrane at the end of the ear canal. On the other side of the eardrum, the middle ear is open to the throat at the back of the mouth through the Eustachian tube, which equalizes air pressure on each side of the eardrum. If the tube gets infected and becomes blocked by mucus, pressure builds up on one side or the other, and the eardrum bulges, causing pain. Airline passengers experience the same painful effect briefly when the air pressure suddenly changes. In this case, swallowing or yawning opens the tube – sometimes with an audible 'popping' – restoring pressure equilibrium.

Sound waves make the eardrum vibrate. The vibrations are transmitted along three tiny bones to the oval window, a membrane-covered hole. The lever action of the bones and the fact that the oval window is smaller than the eardrum amplify the vibrations about twenty-fold. On the inner side of the oval membrane is a long, spiral, fluid-filled tube called the cochlea. The amplified vibrations spread through the fluid and stimulate sensitive hair cells, which convert the vibrations into nerve signals that travel along the auditory nerve to the brain. Short fibres in the first part of

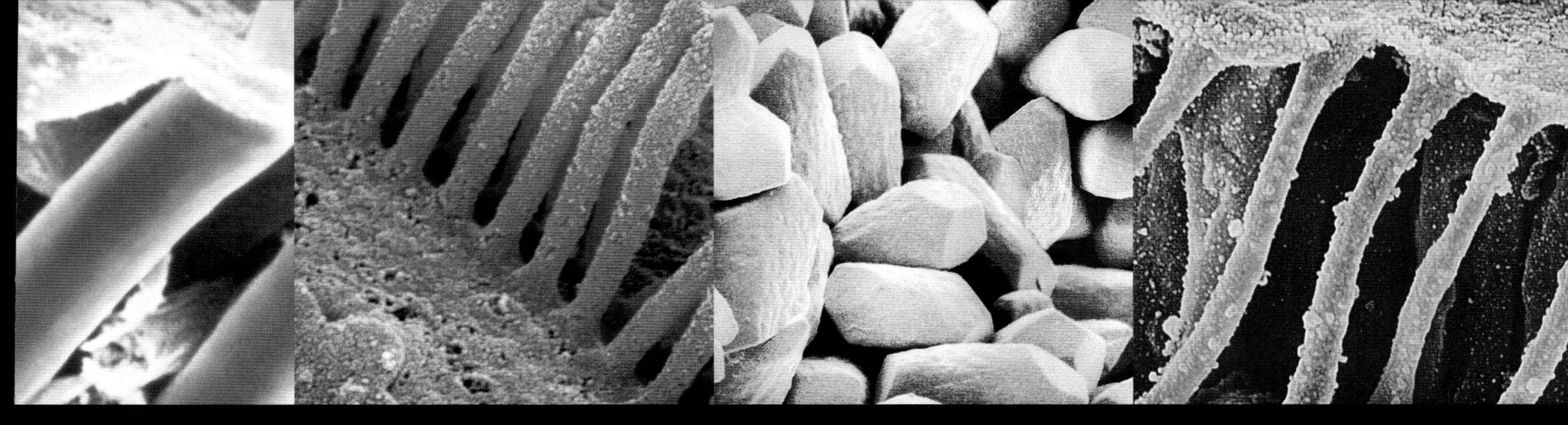

the cochlea respond to high-frequency vibrations, while long fibres in the last part are sensitive to low-frequency vibrations.

The ears also play an important part in maintaining balance. Above the inner ear is a system of fluid-filled sacs and three semicircular canals. The sacs contain chalky granules, called otoliths (balancing stones), which lie on sensory fibres. When your head tilts, the otoliths move under gravity, pulling on the fibres. The nerve signal fired off from the fibres reach the brain and set off a reflex tending to return the body to its normal posture. The semicircular canals contain sense organs that respond to movements of the fluid. When you rotate your head, the fluid lags behind and bends a jelly-like blob called the cupula at one end of the canal. Tiny hairs rooted in nerve cells at the base of the cupula are stimulated and send information to the brain about the rotation of the head. The three canals are at right angles to each other so that the brain can detect movement in any direction.

Hearing and balance

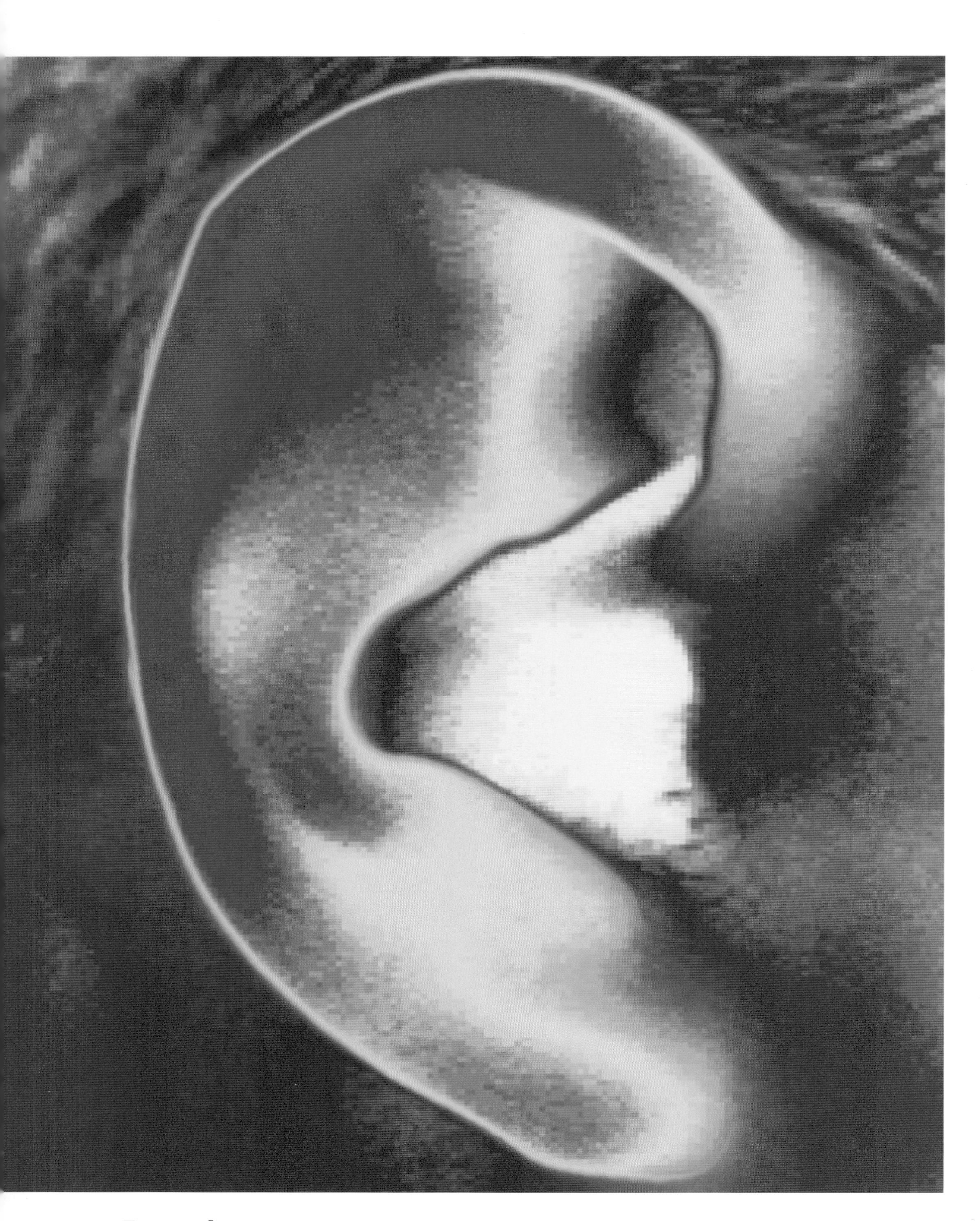

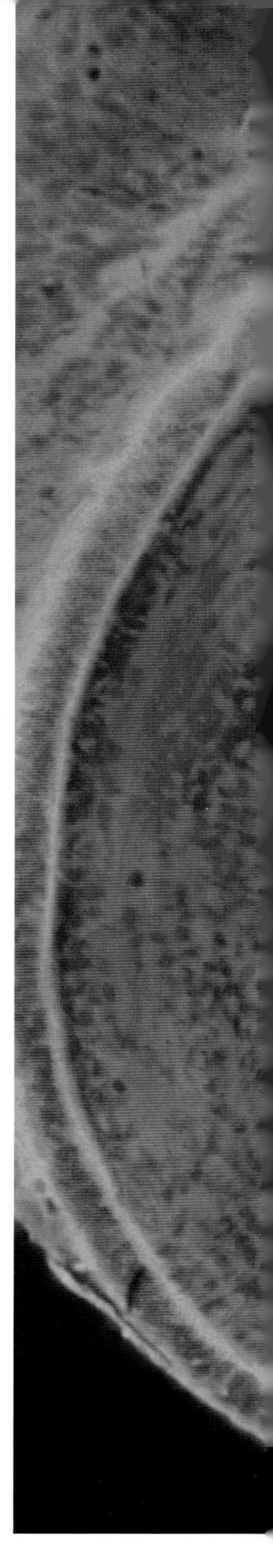

External ear [Thermogram]

THE VARIATIONS IN COLOUR here reflect the different temperatures on the surface of a man's ear – from red (warmest), through yellow, green and blue to mauve (coolest). The fleshy, dished outer ear collects and amplifies sound waves, which then pass along the ear canal to the eardrum.

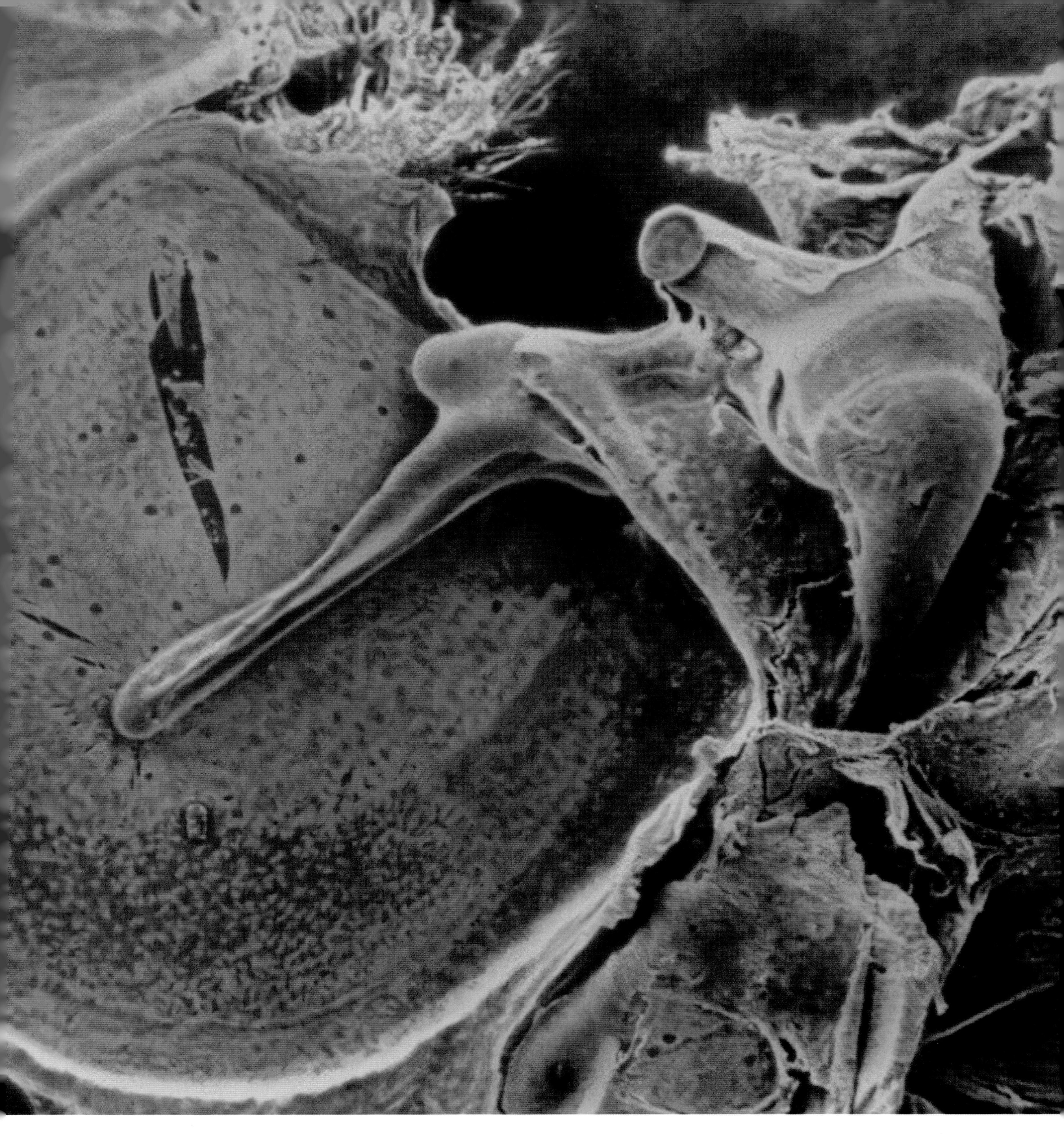

Eardrum and ossicles [SEM]

SOUND WAVES ENTERING the ear travel along the ear canal and strike the eardrum (large semicircular structure), causing the tightly stretched membrane to vibrate. The vibrations are picked up and transmitted by the three ear ossicles, the smallest bones in the human body, two of which can be seen at centre and right.

Organ of Corti [SEM]

A SECTION THROUGH the cochlea in the inner ear shows the organ of Corti, the true organ of hearing. The red, pillar-like cells at centre support four rows of hair cells, each containing about a hundred individual hairs. Sound waves displace the hairs, and this mechanical movement is translated into electrical impulses that are transmitted to the brain via the cochlear nerve (not seen).

cochlea displace the fluid that surrounds the tiny filaments projecting from the hair cells, causing them to bend. The movement is converted to a signal by the cochlear nerve and carried to the brain.

Inner ear sensory cells [SEM]

BUNDLES OF HAIR CELLS (blue) situated within the inner ear detect directional movement, such as the tilting of the head. The cells are immersed in fluid. Currents in this fluid move the cells, resulting in the production of nerve pulses, which are sent to the brain. Also seen are two otoliths (balancing stones), which stimulate the hair cells when the head changes direction.

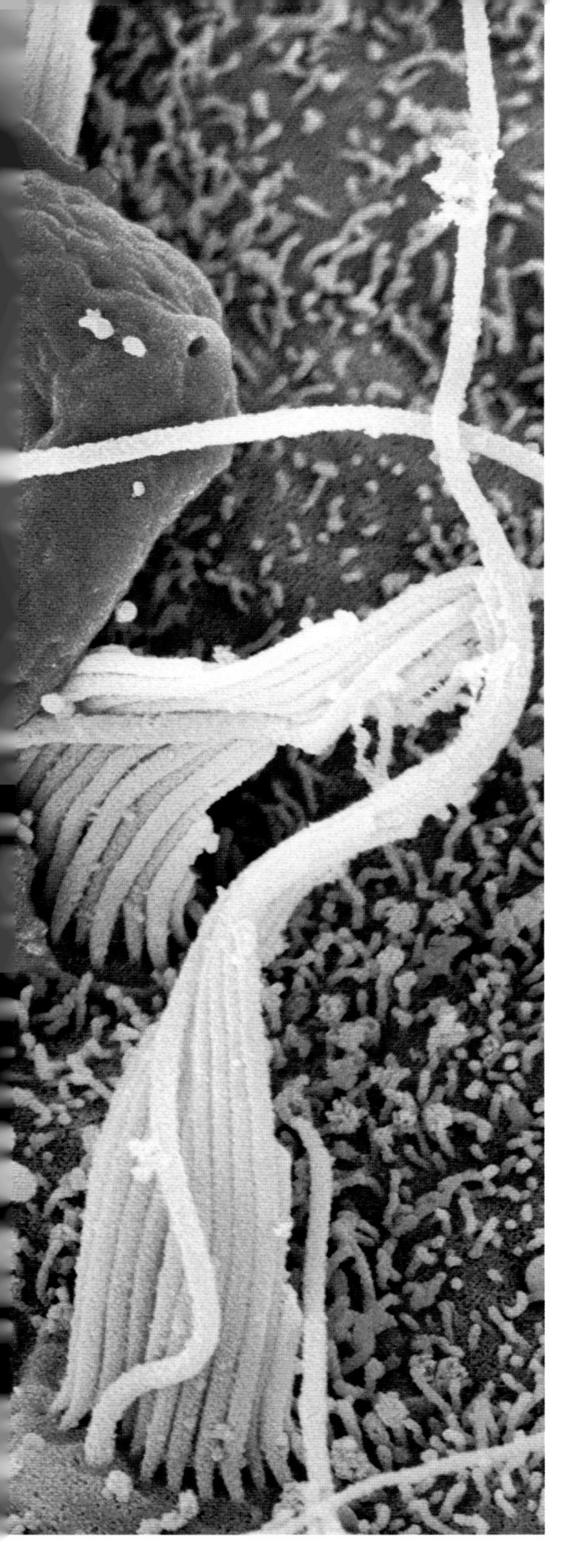

Otolith granules from inner ear [SEM]

THESE GRANULES ARE CRYSTALS of calcium carbonate on the surface of an otolith, a balancing stone attached to sensory hairs in the inner ear. When the head tilts, the movements of the otoliths trigger nerve impulses to the brain, which responds by adjusting the body's balance and orientation. Overstimulation of the otoliths can cause motion sickness.

civilization has blunted our sense of smell, and we no longer use our noses to find food. Smell does, however, still tell us whether food is fit to eat and alert us to other dangers, such as fire. And although we may not be aware of it, subtle and not so subtle body scents helps us to select sexual partners. Smell is the most directly acting of the senses. It is difficult to recall scents, harder still to describe them. At the same time, smell is strongly linked to memory. A fleeting scent can waft us back to a particular day in childhood.

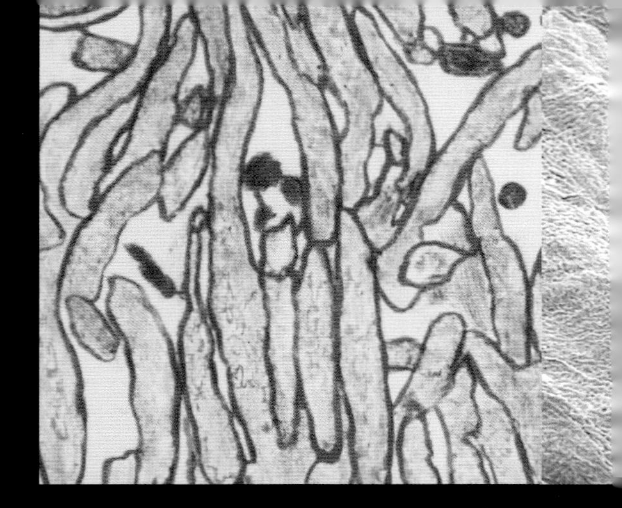

Each time we inhale, we breathe in millions of odour molecules. When we wrinkle up our noses and sniff, it is to move the molecules of smell closer to the odour receptors, which lie at the top of the nasal cavity, behind the bridge of the nose. Some of the molecules get trapped in the mucus membrane containing the smell receptor cells. When the cells are excited, signals are carried by the adjoining nerve fibres to the brain's olfactory centre, where the sensation of smell takes place.

Humans have about ten million smell receptors. Although dogs have about a hundred times as many, our noses are incredibly sensitive to some odours. For example, we can detect one five-millionth of a milligram of vanilla per litre of air (one twenty-five thousand-millionth of an ounce per pint). Some of the smells that we find particularly potent are associated with sex. Significantly, many of the traditional ingredients of perfumes – musk, civet and castor – are obtained from the sex glands of animals.

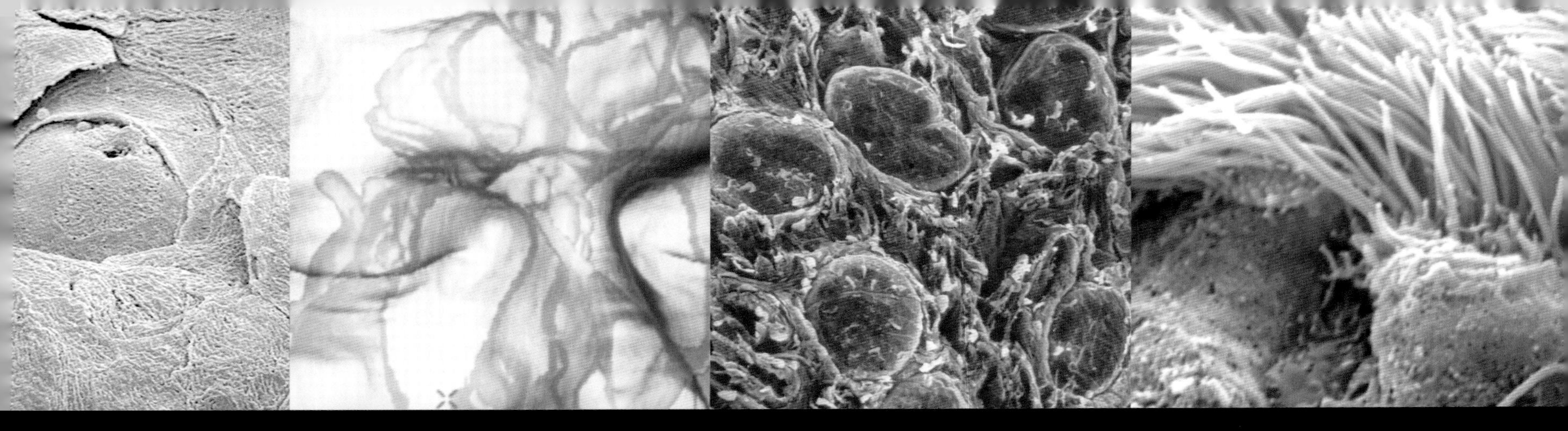

Our sense of taste depends on clusters of cells called taste buds. Humans have about ten thousand taste buds in the mouth, most of them on the tongue. The taste buds are sensitive to only four main tastes: sweet, sour, salty and bitter. We taste sweet things at the tip of the tongue, sour things at the sides, salty things over the surface, and bitter things at the back. Each type of taste bud has a different degree of sensitivity. While we can detect sweetness in one part in two hundred, we can detect bitterness in as little as one part in two million. This is probably an adaptation for survival. In nature, sweet things – fruits and starches – are generally edible, while bitter flavours are associated with poisons.

Our sense of taste is much less acute than our sense of smell. It takes about twenty-five thousand times more molecules of a substance to taste it than to smell it. Much of what we think of as taste is really smell – the reason why we cannot properly 'taste' food when the nose is bunged up with a cold.

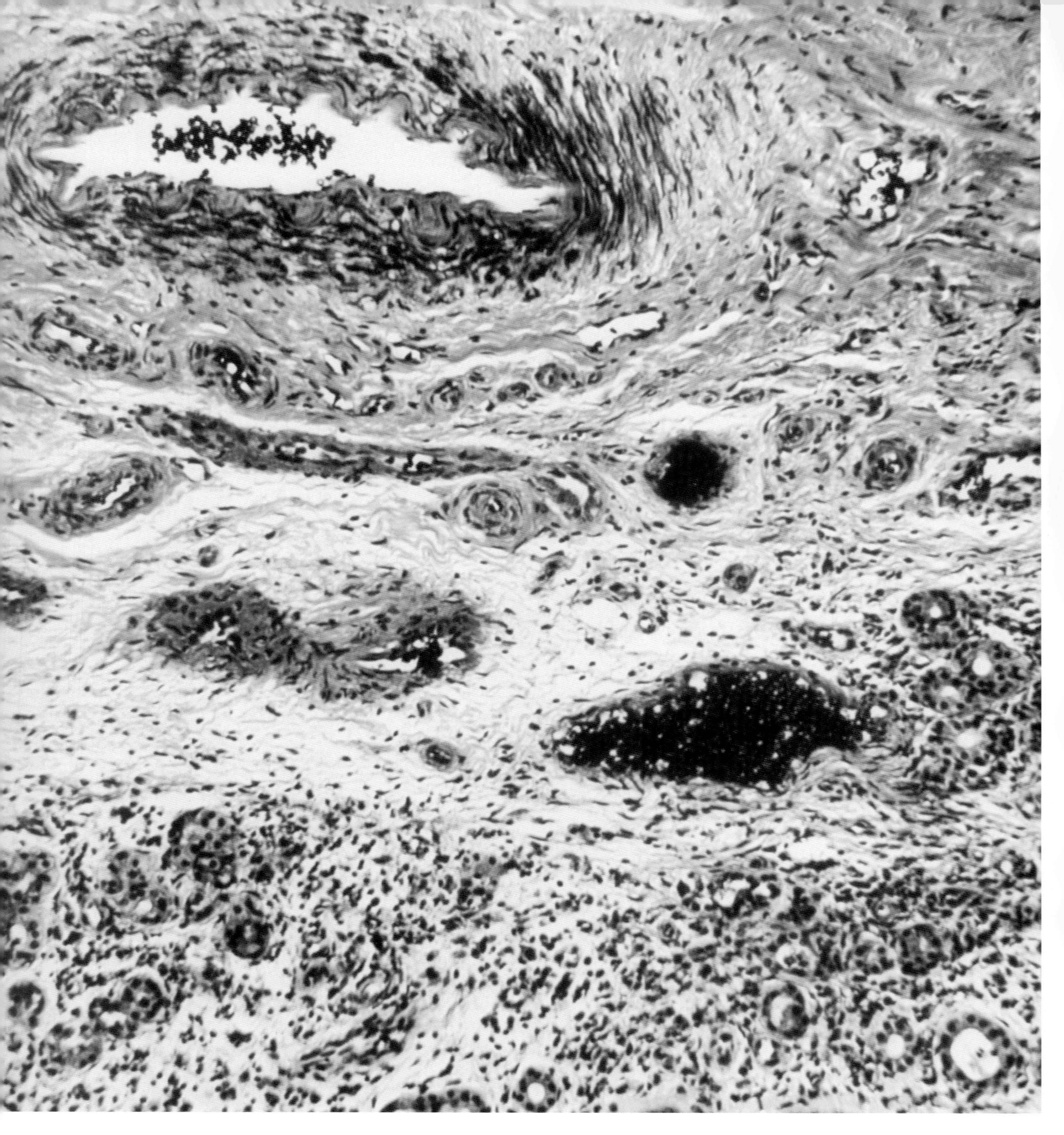

Olfactory epithelium [LM]

A SECTION OF SKIN from the human nose shows part of the olfactory epithelium (purple), an area of tissue in the upper nasal cavity where the smell-sensing cells are located. Although the olfactory epithelium is no larger than a postage stamp, it contains about ten million smell receptors of at least twenty types, each of which is sensitive to a specific range of odour molecules.

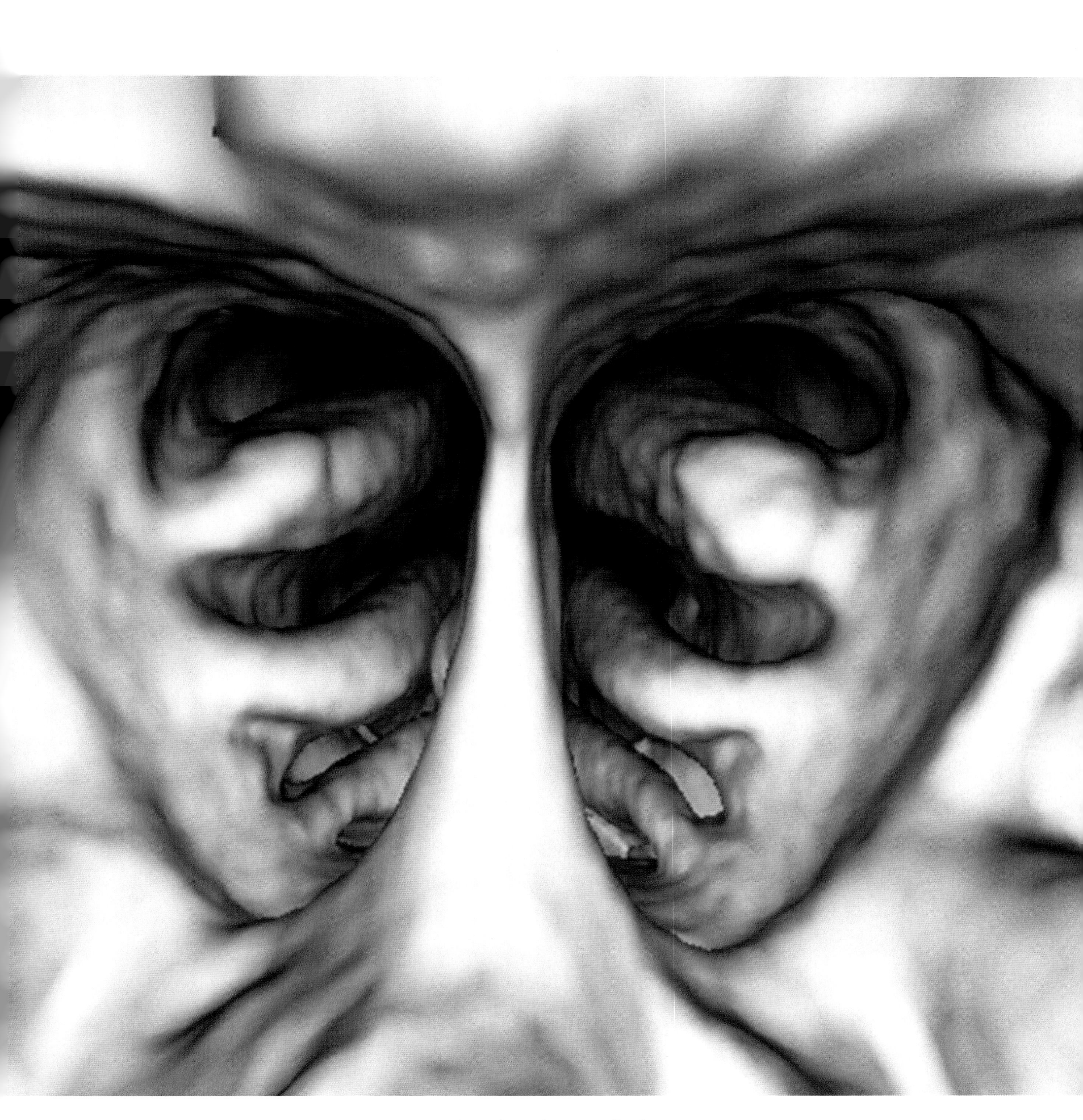

Conchae in nasal sinuses [CT scan]

EXTENDING FROM EACH SIDE of the septum that divides the nasal cavity are three conchae – bony plates covered by a mucous membrane. The two lower conchae help to humidify and warm the air we breathe in, while the upper concha is covered by olfactory epithelium containing smell receptor cells.

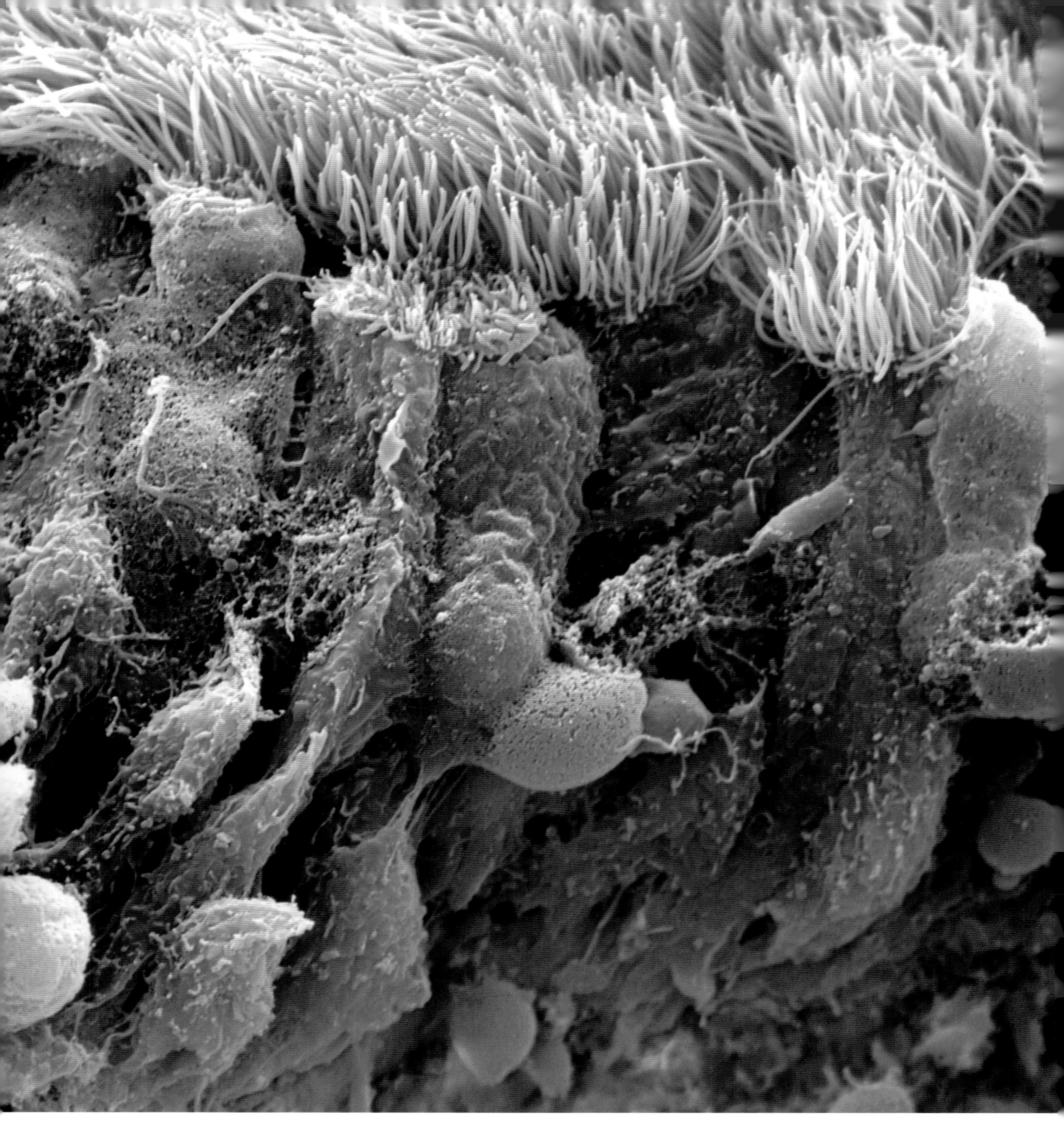

Nasal epithelium [SEM]

THE MUCUS-COATED LINING of the nose traps odoriferous particles, and hair-like structures called cilia (yellow) transport them to the olfactory centre at the top of the nasal cavity. Only about 5 per cent of the nasal epithelium is directly concerned with the sense of smell.

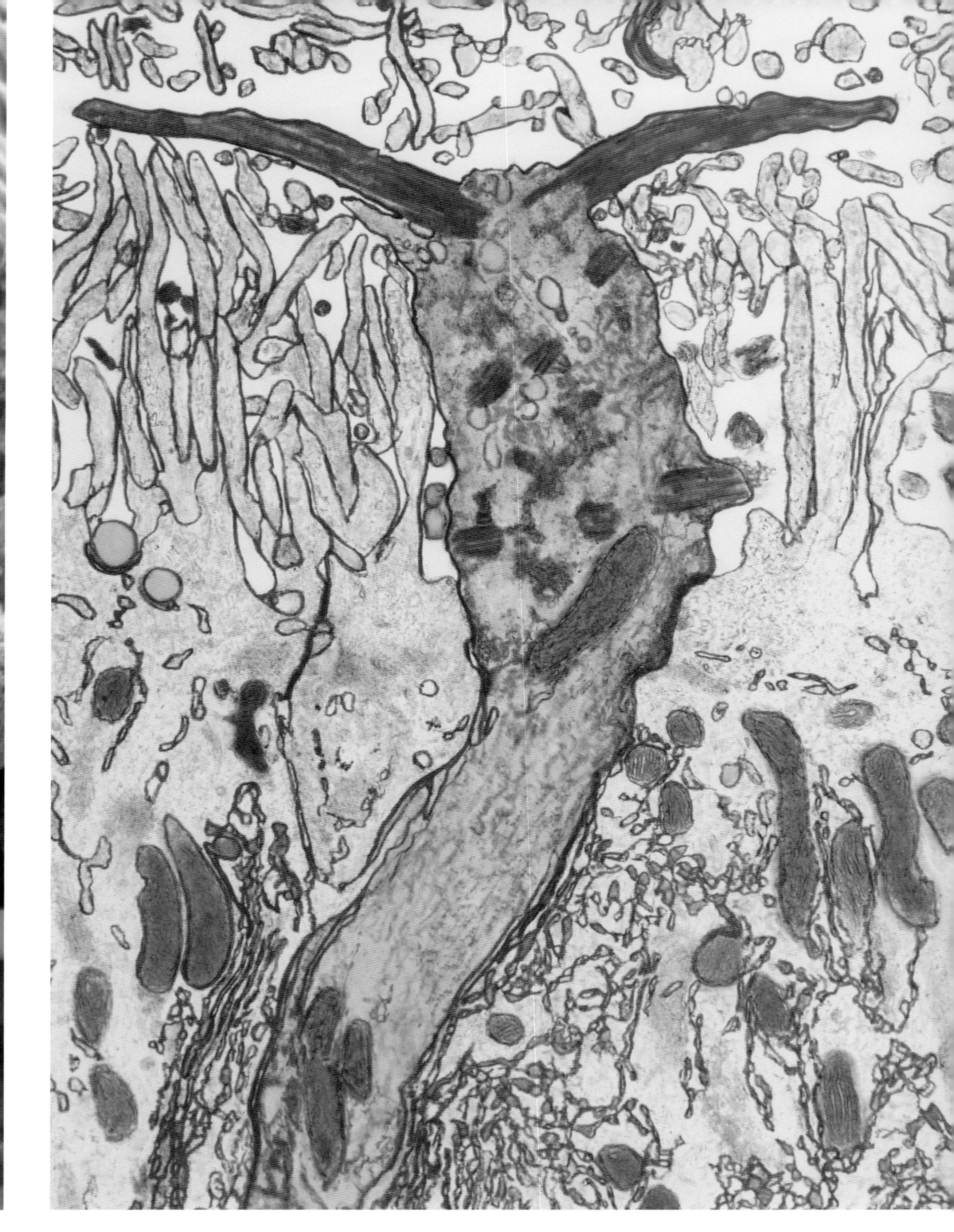

Smell receptor cell [TEM]

A SMELL RECEPTOR CELL (orange) is topped by two long, non-motile cilia projecting into the mucus lining of the nasal cavity. The cilia are thought to be the sites of interaction between odoriferous substances and the receptor cells. The cells are actually neurons (nerve cells), which are shed and replaced every four to eight days.

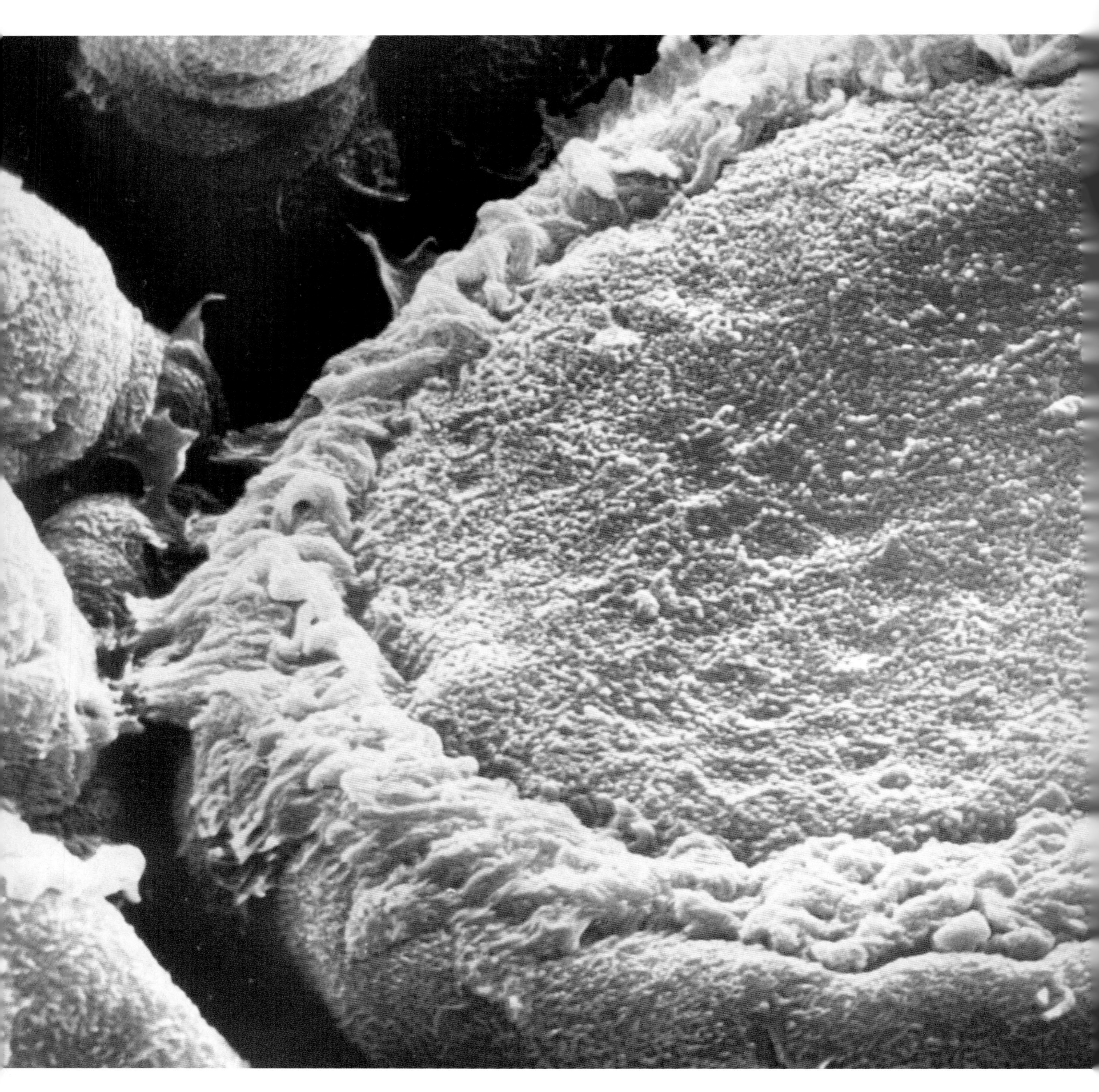

Circumvallate papilla on tongue [SEM]

THIS TYPE OF PAPILLA is associated with taste buds that are sensitive to bitter taste. Called a circumvallate papilla, it is 1–2mm/1/32–1/16in in diameter. There are about ten of these papillae arranged in a V-shaped pattern at the back of the tongue. The taste buds themselves are found in grooves surrounding these papillae.

Papillae on tongue [SEM]

THE UPPER SURFACE OF THE TONGUE is covered with small projections called papillae. Taste buds are found in the centres of many of the large red papillae, which are called fungiform papillae because of their resemblance to mushroom caps, The smaller filiform papillae between them are not involved in taste, but are mechanically adapted to help process food.

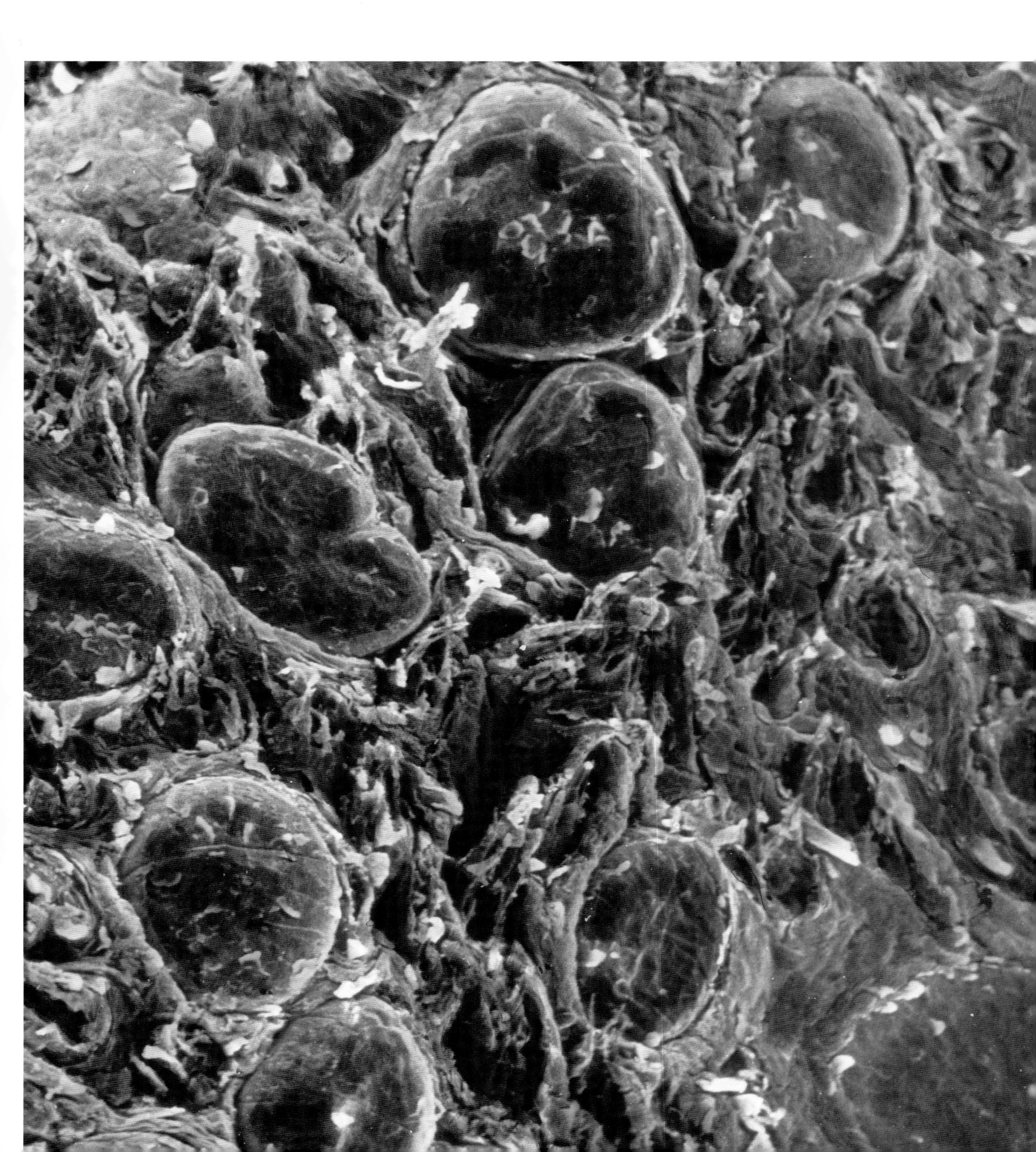

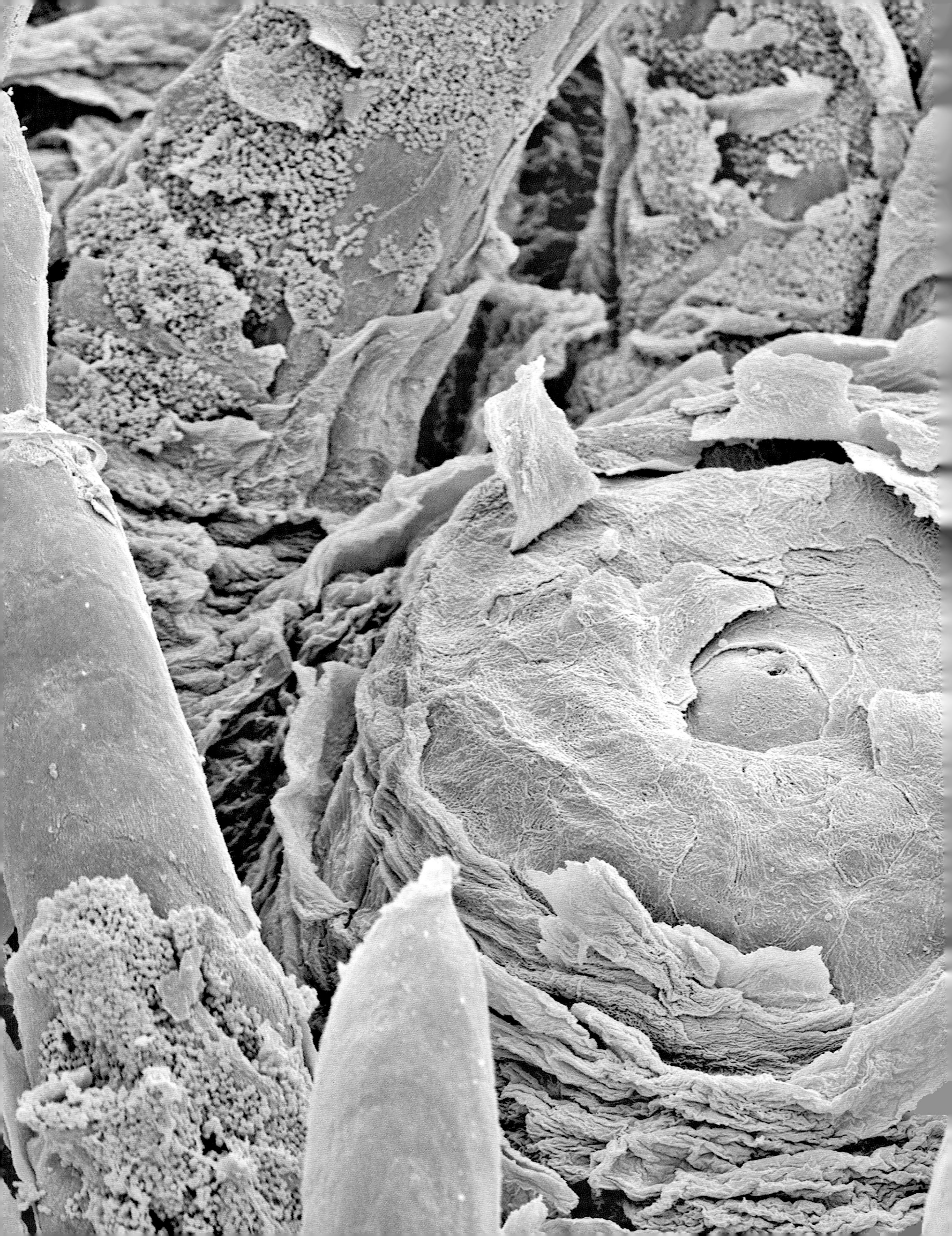

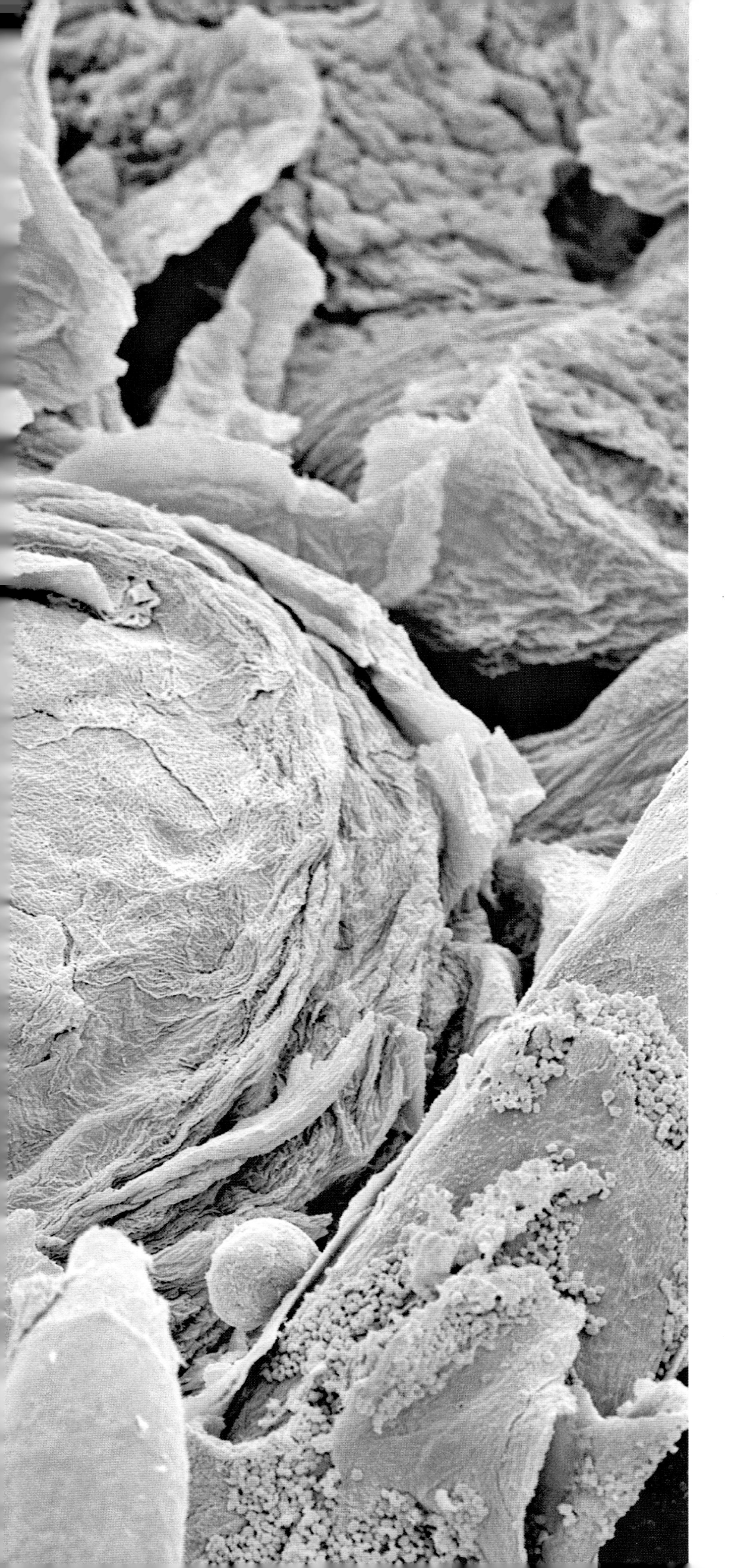

Taste bud [SEM]

A TASTE BUD can be seen in the centre of a fungiform papilla surrounded by filiform papillae. The taste bud itself is a neuro-epithelial cell that reacts to chemicals in food, transmitting nerve impulses to the brain. These impulses are processed to form the sensation of taste.

Vision is the most complex of our senses, and the one we rely on most to interpret the world around us. Seeing is believing, we say.

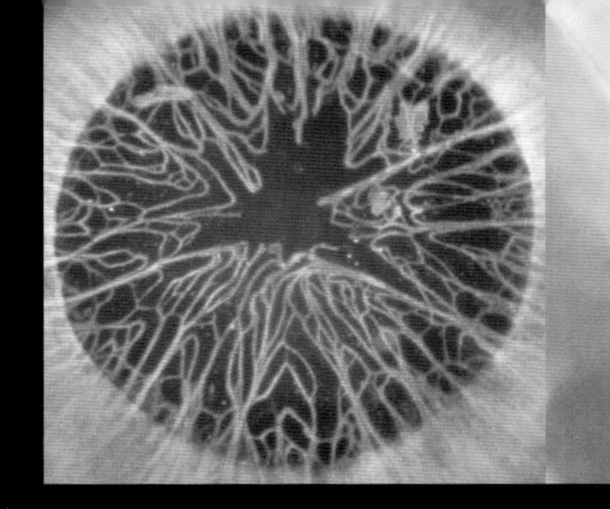

Our eyes work like a camera; or, rather, a camera is a crude mechanical version of an eye. Two eyes give us binocular vision, producing two slightly different images of an object that our brains combine in a detailed 3-D image. Binocular vision, a feature common to other primates and predatory animals, helps us to judge distance and depth.

Light enters the eye through the transparent cornea. Just inside the cornea is the iris, a muscle that regulates the amount of light reaching the lens by constricting or dilating a small hole, the pupil. The iris is what gives our eyes their colour, and its detailed structure is as individual as a fingerprint.

Camera lenses and the eyes of fish and reptiles focus by moving closer or farther away from an object. The drawback to this arrangement is that it takes up a lot of space. Human lenses get over the problem by changing shape, becoming thinner to focus on distant objects, thicker to examine a near object. Six sets of muscle control the shape of the lens, and they work so efficiently that in a split second humans can change focus from an object on the horizon to an object within touching distance. As we age, our lenses become less elastic, which is why elderly adults often need glasses for reading.

Light passing through the iris is focused on to the retina, the light-detecting layer at the back of the eye. It is a tangled mass of cells called rods and cones. The 120 million rods in each eye are

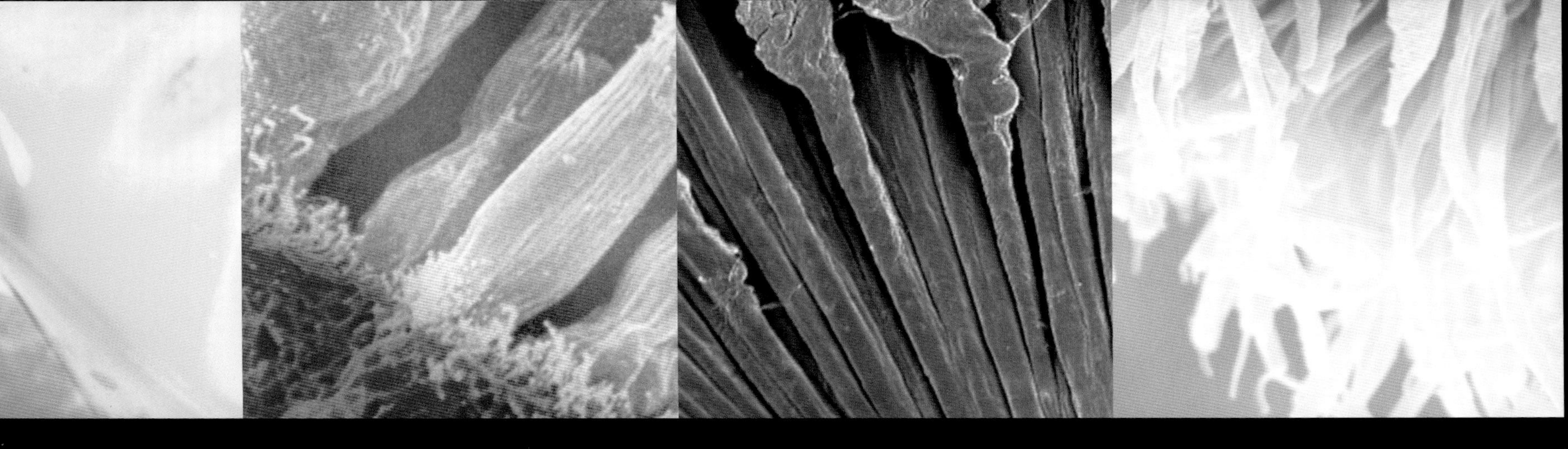

extremely sensitive to light and can see only in shades of black and white, while the seven million cones process colour in bright light.

When light strikes the retina, there is a short time-lag before it takes effect and another time-lag before the effect wears off. This delayed action is what makes us see a movie as a seamless continuum of images rather than as a flickering succession of individual frames. It also explains why we see a moving, two-bladed propeller as a rotating disc.

Signals from the eyes are processed in the brain. The loss of sight is a terrible tragedy, yet such is the power of the brain that we don't need eyes to see. Memory enables us to replay scenes from years ago as clearly as if they had happened yesterday. In our dreams we experience things more vividly than anything that happens in reality. And in our mind's eye, we can picture completely imaginary events.

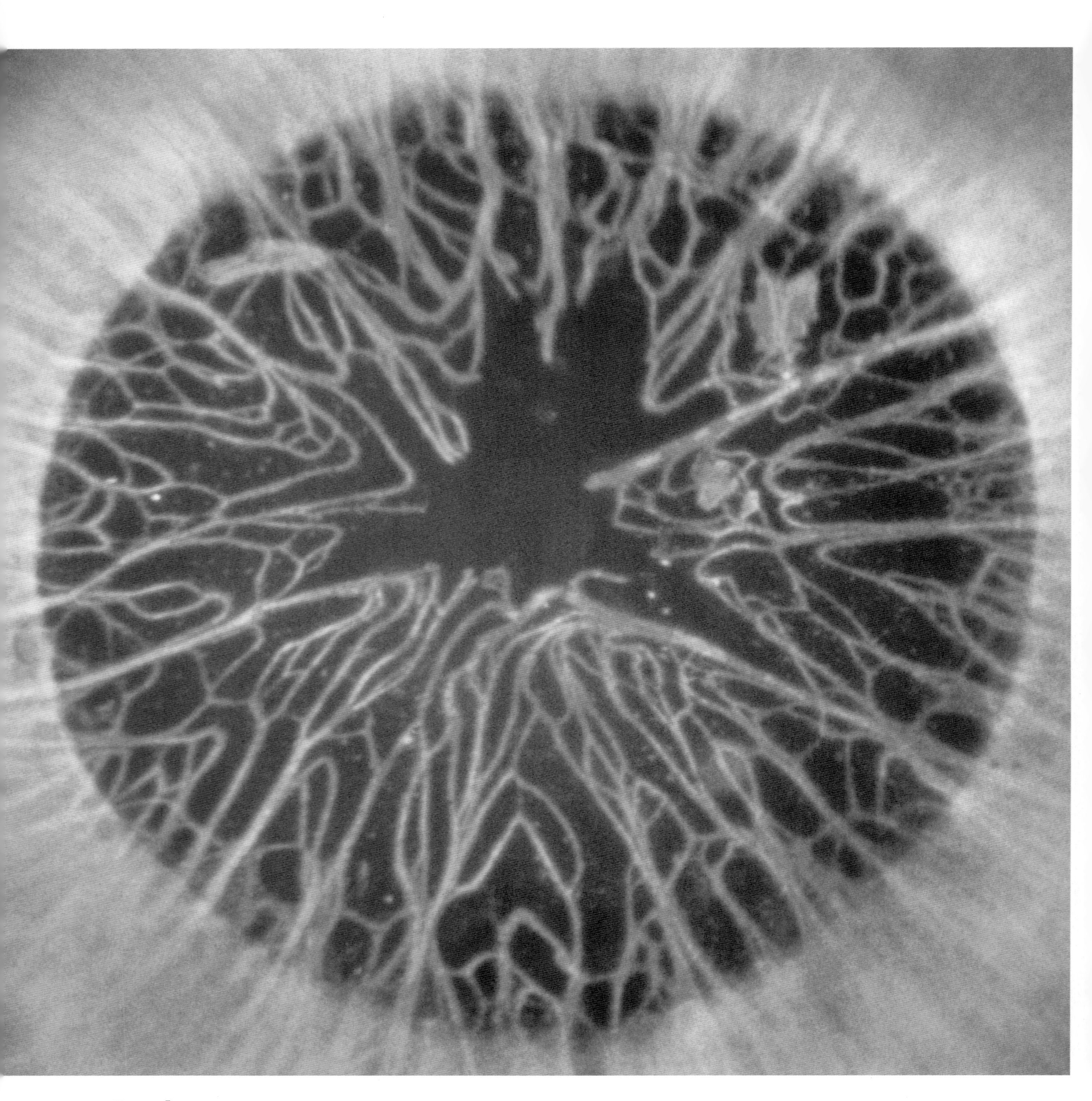

Foetal eye [Macrophoto]

THE LARGE BLUE PUPIL of a human foetal eye is covered by a lace-like membrane. Eye formation begins during the third week of foetal developement, and at five months the retina lens and other major structures are formed. At birth, the baby's eye is already two-thirds the size of an adult's, but growth and developement continue until puberty.

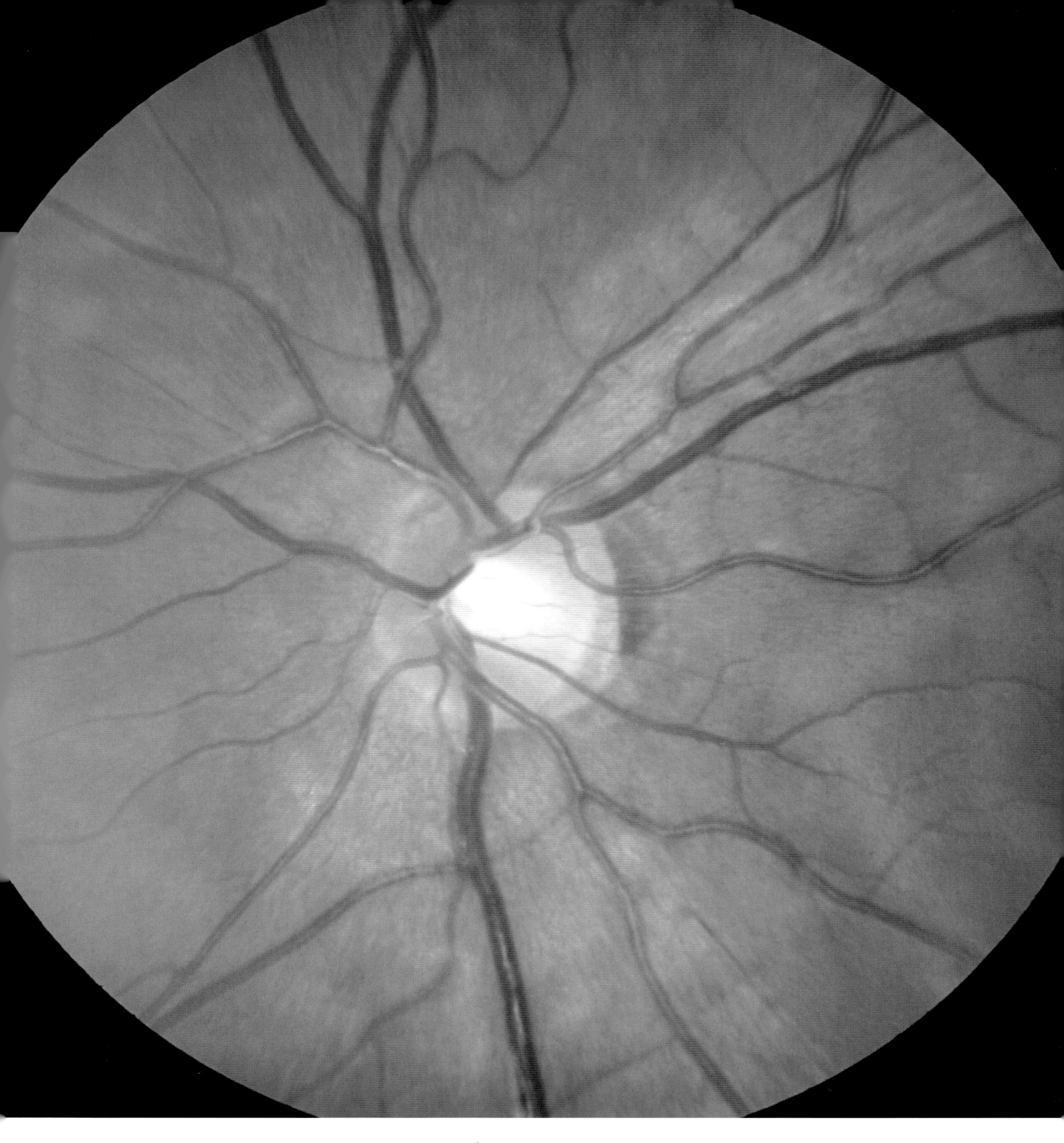

Retina [Fundus camera image]

THIS IMAGE SHOWS the distribution of veins and arteries in the retina, the light-sensitive layer at the back of the eye. The retinal artery emerges from the centre of the optical nerve, which joins the retina at the optic disc, the so-called blind spot (pale central area). This retina has a greenish hue because the subject was Asian; in Caucasians it is reddish.

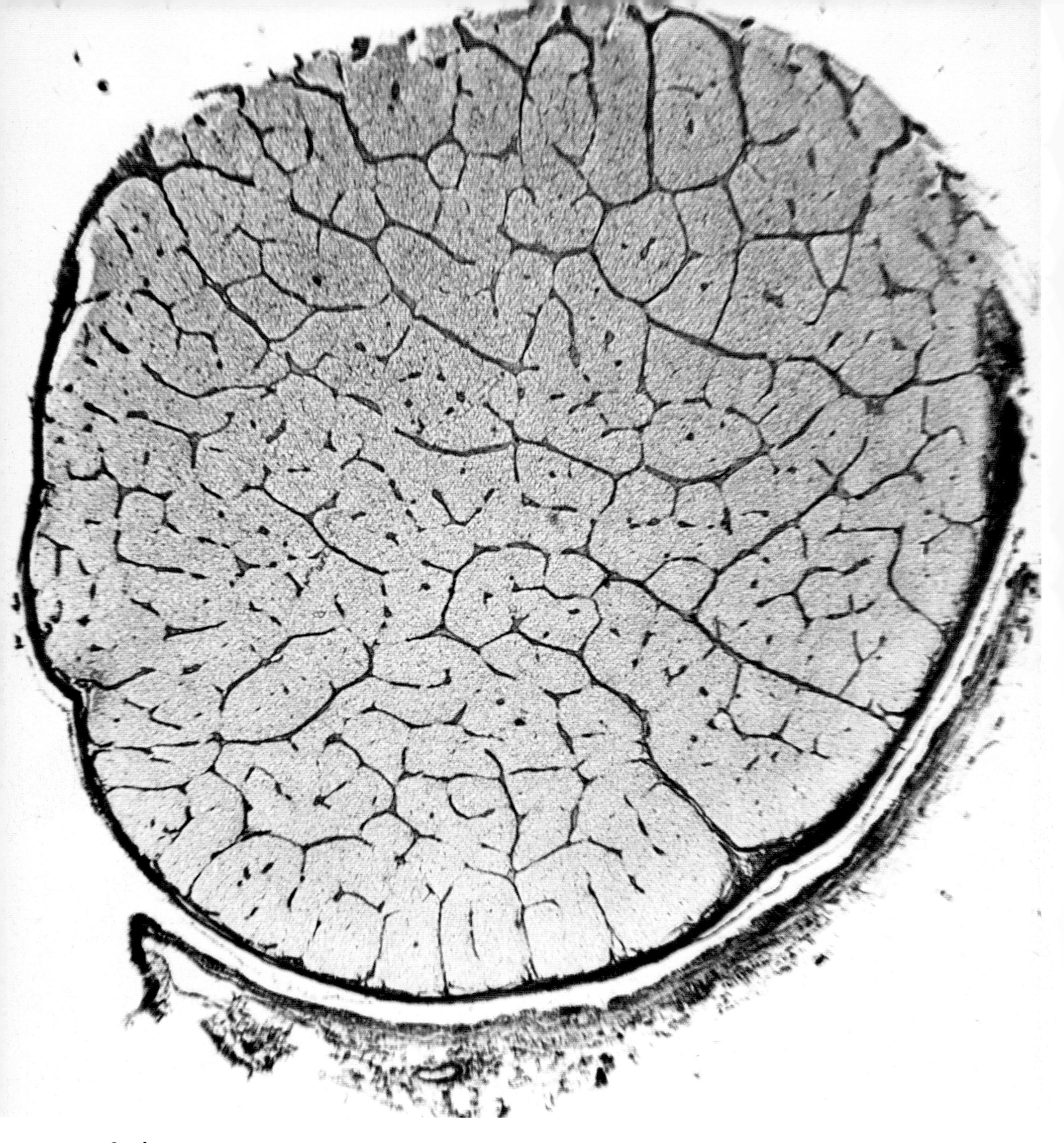

Optic nerve [LM]

SEEN HERE IN SECTION, the optic nerve is responsible for vision. Information from the photosensitive cells (rods and cones) of the retina is carried to the visual cortex in the occipital lobes at the back of the brain. The information is processed, together with other sense data, including memory, to produce the perceived images.

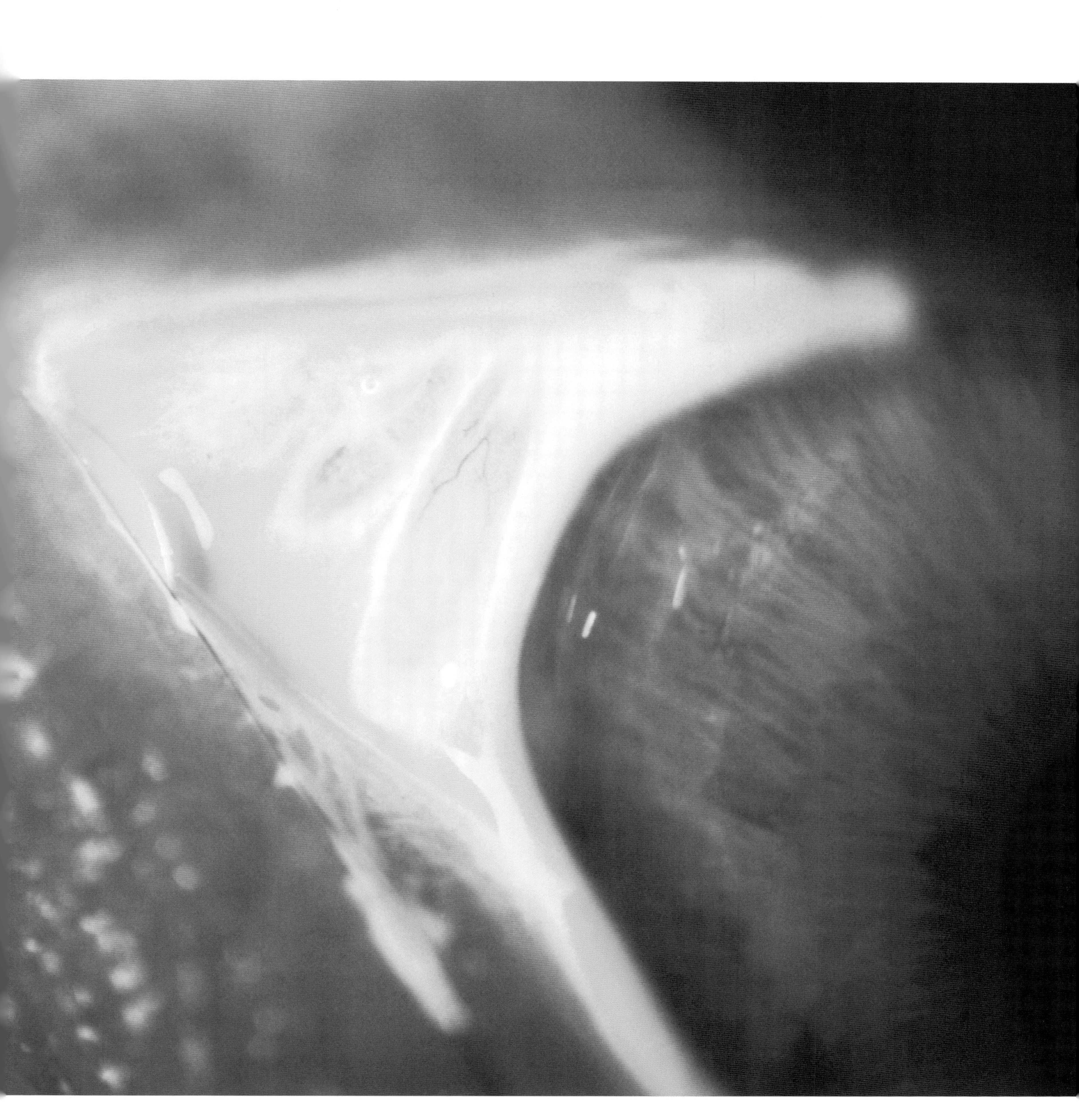

Medial canthus [Macrophoto]

TEARS APPEAR YELLOW in this image of the inner corner of an eye, called the medial canthus. The eye has been treated with fluorescin, a harmless orange dye used to detect damage to the cornea and conjunctiva, the delicate mucus membrane that covers the front of the eye. Tears secreted by the lacrimal gland in the upper and outer part of the eye socket protect the conjunctiva. Tears drain diagonally across the front of the eye into canals at the medial canthus, from where they run down into the nasal cavity.

Lacrimal gland [SEM]

THE LACRIMAL GLANDS (one shown here in section) produce tears that keep the front of the eye moist and lubricated. Lacrimal secretion (one droplet visible, red, centre) also contains lysozome, an antibacterial enzyme that helps to protect the eye from infection.

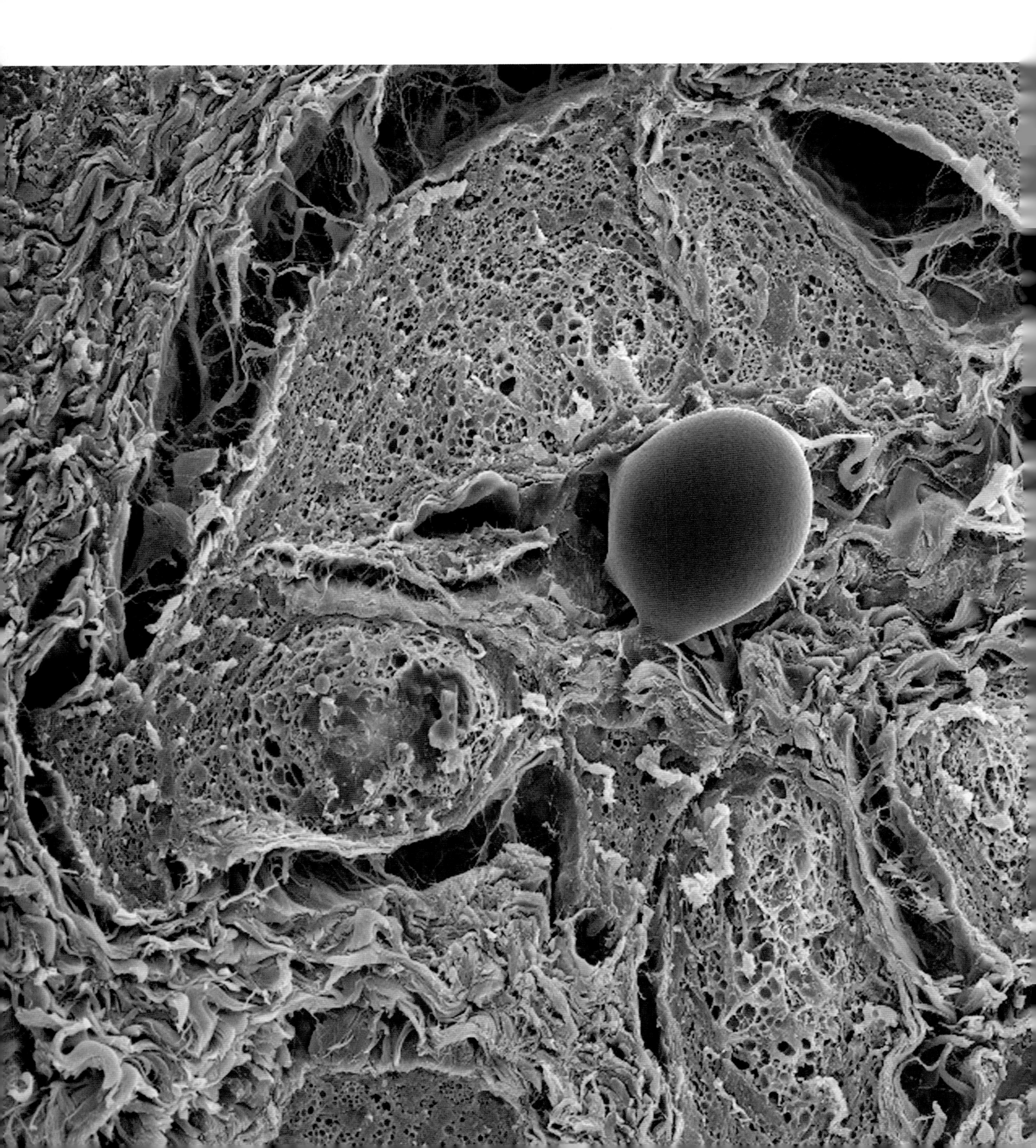

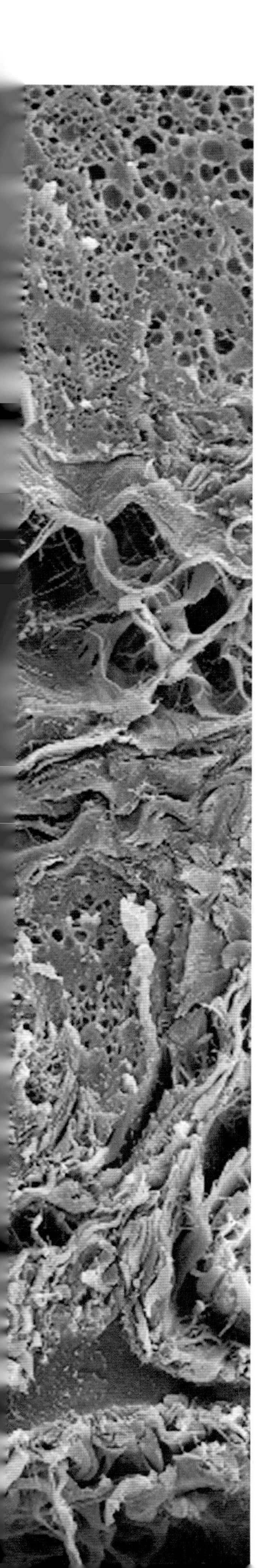

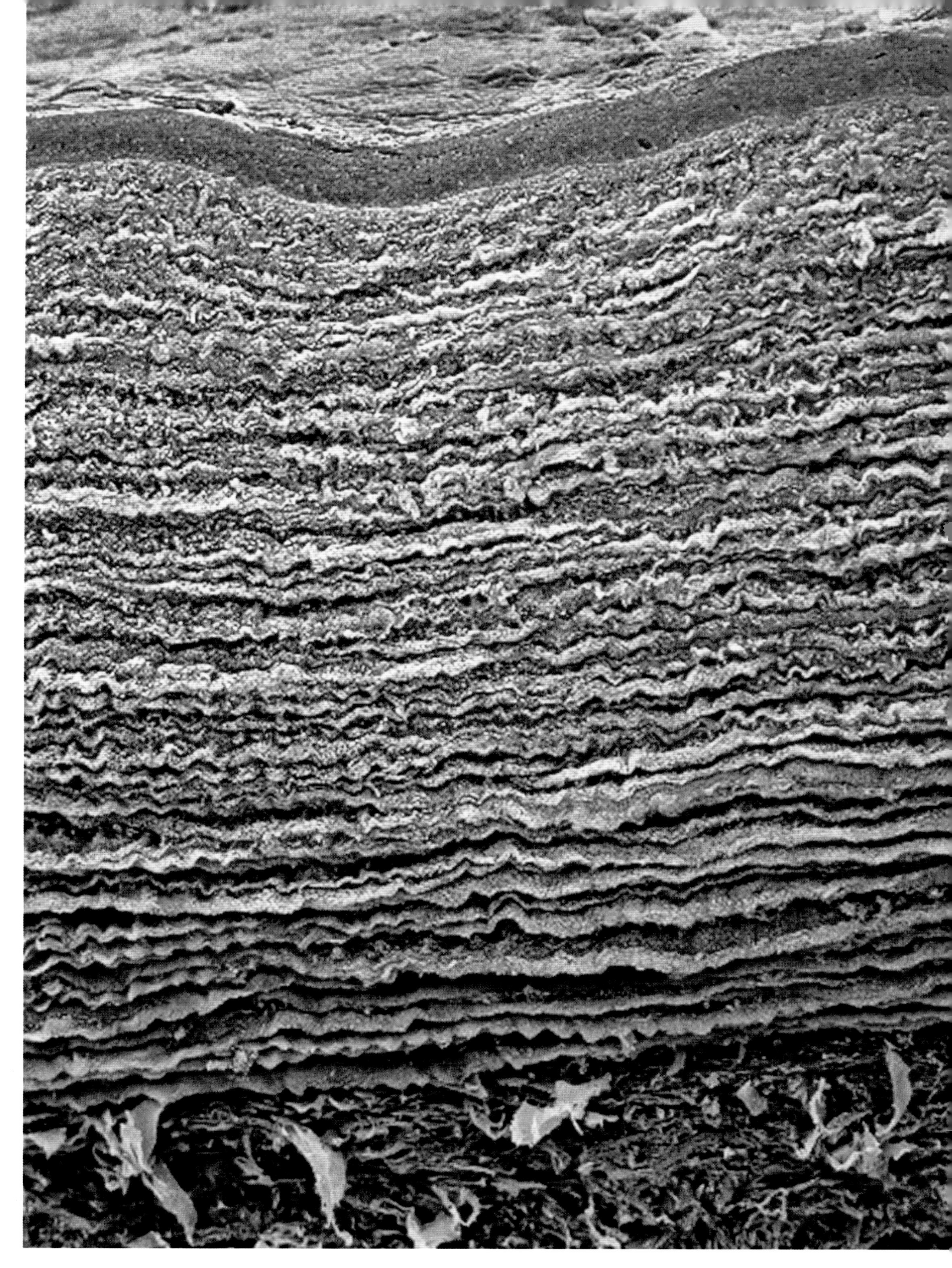

Cornea [SEM]

A SECTION THROUGH A CORNEA, the transparent central region of the eye's outer surface, shows the corneal epithelium (brown) above dense connective tissue (parallel white/blue lines). The cornea acts as a window, but also refracts (bends) light on to the lens, helping to focus images on to the light-sensitive retina at the back of the eye.

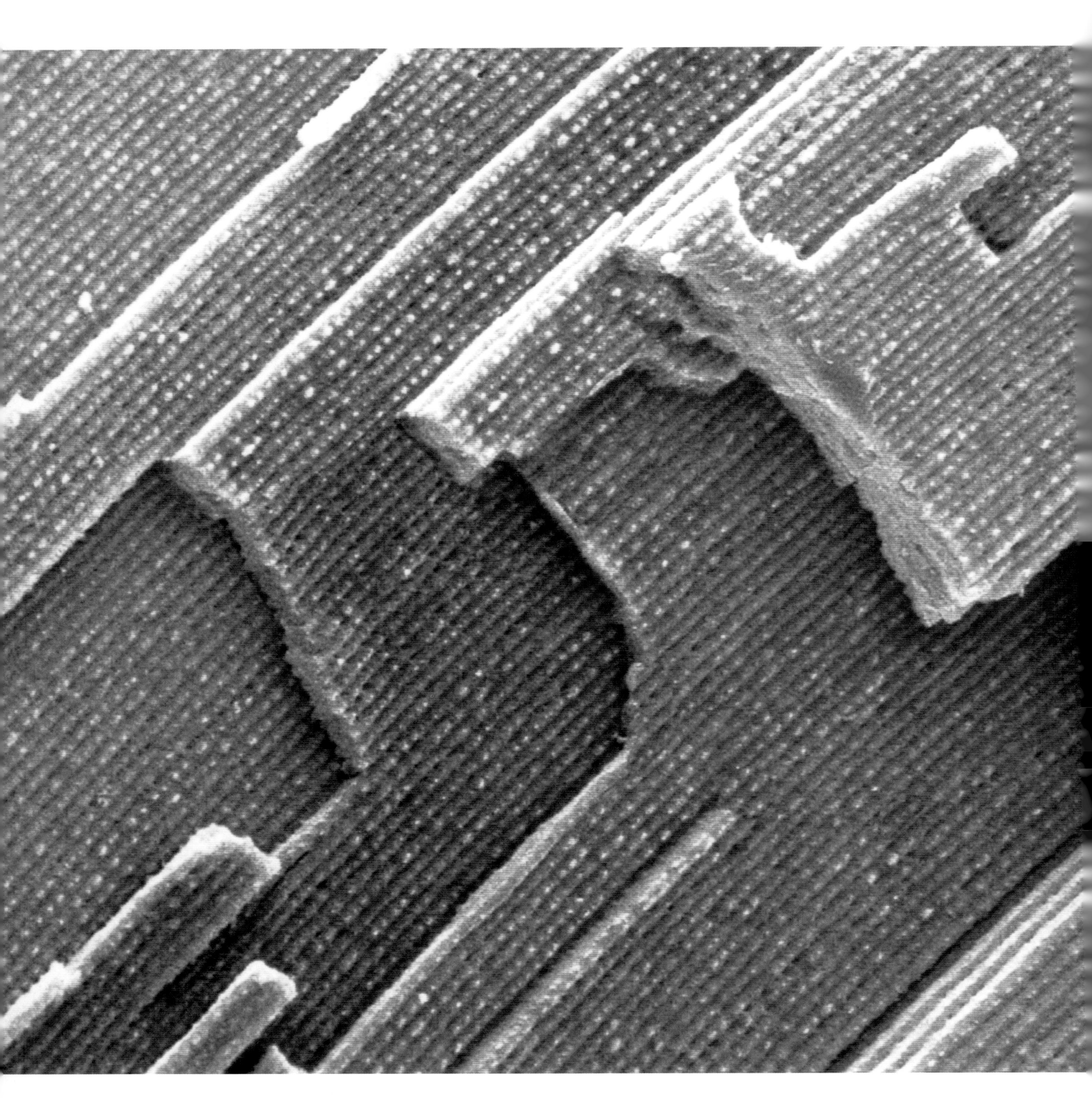

Cells from lens of the eye [SEM]

THESE LONG STACKS OF CELLS are from the lens, the transparent part of the eye, which focuses light on to the retina. The transparency of the lens is due to the absence of nuclei in its cells and to the crystalline precision of their arrangement. The cells are called fibres because they are up to 10mm/⅖in long. With age the lens can become cloudy, but can be replaced by an artificial plastic lens.

Lens fibres [LM]

SEEN END-ON, THE LENS FIBRES – actually cells – are stacked in vertical columns. This arrangement contributes to the lens' transparency. It also ensures that the lens can adjust its shape to focus light rays from all sources, near and far, on to the light-detecting retina at the back of the eye.

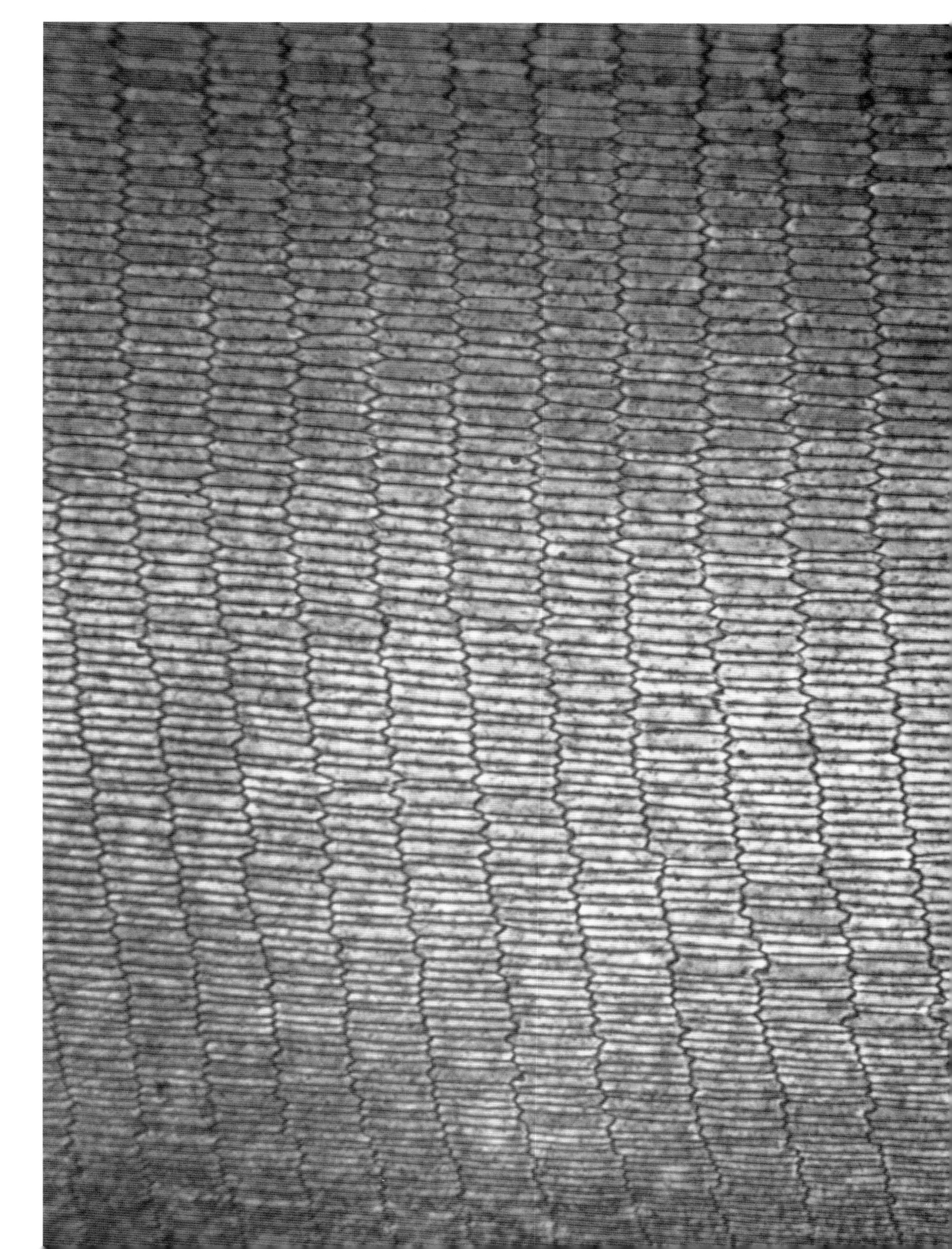

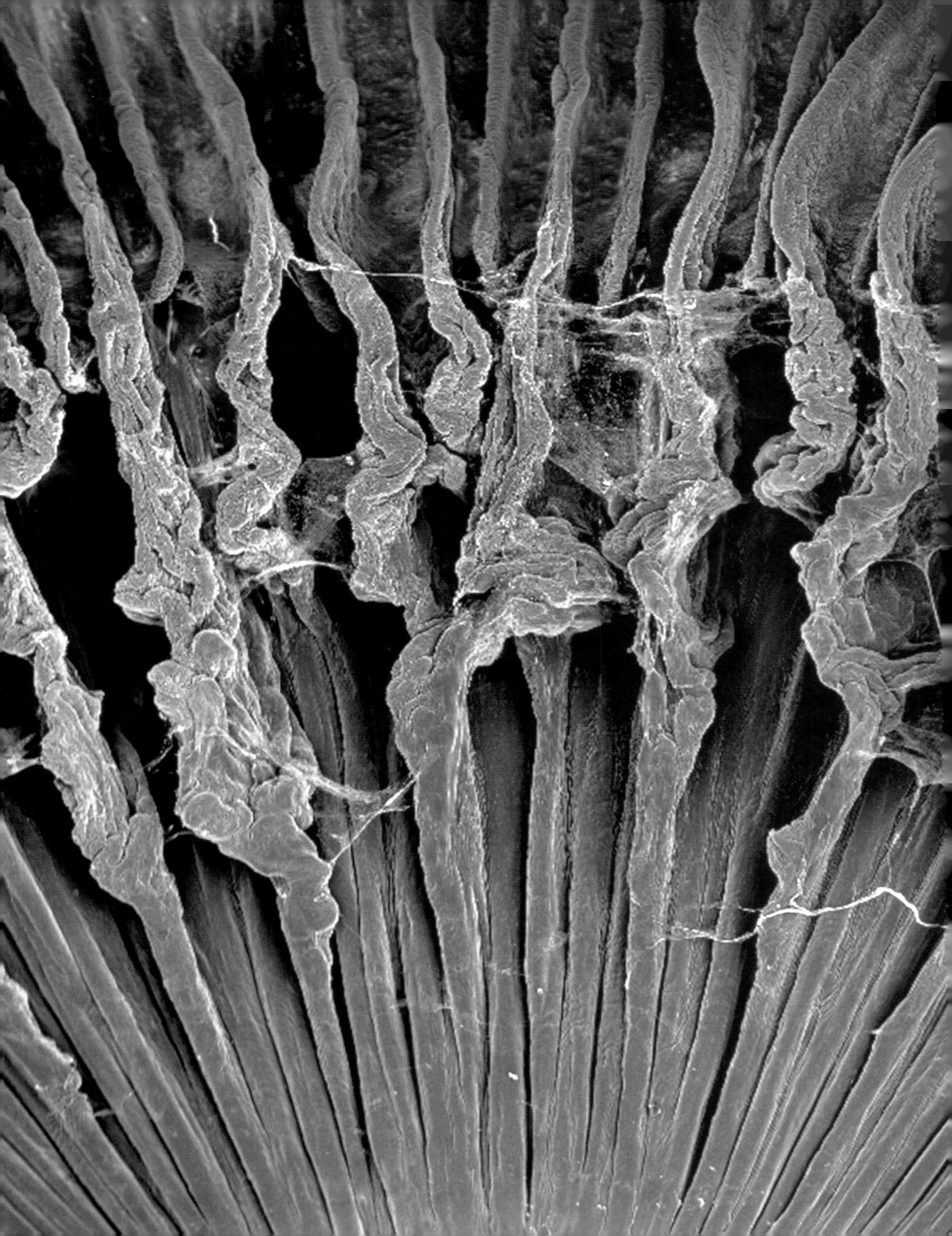

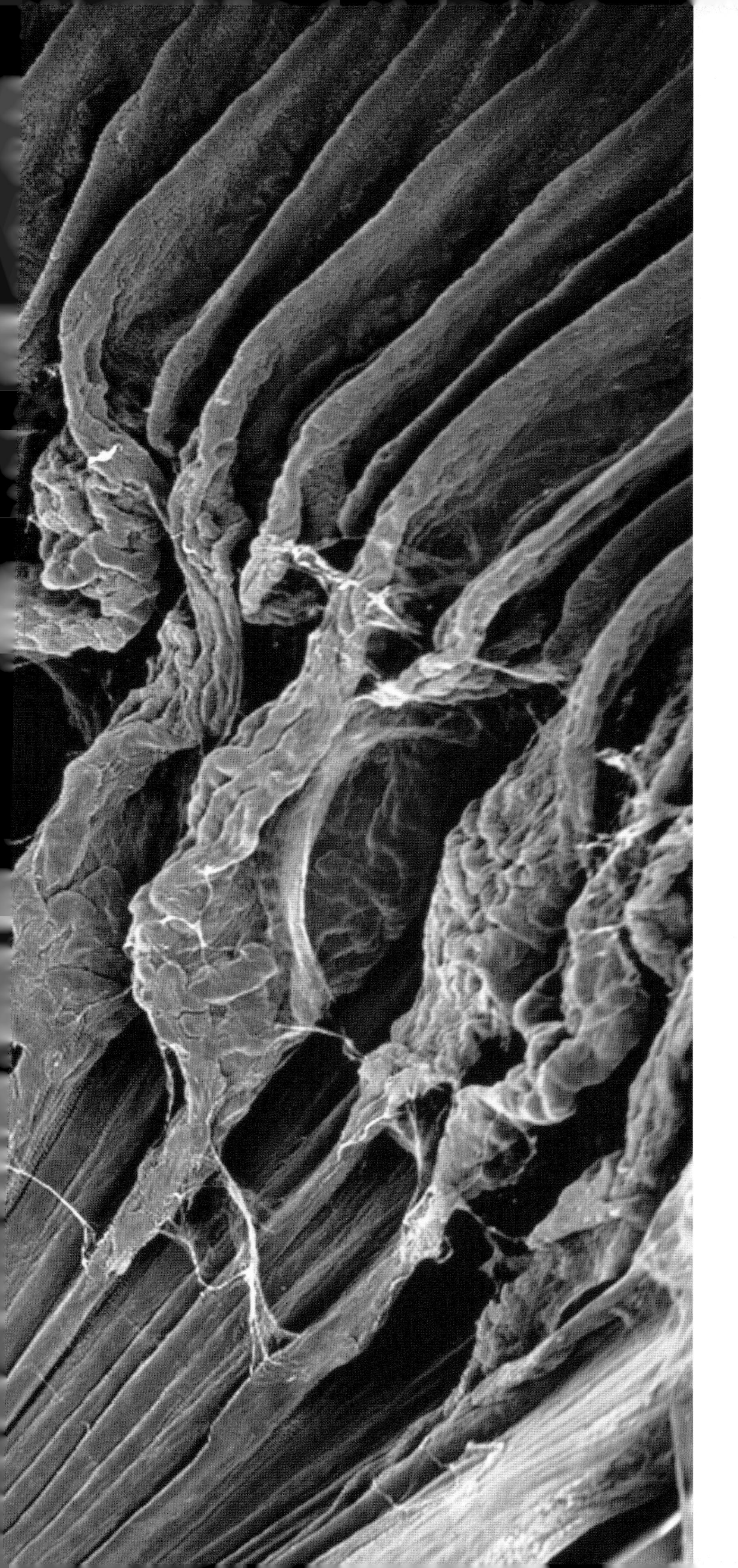

Section through part of eye [SEM]

THIS PARTIAL SECTION through an eye shows the iris (grey-green, lower half of image), the ring of muscle that controls the amount of light that enters the eye. The fibres above it are part of the ciliary body, which contains muscles that contract to alter the curvature of the lens and focus light on the retina. The ciliary body also secretes aqueous humour, a thin, watery fluid that fills the cavity between the cornea and the iris, and the cavity between the iris and the lens.

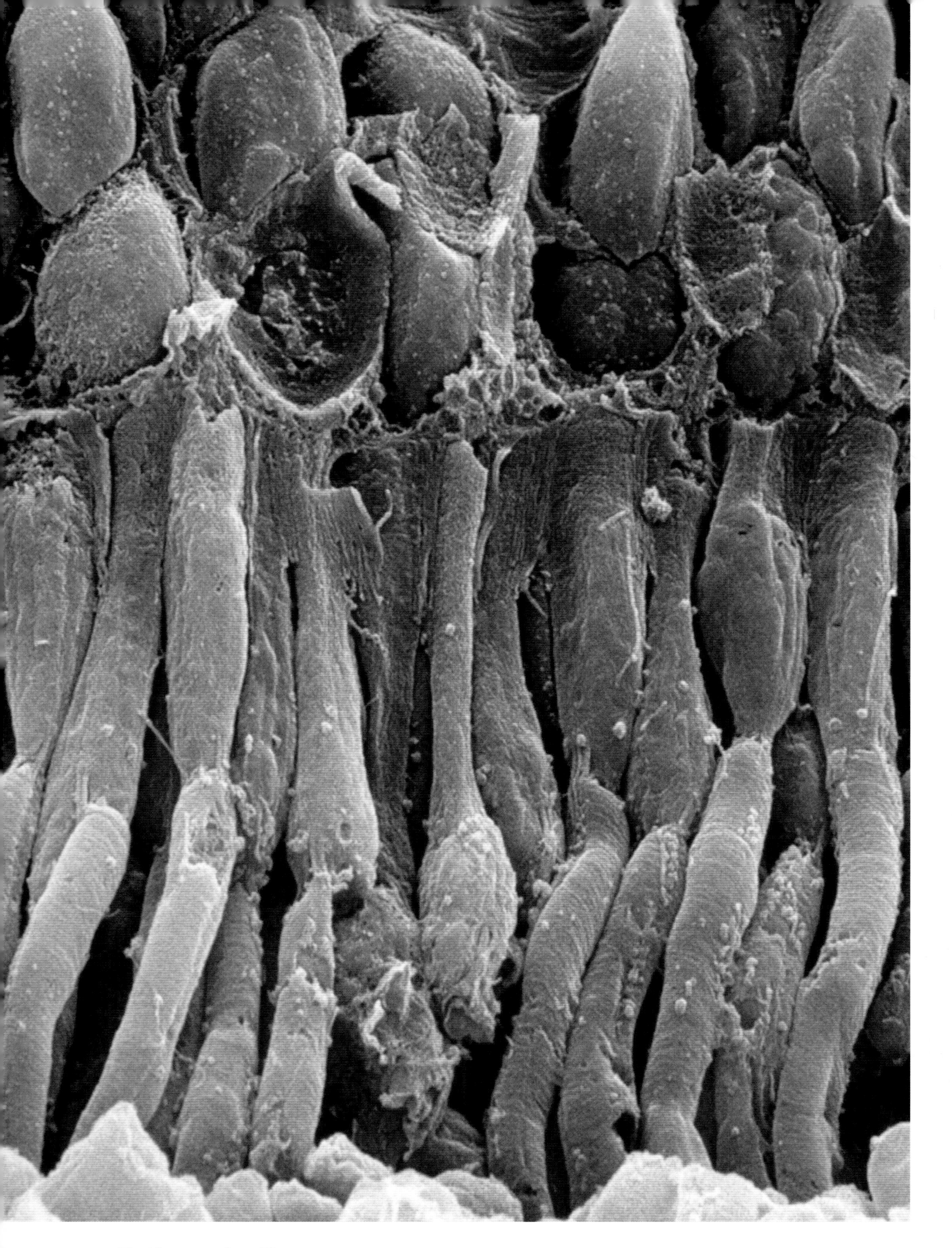

Retina rod cells [SEM]

OUR ABILITY TO SEE in dim light is due to these rod cells on the retina. The rods contain a purple pigment called rhodopsin, which is very sensitive to light. In bright light the rhodopsin breaks down and the rods become inactive. If the light is dimmed, rhodopsin is resynthesized. The pigment does not reform immediately, which is why it takes time for your eyes to become adapted when you step from sunlight into a darkened room.

Cones and rods 1 [SEM]

CONE CELLS (green-blue) are sensitive to coloured light, while rod cells (blue) give black and white vision in dim light. Visual acuity in bright light is directly related to the number of cones on the retina. The human eye has up to 200,000 cones per square millimeter/130 million per square inch, but hawks possess about five times that number and can distinguish details two to three times farther away than humans.

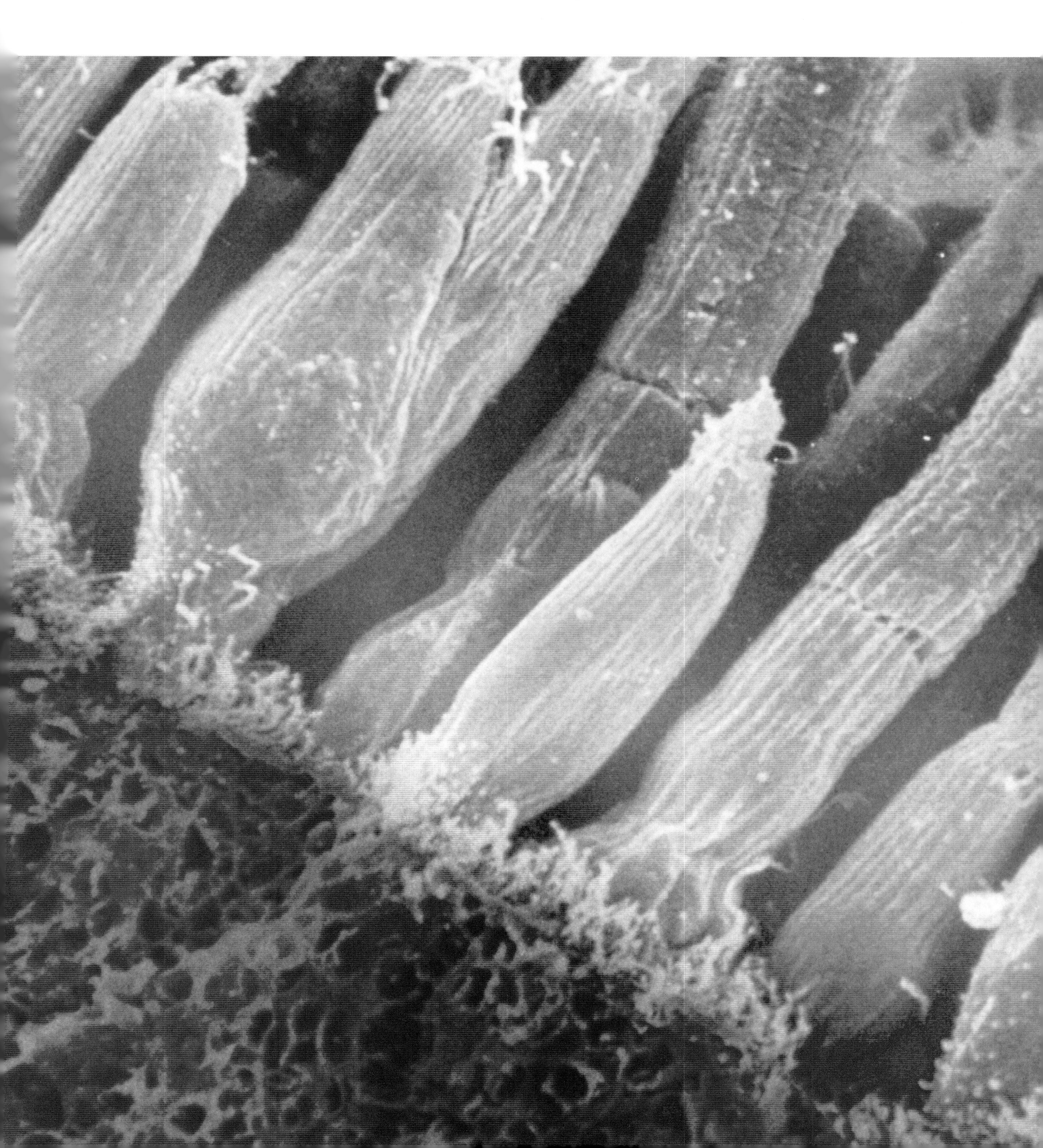

Cones and rods 2 [SEM]

A SECTION THROUGH THE RETINA shows the two types of photoreceptor cells. The yellow cells are cones, which give colour vision in bright light. The white cells at bottom are rods, which give black and white vision in dim light. Rods are connected to nerve cells in groups and produce a general image. Although much less numerous, cones are individually connected to nerve cells and so produce a detailed image.

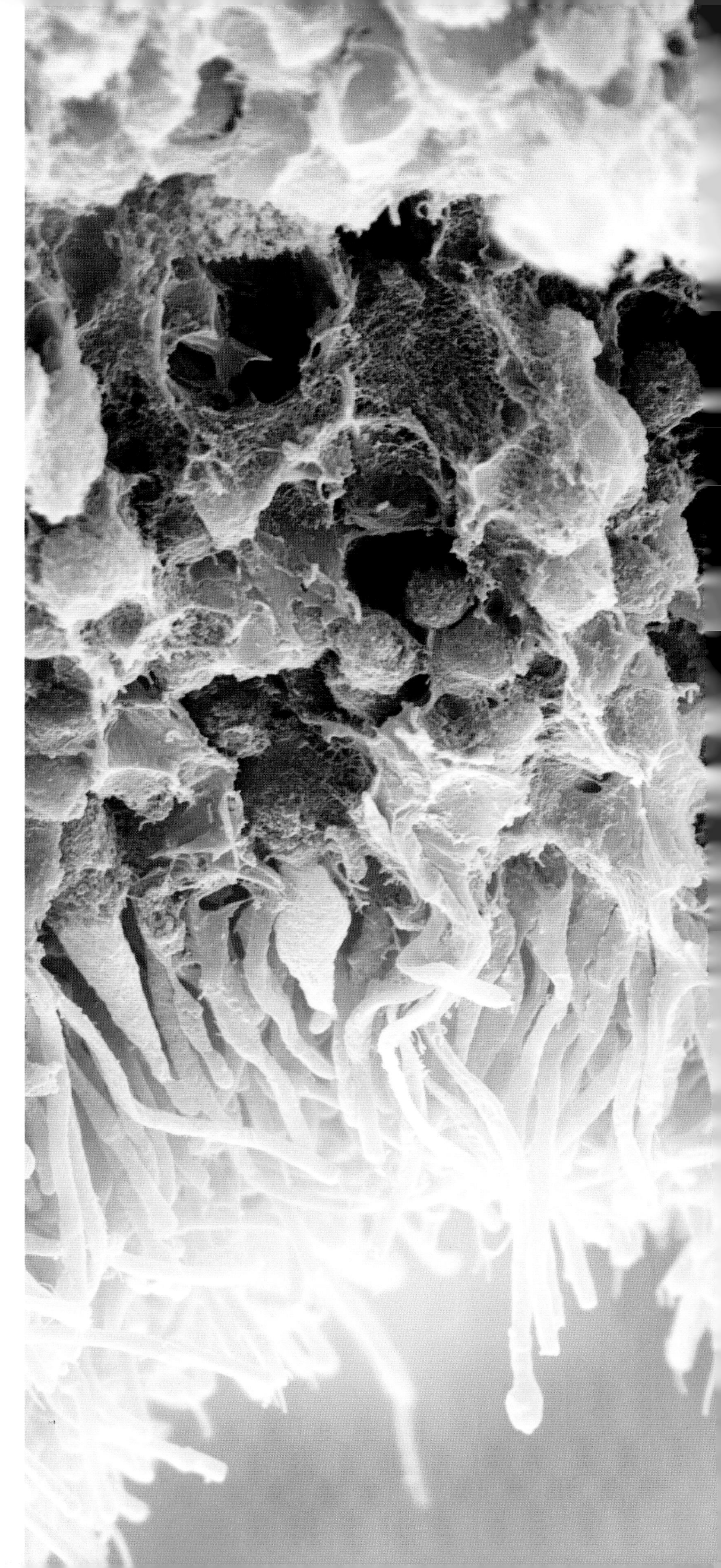

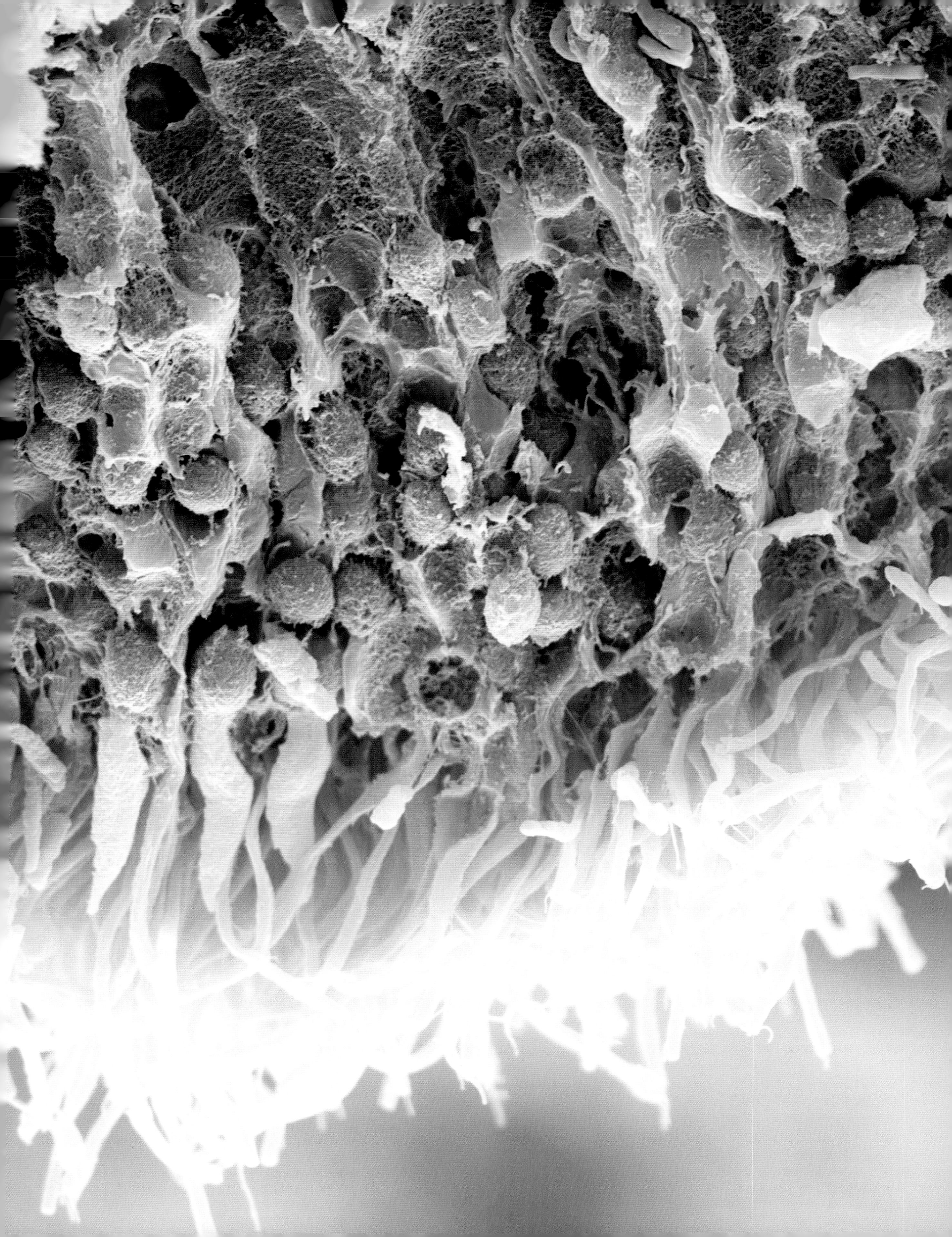

index